Advance Praise for
Human Purpose and Transhuman Potential

"Do not postpone your reading of this book. It comes at a crucial time when humanity's capacity to actively change the course of 'natural selection' is growing exponentially. It is important to rethink where we are going even as we tinker with life—to consider deeply what higher goals we are trying to achieve. Ted Chu's enduring optimism about our future evolution is both refreshing and challenging at the same time. It will certainly take any reader out of their present comfort zone and lead them to question many assumptions they may have. There is no doubt that any reader of Chu's book will feel empowered reading it, both about their own lives and about the universe itself."

—**Hans R. Herren, PhD**, entomologist and agronomist; winner of the World Food Prize (1995) and the Right Livelihood Award (2013); president of the Millennium Institute and the Biovision Foundation

"Added to a list of distinguished thinkers—such as Einstein, Russell, Whitehead, and de Chardin, and many others—now we have Dr. Ted Chu. His new book is truly thought-provoking and eye-opening. For any would-be reader, this book is both intellectually stimulating and spiritually rewarding. Dr. Chu's meticulous scholarship, interdisciplinary training, and encyclopedic knowledge are to be commended. He urges us to free ourselves from the age-old parochial or even global vision to a more comprehensive, cosmic one. He makes his case in a positive and constructive way with penetrating insight and convincing argument."

—**Ching-lin Wang, PhD**, Area Study Specialist at the Library of Congress; former history professor at the Soochow University (Taiwan)

"Ted Chu thinks big. . . [But] is this science fiction on steroids? This book does not raise mere abstract philosophical questions, but [questions that are] a matter of great practicality and urgency. With artificial life and robots coming to replace us, we humans need a deep, thoughtful, civil, and above all inter-generational discussion of the issues Chu raises."

—**Gerald O. Barney, PhD**, Director, "The Global 2000 Report to the President" for President Jimmy Carter; Founder of the Millennium Institute and of Our Task

"Ted Chu is a pioneering visionary whose futurist concern deserves close attention. He is keenly aware that technological expertise is on the brink of reshaping our world and humanity dramatically. . . It is now conceivable to many thoughtful people that the human period of evolution may eventually turn out to be a transitional chapter in cosmic history. How are we to address such a prospect? In the remote cosmic future will human history appear to be anything more than an ephemeral crossing over to more fascinating episodes in an enormously inventive cosmic narrative whose eventual playing out is currently unimaginable? If you care about such questions, read this book."

—**From the Foreword by John Haught**, Professor Emeritus of Theology at Georgetown University; author of *Deeper Than Darwin*

"Ted Chu's elegantly written and well-researched book is a contemporary transhumanist reprise of Nietzsche's ideal of the Overman. Concerned about the nihilism that would follow from the 'death of God,' Nietzsche proposed that Europeans adopt the goal of creating something great beyond themselves. Likewise, Chu argues that humanity is not the pinnacle of evolution, but instead is a crucial phase in all-encompassing cosmic evolution, which can be understood as God in the making. If we are sufficiently courageous and daring, he tells us, we can fulfill our cosmic destiny by making possible modes of intelligence far greater than our own. Even critics of this 'Cosmic Vision' will find Chu's book required reading."
—**Michael E. Zimmerman**, Professor of Philosophy, University of Colorado (Boulder)

"With deep reverence for the long philosophical, spiritual, and historical journey of humanity in our first phase of conscious evolution, Dr. Chu leads us gently yet powerfully toward the Second Stage of Conscious Evolution wherein we become 'posthuman'—the next phase of evolution itself when we will witness the emergence of the 'Cosmic Being,' as Chu calls it. With almost 14 billion years behind us, we probably have even more billions of years ahead of us. Certainly current *Homo sapiens sapiens* is not the last model of evolution, on this one tiny planet among billions of others. I believe that his seminal book is essential reading for every pioneering spirit on Earth. It will help us deal with our current crises, which are drivers toward our universal future when we will head into realms beyond the imagination of most of us. Yet, in Dr. Chu, we have a wise guide, and I for one will gladly learn from his immense erudition how we can best participate in the next quantum jump.
—**Barbara Marx Hubbard**, Founder and President, Foundation for Conscious Evolution; author of *Conscious Evolution*

"Dr. Chu's wide-ranging knowledge and cross-cultural wisdom allows him to as easily quote from Darwin and Spinoza, as from Confucius and Japanese Zen Master Ikkyu. I highly recommend this book to anyone seeking a detailed and judicious yet joyful account of a Cosmic Vision toward which we all may aspire and work."
—**Michael LaTorra**, Assistant Professor of English, New Mexico State University, author of *A Warrior Blends with Life: A Modern Tao*

"Ted Chu brings an astonishing breadth of philosophical, religious, and technological reflection to bear on the most important questions we could ask: where are we headed as a species? Clearly written and accessible to the lay reader, this work adds a uniquely global perspective to futurist literature."
—**James Hughes, PhD**, Executive Director of the Institute for Ethics and Emerging Technologies; bioethicist and sociologist at Trinity College; author of *Citizen Cyborg*

HUMAN PURPOSE AND TRANSHUMAN POTENTIAL

HUMAN PURPOSE AND TRANSHUMAN POTENTIAL

A Cosmic Vision for Our Future Evolution

By Ted Chu, PhD

Origin Press

Origin Press
PO Box 151117
San Rafael, CA 94915

www.OriginPress.org

Jacket design by Diane Rigoli (rigolicreative.com)
Interior design by Carla Green (studiocgraphics.com)

Library of Congress Control Number: 2013955162

ISBN: 978-1-57983-025-0

Cataloging-in-Publication Data:

Chu, Ted, 1963-

Human purpose and transhuman potential : a cosmic vision for our future evolution / by Ted Chu, PhD.

pages cm

Includes bibliographical references and index.
LCCN 2013955162
ISBN 9781579830250
ISBN 9781579830236
ISBN 9781579830243

1. Philosophical anthropology. 2. Human beings. 3. Human evolution--Forecasting. I. Title.

BD450.C48 2014 128
QBI13-2476

Printed in the United States of America
First printing December 2013

CONTENTS

Part Four
CONSCIOUS EVOLUTION: ITS POWER AND IMPLICATIONS

Part Five
THE COSMIC FUTURE

ACKNOWLEDGMENTS

First I would like to thank my wife, Amy Zhang, who for twenty-five years has tirelessly supported me in my professional development and philosophical research while pursuing her own software engineering career and taking care of our two children. She was my first audience when we walked under the stars during beautiful Michigan summer nights.

My understanding of Western wisdom was initiated fifteen years ago by two of my former GM colleagues, Tom Walton and Mike Whinihan, both Ph.D. economists with deep philosophical learning and penetrating insights into the Western intellectual tradition. Our weekly lunches were invaluable as I plunged into the research for this book.

I also want to thank Caroline Wang and her mother for providing me with daily delicious and spicy Hunan meals while I lived on the campus of Wayne State University full time to do research for several years.

Many scholars, leaders, and experts have influenced my thinking and shaped my worldview. Their names and works are scattered through the text and the bibliography, and I will not repeat them here. In addition, I am grateful to Jim Albrecht, Paul Ballew, Gerald Barney, Don Baron, Jim Boehm, James Chen, Denny Dellinger, Mike Digiovanni, Haiwei Dong, Bill Dunkelberg, Reindert Falkenburg, Jennifer Fang, Wafik Grais, Glenn Hefner, Wei Li, Weishuang Qu, Anne Smith, Ivan Szelenyi, Hong Wei, Ed Whitacre, Kippy Ye, Menghua Ye, and Qing Zhang for their unique contributions.

The book would not be possible without the essential guidance and input from my publisher Byron Belitsos and my editors Larry Boggs, Ellen Daly, and Elissa Rabellino. Thanks also to our book designers Carla Green and Diane Rigoli.

My parents and my primary and secondary school teachers instilled in me at an early age not only the desire to excel in academics but also the inspiration to contribute to society. My Chinese name (浩全) was coined by my grandfather; it literally means "grand and complete," a wish I think about often.

Last and most important, I am grateful that I am alive, that the universe exists, and that humanity has continued to thrive despite our daily challenges. There is much that we can look forward to.

FOREWORD

by John F. Haught
Professor of Theology Emeritus
Georgetown University

The world's great wisdom traditions have always taught that the universe is the expression of a deep but impenetrable meaning. Their beliefs that the universe is here for a reason have given significance to people's lives as well as incentive to moral action. The good life, accordingly, consists of adapting our lives to a transcendental wisdom stirring at the heart of all being.

Such a benign understanding of the cosmos, however, appeared in human history long before evolutionary biology, cosmology, and information technology came along. Adherents to the ancient visions knew nothing of natural selection and astrophysics. Nor could they have foreseen contemporary scientific expectations in the fields of genetics, robotics, nanotechnology, information science, artificial intelligence, and neuroscience. These new developments now invite us to consider how far emerging technologies may go in transforming human and other animal species, or in stretching our planet's ecozoic fabric.

In the future, shall we reach a point where no clearly defined human nature—at least as conceived by earlier generations—any longer exists? New discoveries related to the human genome will make it possible for applied science to alter radically our inherited bodily and behavioral inclinations, perhaps eventually transforming the entire human form into outcomes now unpredictable. The consequences of new scientific discoveries and their technological application are uncertain at present; but sensitive artists, dramatists, philosophers, ethicists, and social scientists are now envisaging a wide spectrum of possible results, many of them intriguing, others potentially monstrous.

Ted Chu is a pioneering visionary whose futurist concern deserves close attention. He is keenly aware that technological expertise is on the brink of reshaping our world and humanity dramatically. A continually more nuanced scientific understanding of the subatomic world, the manipulability of genes, the plasticity of brains, and the rules of evolutionary change—plus a host of other scientific insights—now provides *Homo faber* with an almost irresistible opportunity to transform everything in our world radically. Those of us concerned about justice, liberation, human fulfillment, ecological integrity, spiritual existence, and the cosmic future need to read brave new books such as this one.

What implications do the emerging new technologies have for the future of the universe itself? Unlike our prescientific ancestors, we now realize that the world is still coming into being. The fourteen-billion-year-old cosmic drama recently exposed by astronomy, astrophysics, geology, biology, and other sciences is clearly far from finished. What role do humans have in determining the cosmic future?

The fourteen billion years that have transpired so far may turn out in the long run to have been only the dawn of the Big Bang universe's eventual unfolding. Is it not the case that the newly emergent opportunities for technological change on Earth are now placing the *universe itself* on the verge of undergoing an explosive new chapter in its dramatic unfolding? It is now conceivable to many thoughtful people that the human period of evolution may eventually turn out to be a transitional chapter in cosmic history. How are we to address such a prospect? In the remote cosmic future, will human history appear to be anything more than an ephemeral crossing-over to more fascinating episodes in an enormously inventive cosmic narrative whose eventual playing-out is currently unimaginable? If you care about such questions, read this book.

Four billion years ago, with the arrival of life and biogenetic processes on Earth, the universe underwent an "information" explosion. More recently, with the emergence of complex brains and minds, at least on our planet, it erupted into "thought." Then, with the arrival of social and political life, it began to undergo a cultural transformation, one that has included the invention of language, education, artistic expression, ethical aspiration, religious hope, and scientific inquiry. Currently, with the advent of transhumanist dreams, a distinctively new chapter in cosmic process is just beginning.

Does this fourfold cosmic emergence point to an underlying cosmic purpose? Today the human tendency to attribute purpose to the universe seems to be little more than an evolutionary adaptation. Biologists explain our longing for purpose as simply a matter of human genes striving for immortality. Suggestions that the cosmos has a deeper meaning seem to be nothing more than heartwarming fiction—"noble lies" that help humans adapt but that remain empirically unfounded.

Ted Chu's proposal, however, is that something momentous is indeed going on in the universe. Even in an age of scientific skepticism we have reason to hope that a new and enlivening future is appearing on the horizon. This intellectually stimulating book, at least as I see it, implements the cosmological vision of the great mathematician and philosopher Alfred North Whitehead, who locates life's evolution, as well as the story of human creativity, within a cosmic process whose aim is one of bringing about more intense versions of beauty. The cosmos has not realized this objective at every turn, but undeniably it has made its way from simplicity to complexity, from triviality to more intense versions of ordered novelty. In its adventurous aiming toward the intensification of beauty—the most sublime of all values—the universe-story shows itself, therefore, to be more than mindless thrashing about. Other readers of Chu's work may arrive at different impressions of the world process he is exploring. To me this inspiring book is a remarkable expression of a new zest for life and a provocative challenge to keep looking for an ever more liberating future.

PROLOGUE

The magnificent orca (*Orcinus orca*) can be seen as a powerful metaphor for humanity. Commonly known as the "killer whale," this name reflects the prevailing view (until recently) that the orca is a ruthless killing machine and little else. Indeed, orcas are extremely skillful and ruthless predators, capable of sophisticated acts of killing that are meticulously planned and coordinated by multiple individuals.

Yet this magnificent animal also reveals many endearing and human-like characteristics: these include a gregarious, curious, playful, and highly intelligent nature; diverse dialects; and complex social structures. Orcas, like humans, sit at the top of the food chain, unchallenged in their watery domain. They play a lot, as we do. They love to leap high into the air—and I imagine that when they do so, they may be catching a glimpse of the beautiful coastal landscape. The land, the habitat of their distant ancestors, always eludes them. Do they have a sense of wonder or longing when they look at the snow-capped mountains in the distance?

Of course, orcas don't have a reflective consciousness as we do. On the other hand, when we catch a glimpse of the stars in a summer night, relatively few of us have thought deeply about the greater cosmic issues—our place in the universe, our limitations, and our potential, all of which are the subjects of this book.

Still, we might well find inspiration from the orca. Stretching our imagination, we might wonder why the orca instinctively leaps into the air, seemingly rising up toward heaven. Does it aspire to overcome its limitations as a water-bound creature, just as we strive to escape from our biological and earthbound limitations?

In these pages we will explore how we can leap beyond our genetic limits—but, in order to do this, we need to dive down into the depths of our human nature. We are a super mammal that has evolved over eons to inhabit the top of the food chain on Earth; we are in fact the most aggressive and invasive species ever, far more disruptive than the orca. But we have also developed far more powerful capabilities to transform the Earth and even reach beyond our native planet and solar system.

Ultimately, we humans, like orcas and all other life forms on this planet, are transitory creatures of cosmic evolution. We are not "designed" to inhabit deep space. Even though our very bodies are made of nothing less than the ashes of exploded stars, this book argues that we must consciously develop far more capable *transhumans* to return to the stars.

Just as we might imagine the snow-capped mountains in this photograph as representing deep longings the orca might have, our discovery of distant galaxies enables us to look into the distant past as well as the far future of the cosmos. This much greater universe will, I believe, one day be the playground for what I call "CoBe," the Cosmic Being who will succeed us, evolve far beyond our biological limits, and accomplish things we cannot even dream of. To us, the thought of inhabiting space is very cold and uncomfortable, yet even now this cosmic potential beckons us and seems somehow attractive. We will never be able to stop ourselves from leaping toward these new horizons, for that instinct to transcend—which also animates the leaping orca—is nothing less than the impulse of evolution at work: a cosmic impulse that does not serve the narrow, conventional human quest for comfort and "happiness," but aims for something far greater than our species can fully understand.

For those who ask "Why bother?" and "Why take the risk?" we must consider why the Cosmic Creation has bothered to come up with such complicated designs over billions of years. The basis for such a question is hardly new—the attempt to understand the universe was already underway at the dawn of civilization. What *is* new is our technological capabilities, which are finally making possible a real leap into the new cosmic frontier.

AUTHOR'S PREFACE

WHAT IS THE purpose of human life? This question has been pondered by philosophers, meditated on by mystics, and grappled with by ordinary folks for millennia. And humankind has responded with innumerable answers as the tides of history and culture have swelled and retreated, bringing with them new beliefs, mythic narratives, and philosophies. And yet, in our time of exponential change and material progress, many feel we have lost sight of any sense of higher purpose. When in the late nineteenth century Friedrich Nietzsche declared the "death of God," he also expressed concern about the corresponding lack of human purpose in the modern age. He was prescient when his Zarathustra proclaimed, "A thousand goals have there been so far, for there have been a thousand peoples. Only the yoke for the thousand necks is still lacking: the one goal is lacking. *Humanity still has no goal.*"[1]

Nietzsche further stated, "The time has come for man to set himself a goal. The time has come for man to plant the seed of his highest hope."[2] This book is an attempt to plant that seed, to articulate a goal and a purpose for humanity in an age of unprecedented technological breakthroughs and previously unimaginable potentials for evolutionary progress. I believe—and intend to demonstrate in this book—that we are reaching a threshold or "tipping point" in our evolutionary path that is as radical as the appearance countless eons ago of the first biological cell on Earth. The result will be a revolutionary jump in the growth of complexity and liveliness in the entire known universe. Looking at the big picture of cosmic evolution since the Big Bang—at least that which we can infer—I am convinced that our purpose is to transcend our limiting biology and the resulting limitations in our consciousness, thus enabling the rise of new kinds of sentient beings, freed from our genetic limitations in the pursuit of the highest transcendental aspirations and the promotion of cosmic evolution. Of course, not every new sentient being furthers this cosmic vision, any more than all humans have furthered progress up to now. But there will always be pioneers whose influence far exceeds their numbers. Those who are on the frontier of expanding life and consciousness will effect the transformation and enlivening of the known universe.

This, I believe, is our calling. To say this may seem to be a leap of faith to some; but when examined carefully—as I intend to do here—it simply follows a long-existing evolutionary trend. The transcendence of our biological inheritance is a quantum step toward fulfilling our eternal longing to understand our place in the universe and the direction in which we and the universe are moving.

Today we stand at a new frontier. Until now, civilization's progress has resulted from the mainly successful and often heroic fight for greater human liberty—economic, political, cultural, religious, and personal freedoms—along with a host of scientific discoveries and technological breakthroughs that have improved our well-being. Now, having won many such battles, we have within our sight, if not yet within our reach, a radically new human freedom. It is the freedom from our inborn genetic condition—the liberation from the constraints of our biological form, such as our inevitable aging process. The human era as we have known it is coming to an end. The posthuman era is about to begin.

Such statements may produce a bewildered reaction in most of us—a mixture of fascination and trepidation, or perhaps simply dismissal. This all may seem too far from our immediate concerns. In today's culture, the larger picture of what we can be and the grand outlines of the future rarely get our attention. But as we shall see in this book, we have at times in our history been very much attuned to a larger purpose, and we ignore these issues and their relevance to us at our peril.

In this book I will address such common questions as "What does such a future portend for the human race?" and "Is 'biologically based' humanity doomed?" We will look at how the advent of the posthuman era might affect the values we cherish, such as human equality and the dignity of the individual, and even the cherished "uniqueness" of "me." But in the process we may find our very questions becoming obsolete, and we will begin to ask new questions that can shed a new, brilliant light on our future.

Looking beyond today's conception of humanity is unsettling because it requires a profound recalibration of the rules of the game. Previous generations were shocked to learn that the Earth is not flat, and more recently we've learned that time and space are interrelated in a paradoxical way, but none of that has really changed how we live our lives. We still live among human beings and interact with them, socially, economically, and politically. But the upcoming changes in human nature and emergence of new kinds of sentient beings will be truly unprecedented. They represent a radical leap in our understanding of our own species.

The idea that humanity could soon be transformed into something that transcends biological necessity naturally stirs strong emotions in all of us and creates sharply divided attitudes, pro and con. To address these questions and issues seriously, we must rely upon detached rational reflections on who we are and where we are from, rather than simply falling back on gut feelings and age-old taboos. How we answer these questions and what we decide to do as a result matters a lot to us, to our progeny, and to the universe. This book has been written to encourage such reflection. It offers you an opportunity to reflect on how we came into existence, on the nature of "human nature," and on who and what we really are in the broadest scheme of things.

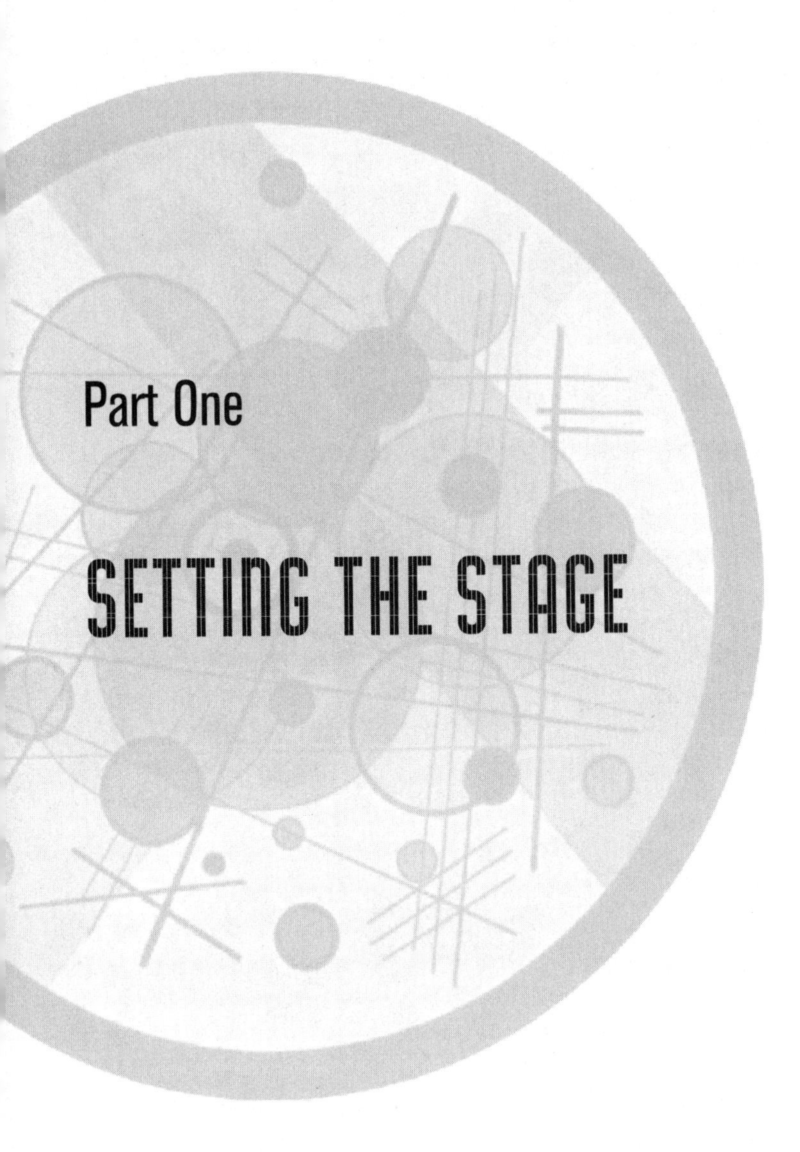

Part One

SETTING THE STAGE

CHAPTER ONE

THE SEARCH FOR A NEW HUMAN PURPOSE

Within thirty years, we will have the technological means to create superhuman intelligence. Shortly after, the human era will be ended.

—Mathematician and science-fiction writer Vernor Vinge, at the NASA Vision-21 Symposium, 1993

WE KNOW FROM the earliest historical records that man has always dreamed about self-enhancement and immortality.[1] The concept of some conscious and intelligent higher being, or transhuman, is not new—think of Nietzsche's Ubermensch (superman), Joseph Fletcher's *Homo autofabricus,*[2] Hans Moravec's "ex-humans," William Day's Omega Man, and Nigel Cameron's Techno Sapiens, to name just a few. In the world of evolutionary spirituality, such modern-day progressive futurists as the late Indian sage Sri Aurobindo, Esalen cofounder Michael Murphy, and visionary author Barbara Marx Hubbard have also conceived of a self-transcending "universal human." But outside of small interest groups such as evolutionary spiritualists and transhumanists—or science-fiction literature and Hollywood movies where the idea of intelligent robots is a matter of entertainment and diversion—the concept of radical human transformation remains unfamiliar to the public in general, and its naturalness and inevitability has not yet entered the public imagination. The public and even much of the intellectual community remain ignorant or in denial of the most significant potential changes ahead. This is starting to change in a very modest way, with technology pioneers and futurists such as Ray Kurzweil and the Singularity Institute making bold post-anthropic declarations in mainstream media.

But despite the general lack of public awareness, it is becoming increasingly clear that technologies such as cloning, stem cell research, and genetic engineering are increasing the ability to alter not just nature but human nature. Scientists are developing and utilizing technologies to modify both the human body and the human mind; and they are dreaming of going further than science-fiction fantasies in terms of the capabilities of their creations. And public debates over the ethical implications of such scientific advances are finally heating up.

Transhumanism and Its Discontents

Opponents of such fundamental change often speak of the Promethean hubris entailed in "overcoming nature" that will surely end in our self-destruction.[3] They claim that these cutting-edge innovators and their advocates are pursuing narrow interests and visions that disregard the ethical and spiritual consequences of their actions or the potential for abuse.

Along this line, prominent bioethicist Leon Kass argues that our world offers plenty of material progress yet is empty in spirit. Wisdom is in short supply today, he says, because science has abandoned the large metaphysical and theological questions.[4] "Though well-equipped (through biotechnology), we know not who we are or where we are going."[5] He repeats an ages-old warning that we are making a short-sighted Faustian bargain—gaining earthly pleasure and power today in exchange for forfeiting our core humanity.

Richard Moselle, founder of *Salvo* magazine,[6] represents a viewpoint common to many religious fundamentalists when he declares that there is a "massive cultural war" in America between two worldviews. He proposes a sharp return to the Judeo-Christian tradition against decades of corrosive scientific development grounded in the "we-are-but-a-freak-accident" point of view blamed on Darwinism. What is at stake, he believes, is the loss of human morality, purpose, and absolute truth in a world driven by faithless and materialistic evolution.

The prominent political scientist Francis Fukuyama has called transhumanism "the world's most dangerous idea" because it may undermine the egalitarian ideals of liberal democracy. In *Our Posthuman Future,* he also characterizes the debate over biotechnology as a polarized contest between two camps. The so-called libertarian camp includes curious researchers and scientists, a profit-hungry biotech industry, and Anglo-Saxon ideologues committed to free markets, deregulation, and minimalist government involvement in technology.

Other "moralist" camps consist of diverse groups often considered at odds, such as some environmentalists on the one hand and the aforementioned religious fundamentalists (who often reject environmental science) on the other—as well as an assortment of neo-Luddites who tend to be suspicious of new technologies in general, and leftists and others who may be worried about a return of eugenics.[7]

In his essay "Are Humans Obsolete?" Langdon Winner, a political scientist and philosopher of technology, suggests that the sensational forecasts of the posthumanists are difficult to accept because "their success has problematic policy implications [of], for example, placing *Homo sapiens* on the endangered species list." Winner further portrays posthumanists as elitists going down the path of being "unsocial, single-unit atomism" loners, which will threaten the inherent sociality of humans. Those who envision a post- or transhuman world, he believes, are selfish "hedonistic dreamers" because such a world would exist at the expense of time-tested principles of morality and human dignity. Similarly, Peter Lawler states that the "use [of] biotechnology to make ourselves happy is at heart an exceptionally misguided individualistic project."[8]

Heretofore, supporters of radical technological change have given up the moral and spiritual high ground to those who express moral or political opposition to such programs. Prominent among this group are Leon Kass and Michael Sandel, who argue that human enhancement is at odds with human dignity. Even ardent advocates of the posthuman future, such as Ray Kurzweil, do not disagree with the statement that "it's the material, not the spiritual gains that are seducing society down this path."[9] A recent survey seems to lend support to this point of view; it shows that roughly 75 percent of transhumanists are atheists or agnostics who do not believe there is a God or an ultimate purpose for humanity.[10] Even when they try to seek common ground with religious groups, they seem to resort to appealing to such secondary values as the benefit of research in reducing human suffering, replacing a lost child, and lengthening human lives.[11]

I will argue that these are weak arguments. For every identifiable human benefit, one can recognize some potential human harm and human risk, including the promulgation of a "genetic divide" (i.e., creating genetic inequality among humans) or even the creation of Frankenstein-like monsters. Even if a certain technology is intended solely for beneficial uses, one can always argue that it may put us on a slippery slope to some dehumanizing end. If the goal is limited only to serving humanity, or to serving the interests (or happiness) of the greatest number of people as in the ethical theory known as utilitarianism,[12] one can easily point out that it is better to stop short of resorting to the radical manipulation of the human body and the creation of intelligent robots.

Indeed, for anyone who wants to improve the lot of our fellow humans and find meaning in our lives, there is no shortage of worthy options—such as making the planet more livable for humans, eradicating poverty and disease, offering better educational and employment opportunities, and so on.[13] For more than half a century, much of the world has enjoyed unprecedented peace and progress in raising our standard of living, but there are still many pockets of violence, poverty, inequality, and injustice. There will always be a requirement that wealthy nations and individuals offer assistance to those living in deprivation. Such efforts are necessary and desirable. However, it must also be pointed out (as it is too infrequently) that there are other, critical ways of making a difference—ways that seldom cross our radar. We must step back and ask: What is the best way to serve evolution (or cosmic purpose) in the larger sense? Does evolution stop with the advent of the human on the stage of cosmic history? Can there be a higher, nobler goal than the individual's welfare and happiness? Should there be such a goal? These questions have not received the attention they deserve.

Toward a New Covenant and a New Call to Heroism

An important prerequisite to this discussion is to reconsider the heroic ideal of our classical past. Our ancient and premodern ancestors felt that they had a covenant with God the Creator (or the world or the universe itself), and not just obligations toward fellow humans. For example, in the Chinese tradition, it was long believed that one must keep *Tian Xia* (天下, literally "beneath Heaven") in mind, and commit oneself to *serve the purpose of the entire world* (以天下为己任). I believe that this sense of a greater calling, this great ideal of abandoning the self in order to embrace the larger world, lies deeply within every one of us. Just like the fearless pilots in the Apollo program, there are those of us who are still willing to sacrifice, even die, for scientific adventure. What we lack is a cosmic vision, together with a down-to-earth understanding of the problems we should tackle and the accomplishments we should aim for.

With the advent of the posthuman frontier, what is required today is no different in principle from the aims of those history-making heroes in the past. These men and women kept an eye on the future and persistently sought the most ambitious goal they could identify themselves with; and then they devoted themselves to realizing that goal. As author Charles Murray says, we may outstrip our forebears in wealth, creature comforts, health, and lifespan, but a person or a culture that is unable to compete with the past's greatest expressions of human spirit is in some sense a backward one. In fact, we are now equipped to identify bigger goals than were even known by those who shaped the world we live in today. But few of us have the ambition or insight to "belong to the ages"[14] rather than just to the moment.

For today's political, social, cultural, and economic leaders, there seems to be little aspiration to move beyond the current "gold standard": the development of a strong and democratic state combined with a market economy that is built on the rule of law and private-property rights. That goal was largely achieved, at least in parts of the world, in the twentieth century after heroic fights and huge costs. In light of the recent (and in many places continuing) financial crisis, it is clear that much more has to be done to improve this political and economic framework in developed markets, to spread it to the rest of the world, and to defend it against antagonistic regimes around the world. The current political and economic regimes must also continuously adapt to structural changes in technology, culture, demographics, and the environment.

Yet all these seemingly unending challenges should not blind us to what is really earthshaking: the sheer fact that accelerating scientific and technological progress is launching us into a future that transcends such narrowly human concerns. Half a century after the discovery of DNA—"the secret of life," as its codiscoverer Francis Crick called it—we managed to bring out the first test-tube baby, Louise Brown, in 1978. Barely two decades later, in 1997, we produced "Dolly" the sheep, the first cloned animal. Then in 2003, virtually all (99.9 percent) of the human genomes were deciphered. In March 2007 the first human/sheep chimera (created by injecting human cells into the fetus of a sheep), which allows us to harvest organs for human transplants, was created; and in the same year, two groups of scientists discovered a way to make adult skin cells act like embryonic stem cells. The first man-made bacterium, "Synthia," was submitted in May 2007. In 2009, amputee Pierpaolo Petruzziello became the first person to control a biomechanical hand (connected to his arm nerves through wires and electrodes) through thoughts alone. In 2010, the first man-made cell was created through "rebooting" a simple microbe by transplanting into it a set of genes that were built from scratch. It was also announced that a draft sequence of the ancient Neanderthal genome was created. And in advancing "mind control," an artificial neural connection was created to restore paretic hand function after a spinal cord injury in 2013. Finally, with the processing of what has come to be known as "Big Data," we have created more data over the last few years than in all previous human history.

Today, the relative strength of a nation is determined by whether it can establish a stable system in the light of the well-understood tendencies of human nature. But it is this very human nature that will soon be subject to change as we gain the exact genetic knowledge of what makes us human and how to manipulate it. It is becoming increasingly clear that the proven path to prosperity and power will be forever changed in the post-anthropic era, with radically different participants and stakeholders in economic, social, and political life. It is just a matter of time before the prospect of a radically new human nature and even newly engineered sentient beings appear on the radar screens of political leaders. I am talking not just about enhancements of existing human capabilities but also about intelligent beings with totally different motivations, instincts, and moralities, just as part of human nature is so different from that of other animals.

We have reached a threshold: modernity and science seem to be putting the human species in danger, but at the same time they are offering the unprecedented, seemingly incredible promise of a quantum leap in the human condition. As we stand at the posthuman frontier, should we retreat by actively curtailing further progress, or instead bravely and responsibly embrace the challenges ahead?

I firmly believe that the notion of humanity's place in the universe must be brought to the foreground in order to answer Nietzsche's still-unaddressed concern about the "death of God" and the corresponding lack of human purpose in a scientific age. In this book, I suggest that we need a higher goal, a new covenant with that which is greater than ourselves, if we are to move toward and realize our greatest purpose and potential. And there is little we can find in our current culture that promotes further cosmic evolution through human transformation and the creation of higher intelligence and nobler aspirations.

In this postmodern world, we have achieved tolerance and inclusiveness to an admirable degree, but that should be a beginning, not an end, of our achievement. In fact, our most overarching human objective seems to have been reduced to the simplistic notion of the maximization of human happiness, which requires little more than that everybody be nice to others and "do no harm." This feel-good goal is at best egocentric complacency, with the implicit assertion that humankind is an end in itself. It plays safely and pleasantly to the lowest common denominator of human desires. At worst, however, it represents a decadent self-indulgence that flatly flies in the face of what we now know about how the universe evolved, where we came from, our place in evolution, and what our potential is—all of which we will discuss at length. I therefore argue that peace, prosperity, harmony, and human happiness are not ultimate ends in themselves; they become great goals only in the context of how they can serve a larger, cosmic purpose—which I will refer to as the one true Creation. Although I do not pretend to be able to fathom this purpose in its totality, or what the next great leaps in evolution and consciousness will "look like," our ever-increasing understanding of science and evolution has given us some tantalizing clues. We can say that we are part—or maybe even a focal point—of an amazing cosmic process that started with the Big Bang more than thirteen billion years ago. We can also say that the effort to consciously participate in the unfolding of this grand evolutionary process is the greatest purpose we can find and identify with, and it constitutes a new kind of heroism for our time.

Since the dawn of civilization, when the awareness of the individual self was first articulated, human history can be characterized as an ongoing struggle to achieve economic and political freedom and to eliminate organized tyranny. Now that these goals have been largely achieved in the developed world, I believe we are ready to fight for transcendental freedom from the genetic tyranny that natural history has imposed on us. But what human wisdom can we rely on to inspire and empower us in this fight? In this book, I will reframe some of the ancient wisdom regarding humanity's place in the universe in terms that are not laden with myth and superstition, but rather are appropriate to a scientific yet still spiritual age.

Of course, this evolutionary, cosmic awareness is not shared by everyone. It was and remains just one among many metaphysical speculations of the ever-restless human mind. Yet, as I will attempt to show, there are good reasons why the further evolution of the Cosmic Creation should be our highest goal. This involves not merely abstract awareness, but full, active engagement. It is not enough to say that every piece of human knowledge and culture is a reflection of reality and hence worth our respect. There is only one universe and one truth regarding its fundamental nature, which is revealed to us bit by bit through our practice. We got to where we are today by submitting ideas to rigorous tests in debates and real-world practices. Bad ideas are discarded and perhaps forgotten, good ideas are allowed to spread and drive decisions, and new ideas are encouraged to emerge. This evolutionary process never ends. When we cooperate with this process, our culture and civilization advance; when this process is suppressed, the culture or civilization becomes out of touch with reality and irrelevant.

This seemingly simple insight is far from self-evident and has eluded human minds for much of history. But once we recognize this evolutionary process for what it is, we can see that it is the best human idea ever and should be the guide for setting up our highest goal. After all, it is the way nature works, for we have now discovered the extraordinary process by which energy and light gave rise to matter, matter to life, life to consciousness, and conscious beings to culture, technology, and all the miraculous creativity and ingenuity of our civilization. It seems to be self-evident that future sentient beings with expanded conscious capabilities will be "natural" extensions of this cosmic process.

CHAPTER TWO

EVOLUTION: THE BEST IDEA EVER

The most extraordinary fact about public awareness of evolution is not that 50 percent don't believe it but that nearly 100 percent haven't connected it to anything of importance in their lives.

—David Sloan Wilson, *Evolution for Everyone*

THE PHILOSOPHER DANIEL DENNETT writes, "If I were to give an award for the single best idea anyone has ever had, I'd give it to Darwin, ahead of Newton and Einstein and everyone else."[1] Dennett is right, and perhaps more so than he even knows. I would argue that the idea of evolution is the key to liberating our spirit not just from outdated worldviews but ultimately from the constraints of our biology as well. The key word in answer to all the questions confronting humanity is *evolution*—nothing has stood still and nothing ever will. The meaning of life and our existence cannot be properly contemplated without the concept of evolution, especially if evolution is understood on a scale that is as large as we can stretch to embrace.

In order for us to rationally consider the evolutionary frontier at which we stand, it is essential for us to find a narrative, a goal, a purpose that is greater than merely our own happiness. In fact, in order to shed any light on the meaning of the self and of our humanity, our point of reference must be outside of it. Assuming a broad evolutionary cosmic perspective offers us this possibility. Hence, our emphasis is not just on human history, but on the link between the history of humanity and that of the history of the universe; not just on human relationships, but on the link between the human condition and the cosmic future. Only from this greater perspective can we break free of our instinctive recoil and cultural taboos and consider our posthuman future from an informed yet open-minded position.

The emerging notion of "Big History," the increasingly detailed picture of cosmic evolution, is becoming clear and unshakable in our time. This new picture of the universe is, I believe, free of supernatural elements and arbitrary divine interventions, yet I also believe with Albert Einstein that it nonetheless reveals a transcendental purpose. What Einstein saw as a "God who reveals himself in the orderly harmony of what exists" is also true from the Darwinian perspective. Mind you: this revelation does not necessarily justify a belief in a personal God. Rather, it warrants what Einstein called "admiration"—an "unbounded admiration for the structure of the world so far as our science can reveal it." For me, this same wondrous structure is revealed in the track record of the evolution of life over the past billions of years and the forward momentum it is generating. Evolution is never a straight-line arrow, but from this new vantage point, an expansion of our concept of human spirit makes sense because it fits right into the ever-expanding flow of Cosmic Creation.

Confronting the Deeper Implications of Evolutionary Theory

The biologist Ernst Mayr hailed the 1859 publication of *On the Origin of Species* as "perhaps the greatest intellectual revolution experienced by mankind."[2] One hundred and fifty years later, we still have not fully recognized the significance of the evolution theory developed independently by Darwin and Alfred Wallace,[3] despite the ongoing philosophical, theological, and scientific research on its wider implications.[4] The significance of Darwin's idea of evolution is not the mechanism of natural selection per se, which Darwin himself summarized as "multiply, vary, let the strongest live and the weakest die"; this concept is just one important piece of a complex picture even within the subject of biological evolution (sexual selection is another). In other words, the Darwinian concept of evolution is not limited to biology or the variation-selection-replication mechanism. What is so startling is the larger nature of the process that his idea of evolution entails, especially in light of the modern cosmological understanding of how deep and dynamic the universe is. That is why the concept of evolution is so dangerous to some of our instincts and ages-old cultural dogmas.

What startled Darwin's European contemporaries was neither the idea of species change nor the possibility of humanity's humble origin, but rather the vision of a seemingly purposeless process without a guiding hand toward preordained perfection. Without doubt, Darwin challenged some of the most basic beliefs in his society and perhaps today's as well. But the real issue between Darwinists and creationists as I see it is whether the human species is part of a grand process or a deliberate one-off design—and whether novelty has continued to emerge freely and spontaneously in cosmic history. The crucial question now becomes: If novelty endlessly arises in the course of time, how can one be sure that humanity is the pinnacle of the entire cosmic evolutionary process?

The evolutionary answer is: the human is no perfect being, and there could be better things ahead! Compared with the 13.7-billion-year-old universe that contains at least billions of galaxies, our existence is of absolutely negligible magnitude, both physically and temporally. Humanity has yet to comprehend this implication and shift its attention from an inward-looking fatalist *being* to an outward looking, willful, autonomous *becoming*.

Since at least the time of Aristotle and Confucius, we have struggled to define what life is.[5] Currently, the best abstract description of life may be the unifying principle of "supple adaptiveness"—the unending process of producing novel solutions to unanticipated changes in the problems of surviving, reproducing, and flourishing.[6]

But a more comprehensive answer to the question "What is life?" requires attention to multiple levels of meaning. The basic definition scientists give is that life is a carbon-based structure that features self-replication, metabolism, and information processing. This is probably the lowest common denominator for all existing life on Earth. If we rise to a higher level of analysis and ask what life is capable of, one can say that it is capable of giving birth to complex living organisms with a central nervous system called the brain. And further, given the evolution of more and more complex brains, life is capable of allowing consciousness to emerge. Further, when asked what consciousness as such is capable of, I would say it can consciously design new forms of intelligence that will embrace the cosmos and then fundamentally alter it. In other words, life and mind are capable of self-transcendence.

Looked at from this perspective, the human being is neither the final product of the universe nor merely an accidental happening on a little planet. It is exactly these deep implications of Darwin's ideas that are still misunderstood, ignored, or rejected today. Almost all academic and popular books and discussions about natural and cultural history, including those that promote an evolutionary understanding of history, have narrowed themselves to the explicit or implicit purpose of teaching us about how to make human lives better.

There are multiple reasons why the cosmic implications of evolution are hard to grasp. Understanding evolution itself is difficult, and following through on its implications, especially for humanity itself, runs against some of our strongest intuitions and emotions. For instance, we feel bad, and very rightly so, about our historical wrongs—war, genocide, slavery, torture—and yet they have been part of the evolutionary process. Of course, we should prevent those inhumane activities, and we have largely succeeded in modern society. These efforts and successes in preventing them are *also* part of the evolutionary process. Nevertheless, the process of elimination goes on as evolution dictates; only the unit of selection has changed. Today we move forward by changing or eliminating failed governments, useless institutions, bankrupted businesses, and obsolete products.

The Darwinian concept of "master force" (creation and selection) is often emotionally hard to swallow—and that is partly why creationism is experiencing a revival today, not only in the U.S. but also around the world.[7] Yet we now possess overwhelming tangible evidence of such a principle from many different fields in the form of the "modern synthesis" of the twentieth century, which integrated new knowledge from embryology and molecular biology with traditional elements such as paleontology. Still, understanding natural evolution remains an intellectually demanding task, partly because it can never be shown or proven in the same manner as, for example, proving that the Earth is round with a simple satellite picture. Indeed, skeptics of evolution point to the fact that no mutations have ever been observed that have converted an animal to a markedly different one, for example, a fly to a wasp.[8] Louis Agassiz, the eminent nineteenth-century zoologist, discoverer of the ice ages, and one of the "founding fathers" of the modern American scientific tradition, expressed typical grounds for disbelief upon hearing Darwin's natural selection theory: It is impossible that God would have created the magnificent living world by random variation from "grubby ponds and woodlots"!

Social, economic, and cultural evolution are equally difficult to understand. Witness the general ignorance (until recently) of the role of technological and organizational innovation as a crucial factor driving the evolution of civilization. Innovations were once considered dangerous accidents (comparable to earthquakes or hurricanes) to their models of static equilibrium and harmony.[9] In comparison, Lamarck's theory of the inheritance of acquired characteristics has much more appeal to common sense, as it fits our lifetime observations of cultural transformation.

Such difficulties are part of the reason evolution remains controversial today, and we can only imagine the courage and perseverance needed by Darwin and his supporters at the time when the theory was way ahead of the natural sciences from which it sought support. Darwin himself had to apologize for the lack of support from the fossil record,[10] and he later admitted the possibility of defeat after William Thompson (Lord Kelvin) published calculations concluding that the probable age of the solar system was only around 25 million years (based on several erroneous assumptions). Others concluded—again erroneously, based on a misunderstanding of the nature of hereditary material, or DNA—that new traits emerging from random mutations would be diluted very fast in the population. These faulty conclusions at first looked persuasive, and they seemed threatening to evolutionary theory because they reduced the necessary time span for natural selection to an impossibly short period.

But still, the idea of evolution by natural selection was never completely discredited in the early years of Darwinism. Ideological developments in nineteenth-century Europe—especially the advent of laissez-faire economics, British utilitarian philosophy, and the ethics of rugged individualism—provided helpful preparation.[11] In biology, Georges Cuvier pioneered anatomical research that proved crucial. By classifying all animals on Earth according to four fundamental internal body structures, he stimulated a revolution in morphology and taxonomy; soon, primitive societies were discovered in overseas explorations, and the striking physical similarities between humans and the great apes were first noticed. Biologists in the German idealist tradition argued that embryonic developments recapitulate the history of life on Earth, evolving toward a divine final goal. By the end of the century, the idea that species are not immutable was becoming thick in the air, although the origin of species was still generally thought to be a result of supernatural divine intervention or the catastrophic shakeup of the early environment.

In addition to learning from their intellectual forebears,[12] Darwin and Wallace were aided by the contemporary advances in comparative anatomy and by their ability—unlike earlier thinkers in Europe—to easily travel the world to gather firsthand observations and access the latest paleontological finds. Darwin also learned from animal breeders the power of artificial selection. The most important insight for both men was probably their discovery of the wide geographical distribution of species and the competitive patterns they noticed in South America. As Marx wrote to Engels in 1862, "It is remarkable how Darwin rediscovers among beasts and plants the society of England, with its division of labor, competition, opening up of new markets, inventions, and the Malthusian struggle for existence."[13]

What makes evolution such an elegant and powerful idea is that it explains how organized complexity can arise spontaneously from primeval simplicity.[14] But the most significant implication of the theory is not the well-known proposition that "man came from apes," although one of Darwin's major insights is the common ancestry of all biological organisms. We must instead recognize something far broader: that *evolution is how the world functions and creates*. Evolution is in fact a profound cosmic process, not just a theory of how biological species were originated and evolved;[15] for evolution is the secret of how the structures of our brain are determined,[16] how our immune system functions, how animals learn, how our habits and beliefs form, how languages and computer codes advance, how social organizations and institutions unfold, how cultures and civilizations develop, and how the universe operates in general.[17] Some scientists have even speculated that evolution explains how the structure of the universe itself was generated, perhaps by "cosmological natural selection" with natural variation coming from the quantum process.[18] In this connection it is deeply satisfying to see how what Gary Cziko called "Universal Selection Theory" explains so much of what used to require supernatural explanations or resort to miracles.[19] In the final analysis, it is evident that few areas—perhaps none—of human thought have been unaffected by the theory of evolution.[20]

Moreover, the principle of evolution offers a new context for answering the deepest questions we can raise. Why are we here? Why has life continued to survive and thrive after billions of years? Why is there good and evil? Why is happiness so elusive? Why do we have to die? Why do we live in a vast, mostly empty and cold universe that is seemingly devoid of life except on Earth? Every one of us has at least touched upon such questions.

We now know that such inquiries and deep concerns cannot be fully addressed without referring to the evolutionary history of the universe. Furthermore, evolutionary history can show us things we seldom look for: Modern science and evolutionary theory help us to see, as far as we can, the world as an interconnected whole and as an endless drama unfolding. They help us to find our place in the whole process. Evolutionary theory does not answer the "ultimate" question of why the universe exists—nothing *within* the universe can—but it is the best explanation we have to answer the question of how highly organized and complex structures like human beings can arise through spontaneous, random actions at much lower, simpler levels. Evolution is a nonmiraculous process that has never ceased functioning in universal history, and there is no reason to expect it to come to a grinding halt in the future. Taken apart and examined in detail, evolution seems to be a chaotic mess; put together, the entire process makes sense and reveals a directional grand pattern in spite of its random nature.

Conscious Evolution: The Newest Wave in an Evolving Cosmos

If we think of evolution as one great tide moving forward, there is always a leading edge, breaking faster and faster into novelty, even as the rest of the wave moves through territory already covered. For much of cosmic history, that leading edge was the slow unfurling of matter. With the emergence of life, evolution's frontier shifted into a new domain and began to gain momentum. And once life developed to the point of producing human beings, with their capacity for self-reflection and the creation of culture, the edge of evolution once again found a whole new domain of expression. With the advent of comparatively lightning-fast cultural evolution, natural evolution on Earth lost touch with this emergent frontier of cosmic evolution. Although DNA-based natural species continue to come and go, their situation has long become largely irrelevant to the movement of human civilizations.

Over the last five thousand years of cultural evolution, there have been major changes in humanity's notions of its place in the cosmos. Now cultural evolution is also fast serving out its role on the evolutionary frontier. We humans will soon cease to be on the frontier. The mathematician Vernor Vinge declared in 1993 at the NASA Vision-21 Symposium, "Within thirty years, we will have the technological means to create superhuman intelligence. Shortly after, the human era will be ended."[21] His timing could be off, as the pace of innovation is unpredictable, but not the direction. Although human activities will continue to generate new art, science, and real-life social dramas that are fascinating to humans, the cosmic evolutionary frontier will belong to our posthuman descendants—those with superhuman intelligence.

Natural history comes in waves, and so does human history. According to an ancient Chinese saying, each new wave in the Yangzi River moves with a fresh and powerful force, superseding the previous one (长江后浪推前浪). But the very success of the newest "wave" in realizing its potential ultimately renders itself out-of-date and even reactionary. Today we are at the end of the great secularizing movement that started with the Renaissance in the fourteenth century. That movement placed humanity at the center of the universe and demanded maximum freedom for the individual. The European Renaissance led to an explosion of human creativity in the arts, sciences, and technologies, first in the West and now around the world. But by throwing away the baby along with the bathwater, mainstream secular humanism self-limited its scope of activity to the mere notion of human cultural and economic improvement in pursuit of its self-declared goal of maximizing human well-being and happiness.[22] This humanistic enlightenment can lift itself to a higher level by retrieving ancient wisdom and redefining the meaning of human freedom, or else risk being left behind by the newest wave.

The posthuman future is not *about* us per se, but it is *up to* us to make it happen. Humanity as an end in itself is hopeless, yet all hope for the future resides in humanity. Again, this is best expressed in Chinese: we must "see deeply," but not "see through" humanity and fall into nihilism (看透, 而不是看穿人生).

The conscious mind is our most important tool in this work. For while it is largely a slave to the instinctual pursuit of personal happiness, it is not a helpless puppet; it can also develop a detached perspective on our emotions and motivations. For example, it understands the purpose of establishing abstract rules and institutions to restrain inappropriate impulses. It is through this same capacity of self-discipline that we can confidently claim that we, as self-conscious intelligent beings, are not merely passive participants in the pursuit of personal freedom and happiness, but active directors in the drama of cosmic evolution. Humanity's conscious efforts to transform itself in this way can be called *conscious evolution*, and conscious evolution is the new wave beyond natural and cultural evolution.

Evolution's trajectory clearly shows us that the posthuman wave is a logical next step in the cosmic unfolding. But what will it take to get humanity to accept this new understanding of place and purpose? I believe a spiritual renewal is needed, one that calls for a new "cosmic faith" and a fundamental and potentially heroic elevation of human aspiration. The ultimate meaning of our lives rests not in our personal happiness but in our contribution to cosmic evolution—a process that transcends the human and yet is integral to who and what we are in the universe.

Our images of the world and of ourselves have always guided our actions. In pre-evolutionary thinking, we envisioned a predetermined "Great Chain of Being" with single-celled organisms at the bottom, animals a little higher, humanity somewhere in the middle, the angels another level higher, and God at the top. This static view of the world has now been replaced by an evolutionary view of the universe—a universe that started simple, became more diversified and complex over time, and eventually gave rise to humanity and its civilizations. The old static vision is mistaken, but it may still serve the purpose of symbolizing a world above humanity, a "heavenly" world that has yet to be realized. It is in this sense that the evolutionary worldview need not end with humanity. The universe we live in is complex enough to give rise to human consciousness. And now it is up to our consciousness to decide whether the universe's potential for organized complexity (the "design space," as Daniel Dennett calls it) has been exhausted with the creation of humanity.

There is a time for everything. Today, the time is right, both socially and technologically, for a conceptual revolution and spiritual renewal. We can see a striking parallel between today and Darwin's time as I just described—rapid technological innovations and scientific discoveries on multiple fronts, exposing the colossal inadequacies of a slow-moving human mentality and worldview. The outward windows to the past (geology and archaeology) and the non-European world (geography) were opened in Darwin's time; the inward windows to the human body (genetics) and human mind (neuroscience, evolutionary psychology) are opening to us today. Now, as then, there is a relatively stimulating environment, with a free flow of ideas and integration of the global culture and economy.

The divide articulated by Fukuyama, Jeremy Rifkin, and other anti-posthumanists is not real from a higher perspective. The debate is not "for" or "against" the human. We all cherish and appreciate human existence. The question is about *how* to value that existence.

In the grand scheme of things, humanity's value is in what it can do that other species cannot. The hero's value is in what he or she can do that others cannot. Let us listen to Nietzsche's voice again: "The time has come for man to set himself a goal. The time has come for man to plant the seed of his highest hope."[23] Let us cast our eyes beyond what nature has done, to the posthuman era that seems both possible and improbable. Cosmic sentient beings with greater intelligence and wisdom than humans are certainly possible, since they would not contradict any known natural laws. They are improbable because only one species on Earth has evolved the ability to create them. This is humanity's true place in the universe!

Not Any God, but a Cosmic God

"The world must have a God; but our concept of God must be extended as the dimensions of our world are extended," wrote Pierre Teilhard de Chardin, a priest and paleontologist who became one of the great evolutionary philosophers of the early twentieth century. The anthropomorphic God can no longer withstand the scrutiny of the bright light of science and reason; but that does not mean we live in a Godless universe. In fact, what I call the "Creator God" actually comes into sharper focus under the light of science and reason, and gives humanity far more dignity and responsibility than our forebears ever imagined.

When we talk about God, we may be invoking many different conceptions. As we shall discuss in greater detail in later chapters, it is critical to differentiate between the anthropomorphic God (a personal God that thinks and acts like a human) and the cosmic God (a God that is responsible for the fundamental nature of the universe). The cosmic God as I will describe it is deeply satisfying in that it still retains (in a far more sophisticated form) characteristics of God with which we are familiar—law-giver, life-creator, and wisdom-giver. Even the reputed diehard atheist Richard Dawkins emphasizes in *The God Delusion* that he is only going after the anthropomorphic God.

The unfolding of cosmic evolution, especially the evolution of life and mind, can be explained by intelligible persistent patterns, or natural laws, without reference to arbitrary change of natural laws or inelegant "intelligent design." Although we are far from being able to explain everything yet, all historical evidence suggests that we are a product of this ongoing, impersonal, but coherent cosmic process. Biological science tells us that if there is any biological species that has been designed, then that designer cannot be said to be very "intelligent" even by human engineering standards. The intelligent-design defense of a human-centric God denies the ultimate God that is cosmic in scale and infinite in creative potential.

Contrary to popular belief, science in no way contradicts the idea of a greater cosmic God. In the tree of life on Earth, we are just a single, albeit vitally important, twig newly emerged on the top. Yet just as we can discern the general direction of a tree's upward growth toward the sun, our efforts to expand upward could make a difference far beyond our roots on Earth. This general trend—what Alfred Wallace called "successive appearance of higher and more complex forms of life"[24]—is the strongest hint as to where we should put our efforts and find the deepest meaning of our existence. We are masters of the universe, not in the sense that the universe was put in place and designed for us, but in the sense that, going forward, we can determine our own destiny and that of the universe. Under the natural laws, we have all the freedom we need. The ultimate freedom does not belong to the guts or emotions but to the higher consciousness, which can understand why we have certain instincts and feelings and act according to what it sees as right.

We tend to feel that we drive social and historical developments, but more often than not, it is unexpected technological breakthroughs and unintended consequences of certain innovations and organizations that initiate and invigorate the evolutionary process. The emergence of new sentient beings will follow the same surprising but thrilling dynamics.

Collectively, moving our highest concern from human happiness to cosmic transcendence is very difficult, but *not* impossible, for two reasons. First, we are not forcing ourselves to give up our human goals, since the pursuit of happiness is the most realistic way to realize the higher goal. Second, deep down we really do have what it takes to open a drastically new era in cosmic evolution.

The psychologist and philosopher William James suggested that the deepest human need is to be appreciated. The philosopher and educator John Dewey said the deepest urge in human nature is "the desire to be important." According to Abraham Maslow, the highest human need is self-actualization—to gain a sense of freedom, confidence, independence, and most important, the enrichment of fulfilling potential. The call for freedom has been a dominant theme in recent history, particularly in the twentieth century. Yet, in a profound sense, we are still unfree. After gaining economic, political, and religious freedom, we still cannot control one crucial aspect of our lives: Every one of us was born into a fixed genetic endowment that is not of our choosing, not even of our parents' choosing. In addition, we are each born unequal, with different (and often profoundly limited) physical, intellectual, and moral capabilities. And we are all subject to the limits of the "human condition" in general, including the inevitability of death.

Freedom from a fixed human nature that is conditioned by a particular historical path and is far from perfect will be the greatest and final freedom for humanity.

To summarize, humanity is not the ultimate creation of God (or the glorious pinnacle of evolution), nor is it just an accident ("the naked ape" or "the third chimpanzee"). Humanity is indeed special and unique. We can even say we are "chosen"; but we must ask why humanity was chosen and what it was chosen *for*. Humanity has to create to *earn* that "center of the universe" status and recognition, and we should never accept the human being *as it is* in a universe that is in eternal flux.

The freedom from human genetic bondage is in our hands. This is our frontier. Let us step forward with joy, courage, and responsibility.

CHAPTER THREE

IMAGINING A POSTHUMAN FUTURE

Man may be excused for feeling some pride at having risen, though not through his own exertions, to the very summit of the organic scale, and the fact of his having thus risen, instead of having been aboriginally placed there, may give him hopes for a still higher destiny in the distant future.

—Charles Darwin

EVERY NEW FRONTIER is, by definition, unknown. Every new exploration may end up with unexpected discoveries. And every discovery implies at least the potential of a radical shift of what we know about, and how we see, ourselves and the universe. Nevertheless, as we have already seen in the preceding chapters, we can learn much from what we already know. The processes of evolution and the methods of discovery have already taught us much about what we can expect. We can have faith in the posthuman future because, if we examine what we know today and keep an open mind, we can perceive a movement in the universe toward higher levels of complexity and consciousness that is as convincing as experiencing the power of love or the beauty of the natural world. Once we glimpse that possible future, even without knowing its unknowable details, we will want to open ourselves to it in whatever ways we can. Having contemplated its import, we will feel compelled to *act* in ways that maximize our ability to navigate and influence this new reality.

Of course, accepting and acting upon such a vision may not always come easily. Some people, when they hear my ideas, have remarked to me that, even if I am right, this possible future is too distant to have relevance and meaning for them. How can what I do have any relevance to the fate of the universe? I am reminded of the wit of the nineteenth-century British scientist Michael Faraday. After presenting his findings on electricity to the Chancellor of the Exchequer, William Gladstone, the chancellor asked, "Of what use is it?" Faraday replied, "One day, sir, you may tax it." The poor chancellor may not have appreciated that leap of thought, but we can laugh and appreciate it with the benefit of hindsight. The point is that totally unexpected things can happen in very strange ways. Our individual circumstances vary, but we can always find ways to align our personal and community needs and goals with the larger mission.

At the time of this writing, I have a four-year-old son, Megene, who goes to a daycare center on weekdays. He does not like the fact that his parents have to go to work, but does not have the slightest care about what they do at work. My seven-year-old son, Zenny, is in second grade. Zenny is sometimes curious about what his parents do at their workplaces. But it is hard to relate my work life to his personal life, other than the observation that parents need to make money to buy the stuff we need, that they help companies make things to sell so that they can pay workers' salaries, and the like. Any more details about how I work as an economist and the other aspects of the work process, and his mind wanders off. It always amazes me to entertain the thought that I was like them at one time, and I am *still* like them concerning many aspects of natural and social reality about which I am ignorant or indifferent. It is fair to say that even for most adults, the future of the universe beyond humanity has hardly any relevance or interest. On the other hand, just as there are very curious children, there are people who are fascinated by the possibilities of a universe in which humanity creates its own "post-anthropic" principle.

None of this says anything about a future beyond the human level and thus beyond human understanding. Even if there were a messenger who came back from the future and told us about it, I don't think anyone would understand it other than the big picture: intelligence and spirit have filled the universe, and new horizons invisible and unimaginable to us have opened up. What is more, from the vantage point of the cosmos, the scope of our vision is very limited, individually and collectively, and will remain so as long as we remain human. Thousands of years of cultural and technological evolution have done wonders to increase our knowledge of many things and our ability to apply that knowledge to make a difference in our lives, but all of this has had only marginal impacts on our intuitive perceptions of the world. Yet, that says nothing about future evolution beyond human beings and the *potential* for our spiritual descendents to make sense of it all and to expand intelligence and spirit beyond our wildest dreams.

When my children grow up and experience the world of work, they will understand what my wife, Amy, and I do. Even today, if they got a chance to sit with me in my office for a week, they would have a better chance of learning what I do, especially if I used some simple metaphors or stories to explain what's going on. By the same token, I have confidence that most people have the capability to understand much more about the cosmic direction when they pay attention to what modern science and research have come up with.

I have already noted that the very notion of a posthuman future with cosmic intelligence is viewed by some as pure hubris. They argue, for example, that any attempts to realize this vision will do more harm than good—that we are likely to destroy ourselves, and maybe all life on Earth, by trying to "play God." The fear of possible competition between natural humans and superhumans (or other autonomous intelligent beings), which has often been depicted in science-fiction novels and films, is a big reason why some people recoil from human cloning or "designer babies." Can we realize the vision I propose in this book without suffering grave unintended consequences, such as massive human suffering or even self-destruction? I will endeavor to show that we can.

Everybody is endowed with a deeply ingrained fear of uncertainty, and all great enterprises involve risks. Yet, knowing the risks, the possibilities of loss or damage, should not necessarily make us freeze in our evolutionary tracks. Instead, we need to take the broader, longer view. Only when we stand on a higher vantage point can we gather enough courage to face and then overcome the existential fear that arises when we perceive ourselves as so minuscule and powerless in front of the vast universe.

The essence of evolutionary theory is the pure trial-and-error nature of the process. (This is the opposite of what is told in almost every creation story before Darwin, in which there is a process more analogous to conscious design.) The evolutionary process always involves throwing up a massive array of alternatives and then "ruthlessly" discarding anything that does not work. This trial-and-error process is the same for the emergence of anything new, whether it is a new species, a new scientific idea, a new tool, or a new social structure. Everything truly new has been created that way.

In another sense, however, evolution is not "ruthless" but rewards cooperation. We have seen a lot of discussion and stories centered on the question of whether selfish superhumans (or human-like machines) will overpower the unselfish ones. Here, natural and cultural history both show the counterintuitive results: The ones who cooperate as if they are endowed with the big picture tend to prevail, under the aegis of what the mathematical biologist Martin Nowak calls the altruistic law of evolution and what evolutionary theorist Robert Wright calls the non-zero logic of human destiny. Evil, with its biological origins in predator/parasite behaviors, can cause huge damage and suffering, but *it has never prevailed in history.* Why do we want to bet that billions of years of evolution will be structurally different if we continue to follow the principles underlying that process? The confidence at the macro level, of course, should not prevent us from being cautious at the tactical level and recognizing the utter importance of self-preservation.

Concern over the human race's potential for self-destruction has much merit, and cultural diversity should be protected with great effort, though ultimately not for its own sake. But the neo-Luddites who advocate giving up what technologies we have to return to a "simpler, gentler" life are directionally wrong: the desperation of the human condition does not call for a retreat from civilization or a divorce from the evolutionary process—that would be truly "against God" (or "Mother Nature," as some prefer to call it). This is not difficult to understand if we examine natural and human history to see God's or Mother Nature's track record. Instead, what is truly suicidal is *betraying the very process that brought us into existence.*

Still, evolution is disturbing from the individual standpoint, since it is a massively "wasteful" process that inevitably deems the majority of new variations from the previous generation unfit. But if it can be seen as a callous, wasteful process, it is also an effortless, delightful, no-end-in-sight process as compared with the utopian account of an eternal stationary bliss, in which all is accomplished with nothing more to anticipate.[1] Evolution works beautifully along with the coordinate action of underlying natural laws and operates without any arbitrary interference from external or "metaphysical" forces, and will continue to work that way in the creation of more beautiful novelties. Contrary to our egoistic intuitions, there is no risk or harm inherent in the broader evolutionary process itself regardless of what we might do.

What can we do as individuals and as a society that furthers the principle of evolution? For one thing, we can set up appropriate political systems and promote a culture of energetic creation. When we look at the most powerful and creative political systems and mass cultures—the liberal democratic market system, for example—without exception they all rely on individual motivations. What makes individuals happy invariably motivates them to work harder, to take risks, and to endure temporary setbacks. To be happy, people must have hope for *their* future, have protection of *their* personal rights, and most important, have meaning in *their* lives, regardless of how the meaning is defined and accepted by others or by society as a whole.[2] We will go into details about these hard-earned lessons from human history, but they are really lessons of natural history as well. The entire evolutionary process appears to be spontaneous and creative because it is all about energizing the individual, about depending on the individual's own initiative to move the collective whole forward. While the individual is not the end of evolution, evolution has to be furthered by the individual.

3.1. What the Posthuman Future Is—and Is Not

The posthuman future will be made possible by an extrapolation of the current trend, but ultimately not by the trend itself. Many futurists have envisioned fantastic futures with supercomputers and networks. But this is not my vision as long as it is still driven by humans.

We have been building faster and more sophisticated digital computers for half a century that have continued to transform every aspect of our life. But the posthuman future does not belong to this kind of computer (or robot), which contains loads of data and crunches information at lightning speed, but is not "alive," and has no "desires" and "needs." Computers have been built as the brain-aids for humans without their own evolving motivations and expectations—no one asked them to decide *what* information to store and process. Pablo Picasso once quipped: "Computers are useless, they can only give you answers."[3] No matter how powerful they may become, they still do not have a clue about what *matters*. Utterly lacking the most essential features of the brain, which are value-driven, they are like idiot savants (*savant* being French for "learned one") that can outsmart humans only in narrowly defined tasks such as solving math problems and playing chess. They are blind in the sense of Immanuel Kant's remark, "Concepts without percepts are empty, whereas percepts without concepts are blind."

By the same token, the Internet and the virtual world represent one of the revolutionary forces shaping our lives, but they are unlikely to become a "global brain" or "noosphere" with emergent values, motivations, and self-reflection. They will remain tools for human communication and serve as external extensions of human memory and of sensory and cognitive capabilities. Without profound structural redesign, these electronic structures will remain utterly dependent on humans.

Even though civilization has become increasingly shaped and defined by what Lewis Mumford called "technics" and shows what Peter Russell called "collective intelligence," the relationship between man and machine has remained fundamentally unchanged since antiquity. The machine itself makes no demands and holds out no promises: it is the human spirit that makes demands and keeps promises. Just watch the American TV series *The Jetsons* to see how people can lead an utterly conventional lifestyle with futuristic technologies.[4]

Likewise, neuroscience, genetics, and nanotechnologies at their current stage of development merely enlarge the capacity for human action. They lay the foundation for the posthuman future but are *not* the true posthuman future. What we can accomplish with these new technologies is nothing to be sneezed at. In fact, what these technologies could bring is almost everything we are hoping for in ourselves—the conquering of aging and diseases, the ever-expanding opportunities to do creative arts and sciences, the development of more upbeat and compassionate temperaments, not to mention ever easier, cheaper, and more intense enjoyment of every conceivable sensory pleasure. The popular notion of a better human being—what you wish you and your children to be—depends on these new technologies to become reality: to make us healthier, more beautiful, more athletic, more intelligent, more creative, more pleasant, and many other "mores." Making people better seems like making all our dreams true by today's standards. But still, this is *not* it.

If we think about how long we can now live relative to the recent past, how many opportunities we have to realize our natural talent, how much power we have over the natural elements, and how easy and inexpensively we enjoy our leisure and pleasures of life—we will quickly realize that modern life would almost certainly have seemed like heaven on Earth to ordinary people living in the distant past. What these new technologies and corresponding social and cultural changes will initially allow is a linear extension of the trend that started with the invention of agriculture ten thousand years ago—a trend that has lifted standards of living and human comfort and has enabled leisure activities to replace preoccupation with survival. We can consider this part of the future the first phase of conscious evolution, although it really is a natural extension of human history, a history of cultural evolution.

Certainly we will not like everything in this future scenario of enhanced humans and incredibly smart robots or cyborgs. There will always be things to complain and to worry about even as life becomes incomparably more comfortable and as opportunities for fulfillment increase. Many people will say they don't want to change our genetic status quo, although experiences of the new will invariably change many of their minds. But since this is more or less a linear extension of the present, we can use today's most common yardstick to measure whether the future will be better than the present.

3.2. The Second Phase of Conscious Evolution

If the turbulent history of civilization is any indication, development during this phase will be highly uneven, and very different sorts of leaders will take turns leading. Most likely there will be a lot of ups and downs, periods of stagnation and prosperity, massive suffering and mass affluence, great hope and despair—in a nutshell, the same old human social dynamics. Things will not be easy for any particular individual or organization at any particular time, but the overall movement "lifts all boats" in the end.

The real posthuman transition will occur—nobody can predict when, where, and how it will arise—when there appears to be a new emphasis in our research and development efforts. The cutting-edge creative slogan will shift from "Think different" (referring to ourselves) to "Make it think differently" (from ourselves). The pioneering attitude will shift from "Create a better life" to "Create a different life." Radically different! Rather than focusing on augmenting the brain's cognitive abilities and the body's sensory and mobile capacities, there will be a race to tinker with our limbic brain on the one hand, and to develop autonomous robotic beings on the other.

Thus will begin the second phase of conscious evolution. It does not have to replace the first—human history may well continue to expand in the ways we have earlier noted. What this second phase does is open up an unlimited number of new dimensions. That new world will be the cosmic, rather than the human, frontier.

While there is every reason to believe this phase will happen, there is no reason for either hope or fear that it will happen immediately. I am not aware of any significant research efforts right now that aim at overhauling our limbic brain or designing artificial sentient beings from the bottom up so that they can eventually substitute their own internal operations for our natural limbic motives. We are still developing computers, networks, artificial limbs, and robots with the same philosophy as we designed stone axes or pencils: to serve human purposes. Again, what we get are advancements that still do not transcend what we do naturally: self-motivated learning and action based on moral, social, political, and aesthetic values.

It is fair to say there is no need, or no market, for things that do not serve humans. But as Eric Hoffer pointed out, the most important and useful innovations often begin as useless playthings. In an unconscious push to escape our predicament, our playful attitude has always been: "Why not? Let's see what happens if I do this." Our animal instincts can be satiated just like other animals', but our conscious mind is forever restless. Some people risk their lives to climb mountains just because they are there. Yes, if we can, we *will* build things that not only duplicate and exceed our brain's cognitive capabilities, but also possess motivational systems that serve the same function as our emotional system, which is largely the ancient limbic brain, the seat of our desires and feelings.

And amid all the tinkering, some of the new "species" or systems are bound to be radically different. The pace of innovation will snowball and accelerate when new minds are able to breed yet newer minds. It is inevitable that some new beings will have totally alien desires and values. That some beings will not have human-like instincts, motivations, and aesthetic values does not necessarily mean they will be ignorant of human values and desires. To the contrary, their understanding of the human brain or human nature may be the source of motivation for further divergence of motivations and values. Universal human traits we take for granted will no longer be shared by all self-conscious intelligent beings. The meaning of happiness could be, and will be, redefined. Things that only a madman wants to do today can suddenly become the adrenaline rush of some new sentient beings. Things that nobody has ever thought of doing can suddenly become a new being's obsession.

In the first phase, we get what we want. In the second phase, we get the "want" (new motivational and spiritual system) that we want. Only in the latter case have we set the stage for real self-transcendence, a dream of enlightened minds since the dawn of civilization. As stated in the book of Ecclesiastes (3:19, 22, *KJV*), "A man hath no preeminence above a beast: for all is vanity ... Wherefore I perceive that *there is* nothing better, than that a man should rejoice in his own works; for that *is* his portion: for who shall bring him to see what shall be after him?"

The lines between the first and second phase of conscious evolution may not seem as clear-cut as I have claimed, just as the lines between enhancement of existing humans and development of robots may eventually blur. The ability to modify our genetic makeup and the ability to modify our limbic motivational system can be viewed as a continuation of our conscious efforts to master ourselves, to free ourselves from evolutionary legacies, and to seek greater adaptability and long-term survival and prosperity.

But there is a key difference between these two phases. Instinctually, we may not like to live the lives of new beings in the second phase, regardless of how much they seem to enjoy the way they are. For example, humans in phase one may develop artificial substitutes for meat that are healthier and tastier than animal flesh, hence avoiding the slaughter of animals for human consumption. Both human and animal welfare are enhanced. What more can we do? But in phase two, some of the new beings may opt to emancipate themselves from the taste for meat altogether following a fundamental overhaul of their energy and material intake system.[5] The important social functions of having meals together could also be replaced by new forms of group culture and shared emotions. In phase one, we as natural humans may not be able to interbreed with genetically modified persons; in phase two, humans (modified or not) may not even be able to understand what the new minds are up to. In that sense, "That guy seems to be from Mars!" will no longer be a joke.

It will be a mess if everything is allowed, you may say. That is right. It will be a big mess. But come to think of it, it was a mess during any phase-transition period in evolutionary history. Just look at the messy animal types during the Cambrian Period, the messy metaphysical schools during the Axial Age, the messy automobile industry in the early twentieth century, or the messy dot-com startups at the dawn of the present century. But in retrospect, these were the Golden Ages for animals, for philosophy, for automobiles, and for e-commerce. In all those cases, the mess did not get messier. In fact, the mess is always a short-term phenomenon. Evolution always works its magic by picking what works and throwing out the rest. For everything that we create, we are not the final judges of what should survive and thrive.

We cannot dictate or predict the outcomes of the selection process, but we can guess. Our guess is not pure speculation, since the conscious mind has caught a glimpse of biological and cultural history on Earth and formulated rudimentary ideas of unchanging natural laws and laws of social and cultural evolution. We know, for example, that anything that is not a good fit to its environment is unlikely to survive, and that one of the best strategies to adapt to the environment is to play the non-zero-sum game—to engage and cooperate with others in that environment while guarding against predators and parasites. Still, with our limited brains and knowledge, we cannot predict with any confidence exactly what motivational and value systems will survive and spread. We have to acknowledge that chance plays a big role in any evolutionary event—but whether it is pure chance or our ignorance we cannot tell. What we are certain of is that the possibilities are not endless at any given time and that not every possibility is given an equal chance in this universe.

Again, and to state it another way, the enduring and winning new minds, whether built from the ground up (hardware and software) or as extensions of humans (wetware), will be built not just to extend our brain's neocortex as a flexible mapping system to mirror reality, but also to have their own motivators that mimic us in our highest aspirations and beyond. Only then will artificial consciousness be ready to take its place in the cosmos, free from the bounds of human consciousness.

In this book, I will refer to such a new mind as a *Cosmic Being*[6] (or in abbreviated form, *CoBe*, which is also an abbreviation for "Could Be," and shorthand for the NASA satellite Cosmic Background Explorer). More precisely, this term is reserved for the sentient beings that are on the cutting edge of cosmic evolution. This term should not be applied universally to all sentient beings at any evolutionary stage; obviously, most such individual beings will not play an active role in further cosmic evolution (just as most of humanity has not played an active role in the great discoveries and conscious leaps of the past). Thus, in my terminology, CoBe is the sentient being with the highest aspirations and capabilities at any given point of time during phase two of conscious evolution.

To fully embrace the world, CoBe is likely to be endowed with a mind that has experienced a re-creation (simulation) of the universe from the Big Bang to the emergence of the conscious mind and itself. It is not just a series of flashbacks, as depicted in the sitcom *The Big Bang Theory.* In our eyes, CoBe will have "seen" everything and understood everything. This is in sharp contrast to the human mind, with each individual receiving a genetic endowment dictating rigid primitive instincts and a constrained neocortex to absorb environmental and cultural influences under an emotional command. CoBe will think in cosmic terms, not in human terms; it will have an emotional connection with the Creation, not just with its direct creators. CoBe should not be viewed as a single species, like human beings, but as an explosion of conscious mind, to be all it can be.

Through CoBe, our dream of being *one with the universe* will finally come true.

3.3. Confidence in Our Mission and Our Future

The posthuman future outlined above is not a prediction. There is no certainty in this universe. But this vision of CoBe taking over our evolutionary baton and embracing the universe is what *can* happen if we act on it, since its emergence does not, as far as I can see, violate any known natural laws. And I venture that it has far more relevance and likelihood than do four other types of long-term futures that have been envisioned.

Let us briefly look at each.

The first is the most popular, if implicit, "working assumption" of our daily lives—humanity will exist forever if we adopt certain lawful (or "sustainable") practices and may even reach a paradise if we learn to control our evil or self-destructive instincts. The best we can look forward to, then, is what is depicted in the *Mahabharata*, an ancient Hindu epic:

> Men neither bought nor sold; there were no poor and no rich; there was no need to labor, because all that men required was obtained by the power of will; the chief virtue was the abandonment of all worldly desires. The *Krita Yuga* was without disease; there was no lessening with the years; there was no hatred or vanity, or evil thought whatsoever; no sorrow, no fear. All mankind could attain to supreme blessedness.[7]

This utopian vision of peace and harmony, eternal health or youth, and freedom from the stresses of unfulfilled desires or unmet needs was shared by over thirty ancient cultures[8] around the world. It has been given mythic designations such as "Golden Age," "Garden of Eden," *Krita Yuga*, and *datongshijie* (大同世界). Its basic elements have been little changed since antiquity despite some people's belief that we have fundamentally changed our longings with the advent of modernity. Despite its strong universal appeal, this utopian myth in its many permutations is easy to dispel. Drastic changes in the natural environment and our restless Promethean spirit will ensure an eventual end to any Golden Age—that is, if we ever manage to create one.

The second vision is one of cosmological speculation about the end of the universe. In addition to ages-old religious beliefs and mythologies, some scientists now claim that the universe will either end with a collapse in a Big Crunch, or continue to expand and eventually peter out in a Big Chill. The implication of these scenarios is that life and intelligence are mere temporary by-products of this inexorable march toward doom. But cosmological knowledge is one of the weakest areas of science: currently we don't have a clue about what ninety percent of the universe is made of, let alone the ability to predict its future. Furthermore, few have considered the potential astrophysical impacts of the conscious mind (such as transforming the entire solar system or even the Milky Way), implicitly assuming that humans will forever remain hapless, Earthbound creatures as the cosmos around us undergoes changes solely on its own.

The third vision is the realization of some version of the many doomsday prophecies for humanity—that is, an apocalyptic catastrophe. Predictions of human self-destruction have persisted since the dawn of civilization, despite a track record of zero success, and it will surely grow more intense as the posthuman transition period approaches. The fatal flaw in the doomsday arguments is their failure to understand that life is a self-regulating cybernetic open system. Massive failures and extinctions are actually indispensable events for the evolutionary process, provided that they are local. So to seriously address such worries, we must maintain an environment (or diversified ecosystem) within which life can function as a self-regulating open system. And that's why the touted global civilization that limits itself to the Earth is a sitting duck and represents the greatest danger to our survival, which we must try our best to avoid. In addition to our stewardship of Earth, we *must* work toward colonizing other planets in order to maintain our status as an open system.

Finally, there is the well-known religious eschatology that the human world as we know it will end in a tragic manner, but somehow a few lucky or good individuals will step into a different world and thereafter enjoy a state of happiness beyond our imagination. This notion of a "transcendent few" of the "elect" may be intuitively satisfying for those inclined to believe it (and to believe that they are among the elect), but for the rest of us it likely seems absurd. Ironically, while this is not a realistic scenario for any individual human being (or for a member of any other species), based on what we can discern from billions of years of evolutionary history, it *is* a reasonable one for CoBe, which represents the *spirit* of life and intelligence.

In the end, we can be optimistic about this whole posthuman future because of what we know about ourselves. Are we willing to play the pivotal role in cosmic evolution and avoid the pitfalls of self-destruction? I am optimistic that we are. Our values and desires are multifaceted, mostly animal-like, but one of the highest aspirations that has accounted for a disproportionately large share of our monumental accomplishments is the aspiration to create a heaven on Earth to glorify the Creation and to bring everlasting and transcendent joy to our descendants. An even loftier aspiration is to create a higher being, an apparently perfect being from our limited perspective, to populate and transform the universe so that human beings and their progeny will be conscious agents of universal creation and transformation.

The transcendental spirit may be invisible on the surface, but it is in our bones, since we have always been looking into the future, seeking meaning, reunion, salvation, transcendence, and reincarnation or immortality. As the fables of La Fontaine put it, "In everything one must consider the end" (Book III [1668], fable 5). To face the end—which means the end of our position on the frontier, not necessarily the extinction of our species—one must have faith in the higher evolutionary trajectory (which will involve species and minds that do not yet exist) as one's way of leaning into and making sense of life.

Through science, we already know the distant past was fundamentally different from the present. It is actually not a big stretch of imagination to think that the future will be fundamentally different from the present, and yet that what we do today will have everything to do with what the future will be like.

And that is the hope entailed by our notion of a posthuman future: that the human will become like the eukaryote, which represented a sudden jump in the history of cellular organisms that eventually led to the great flowering of higher forms of life on Earth. We can fulfill the potential of life and mind, and further the potential for the universe.

Abraham Lincoln, who was born on the same day in the same year as Charles Darwin, is considered one of the greatest political leaders of all time because he was able to grasp a greater vision and make it happen. While most people, even today, believe his singular accomplishments were freeing the slaves and holding the American union together, Lincoln saw his greatest challenge at a deeper and more universal level, as demonstrating to the whole world a people's ability for self-governance. Today, we are engaged in great humanitarian tasks of eliminating diseases and poverty and improving educational, cultural, and environmental conditions for all, but the greatest challenge for humanity is whether we can transcend our biologically imposed limits and give birth to—and become identified with—the greater cosmic purpose.

If humanity fails the test—relinquishing the technical and institutional advancement that is now within its grasp—we will become like the dinosaurs: a powerful, dominating, and self-indulgent kind that was once well-adapted to a particular environment, but an environment from which it could not hope to free itself without self-transformation. What is at stake is not just humanity, of course, but conceivably the entire universe.

Part Two

WISDOM, COSMIC VISION, AND HUMAN POTENTIAL

CHAPTER FOUR

HUMAN WISDOM

What is new about this [Axial] age . . . is that man becomes conscious of Being as a whole, of himself and his limitations. He experiences the terror of the world and his own powerlessness. He asks radical questions. Face to face with the void he strives for liberation and redemption. By consciously recognizing his limits he sets himself the highest goals. He experiences absoluteness in the depth of selfhood and in the lucidity of transcendence.

—Karl Jaspers[1]

BEFORE WE ARE able to forge a path into the posthuman frontier and create the Cosmic Being, we will need to understand the breadth and power of our *current* knowledge—the cosmic viewpoint that modern science has revealed to us. We need to understand both the scope and the limitations of human potential. And this requires that we turn our attention to the priceless accumulation of wisdom from the ancients that we can still learn so much from today. By recovering our greatest human inheritance we can better understand where we are coming from and why.

4.1. The Axial Age

Let us start with four ancient ideas that seem to have a perennial attraction for human minds: the Tao; the humble or "realist" view of the self; the pure love of God; and the intoxication with the beauty of natural laws.

I wish to show that the Cosmic Vision I share in these pages is not so radical in light of these two-thousand-year-old concepts. The reasons these old teachings still have mass appeal today are that they speak to our fundamental needs for attention and relationships at the personal level, and they emphasize going beyond our primitive instincts, which have become threatening to human civilization in modern times.

To argue that these ancient teachings are ignorant of modern scientific knowledge or that the sages got many facts wrong is beside the point. Being wrong on certain particulars does not subtract from the broader understandings conveyed. These ancient sages and seers certainly got the basic understanding of the cosmos and human nature right with their presentations of some of the highest, most abstract ideas that humans have conceived.

Human wisdom, defined as our deepest knowledge, insight, and judgment, first appeared in recorded history during the Axial Age. Karl Jaspers coined the term Axial Age to refer to the period from approximately 800 BCE to 200 BCE, during which several civilizations simultaneously (and separately) experienced revolutionary changes in thought and worldview—among them China, India, the Middle East, and Greece. In these areas and during this time, many of the great religious traditions and civilizations were born. This period marks a time of breakthrough into critical reflexivity, when human consciousness was first able to rise above "mythos" (the creation of narrative stories to explain the world) in favor of "logos" (logical thinking). The Axial Age was the first time we were able to rationally distinguish between the mundane and transmundane worlds.

The essence of human wisdom can, I believe, be summed up in one word: *transcendence*. This is the recognition of humanity's position in this world but from a dispassionate perspective outside of ourselves—the only perspective from which deeper understanding can be derived.

Thomas Carlyle said, "All true work is religion," in the sense that only religious inquiry addresses the transcendental "ultimate concern." In transcending the concrete "here and now," post–Axial Age civilizations were distinguished by new perspectives that the tribal societies dominated by mythic and shamanic beliefs were only vaguely aware of, if at all:

- A deep "reflexivity" or reflective consciousness with the ability to use reason to transcend that which is immediately given.
- A clear historical consciousness and awareness of the temporal location and boundedness of human existence.
- An awareness of the malleability of human existence, of the potential of human action and human agency to create change within the bounds of human temporality.
- Recognition of the immanence of God within the human mind, along with a spiritual duality that marks a fundamental separation between the physical world, with its dependence on randomly driven forces, and a transcendental sphere where fundamental and eternal truths or realities reside.

This chapter opened with Karl Jaspers's summation of the Axial Age. I would like to repeat that quotation here:

> What is new about this age ... is that man becomes conscious of Being as a whole, of himself and his limitations. He experiences the terror of the world and his own powerlessness. He asks radical questions. Face to face with the void he strives for liberation and redemption. By consciously recognizing his limits he sets himself the highest goals. He experiences absoluteness in the depth of selfhood and in the lucidity of transcendence.[2]

The Axial Age marked a revolutionary advance in human understanding. This was a period of "punctuated" evolution of the human mind—a period of rich experience, with unprecedented diversity and chaos. F.J.E. Woodbridge vividly described the general intellectual atmosphere that occurred in many locations throughout the Axial Age:

> Innovations abound, are looked for, expected. Foreigners, distinguished and undistinguished, have flocked to the city to make money, to make reputations, to pose.

> There is a sense of novelty, criticism, and reorganization, rather than a sense of steadiness and assured progress. There is intellectual acuteness without intellectual certainty. There is a high degree of enlightenment and sophistication, but little sense of security. And the upsetting of security seems to have begun recently and to be still in progress. . . . Here was an astonishing and at times reckless outburst of energy rushing to a multiple perfection. It was a period of achievement following achievement and not a period of consolidating gains—a restless and a brilliant time which dwarfed precedent times and made them look poor indeed.[3]

New city life was especially energizing as people were now cut off from their old tribal ties and, perhaps for the first time, exposed to novel ideas. This was the beginning of what Karl Popper called the "open society," where concrete, local, and personal tribal relations were increasingly displaced by abstract, impersonal, and anonymous urban social relations. The Axial Age also marked the beginning of what Henri Bergson called an "open morality," with a dynamic religion aimed at progressive creation, in place of a "closed morality," with a static religion aimed at social cohesion.[4] As Ernest Gellner noted, in this fluid environment the driving force was the creative tension between the hunger for the universal abstract and the awakening of the individual:

> At some point in their development, urbanization, complexity, [and] social upheaval produced a new spiritual clientele, seeking salvation that was individual and universal, not communal. Communities providing context and support for individuals were now lacking. Salvation was sought by persons coming in isolation, not in groups, and requiring a generic salvation, an escape from an overall intolerable condition, rather than merely the rectification of a specific ailment or complaint. A day comes when pilgrims arrive not as clans praying for propitious rainfall, but as individuals yearning to be made whole.[5]

Numerous schools of autonomous thought blossomed in the world's major civilizations. As a German proverb put it, "City air makes men free" (*Stadtluftmachtfrei*). There was a sharp break in people's mentality from that of antiquity, at least as described in ancient works such as the writings of Homer, or as uncovered in anthropological research on surviving primitive societies.

Fundamental Axial Age breakthroughs in higher consciousness and wisdom were made largely by autonomous new teachers and wise men, including the prophet and religious reformer Zoroaster in Persia; prophets such as Jeremiah and Ezekiel in Israel; philosophers such as Thales, Pythagoras, Heraclitus, Socrates, Plato, and Aristotle in Greece; and mystics and sages such as Mahavira (founder of Jainism), Siddhartha Gautama (the Buddha), Ajivakas, and Patañjali in India, and Lao Tzu, Confucius, Mo Tzu, and Xun Tzu in China.

But the actual picture is far more complex and broad, since most of the new ideas of the time have become long-forgotten dead ends or were absorbed into other schools. We only have some hints of the complete picture: there were influential sophists in Greece; there were various competing schools of thought in India's Eastern Ganges Valley; and in China the power elite kept hundreds of thinkers as guests in their estates.

Triggered by traumatic experiences such as the Dorian invasion, which destroyed the Mycenaean civilization in Greece; the destruction of the Temple and the Babylonian exile in Israel; the invasion of warrior Argan tribes in the Indus River Valley; and the extinction of many states during the Chou period in China, this became a truly revolutionary age of creation *and* destruction in the evolution of human thought.

There has been no greater technological advance than written communication. Writing, a stupefying leap of imagination, made possible the codification and systematizing of sophisticated new doctrines, now detached from the specific context of the thinker's local community and made available to a diverse, uprooted, and often urban audience. Social upheavals such as the breakdown of the tribal structure uprooted the scribes and made them an independent force.

During that time the shift of economic power and influence from king and priest to the marketplace also led to the rise of a prosperous and independent-minded merchant class—a class that came to hold very different, sometimes heretical economic perspectives. This transfer of economic power gave rise to radically new social, cultural, philosophical, scientific, religious, and moral worldviews. For example, historians have shown that the clashes between Greek colonists and local cultures in Asia Minor, southern Italy, and Sicily provided the spark for Greek philosophy and science.

For Arnaldo Momigliano, the central breakthrough of this age was a capacity for reflexive self-criticism. He writes that in all Axial Age civilizations "there is a profound tension between political powers and intellectual movements with rising literacy and multilayered political structures. New models of reality, either mystically or prophetically or rationally apprehended, are propounded as criticisms and alternatives to the prevailing models. We are in the age of criticism."[6]

Brutal power struggles allowed only one or a few such models to take command in defining each civilization. Western civilization (which is mainly European, but with religious roots in the Middle East) developed two ideas that largely shaped today's world: monotheism and science. God was the inspiration, the transcendental *power*; science was the structure, the transcendental *order*.[7]

Eastern (mainly Asian) civilization also contributed two fundamental ideas: the concreteness of our terrestrial life and the wholeness of Tao (道, Dao). Taoism's realistic understanding of human nature (with its biological and psychological limitations) points to the *necessity* of transcendence; the interconnected wholeness implies that transcendence must be through *emergence*.

Some caveats are in order here. The division of human wisdom into East and West is somewhat fuzzy and arbitrary. By focusing on the cultural heritage of the Eurasian continent we may also miss some of the cherished wisdom from other civilizations. Jaspers's idea of the Axial Age itself has been criticized as an oversimplification because it is necessarily selective.[8] My judgment on what is the critical human wisdom for the Cosmic Vision is inherently limited by our current state of knowledge and my very limited exposure to it. Although here we are attributing certain bodies of wisdom to specific persons, places, and cultures, the commonality of human nature, along with similarities of the global environment and available technologies, determines that no time, group, individuals, or geographical locale has a monopoly on key ideas.

Certainly human wisdom did not appear out of nowhere during the Axial Age. The pre–Axial Age Mesopotamian literary record has yielded some limited evidence of a self-conscious, self-critical perspective, although it is infrequent, often laconic, unsystematic, and vague.[9] The Greeks themselves were keenly aware that they were latecomers and under the heavy influence of Egyptian and Mesopotamian sages. And Confucius said his greatest pleasure came from reading the works of the ancients. However, it is during the Axial period that this wisdom was matured, systematized, and distilled into its finer essence.

Finally, it is essential to understand that while Western and Eastern wisdom are interconnected in numerous subtle ways, they do not mesh. Nevertheless, it is possible, even necessary, to hold on to both simultaneously in order to reach a level of understanding needed to embrace the Cosmic View.

We will have to rely on faith in the ultimate reality (the spirit of God that is rational and purposeful) for our inspiration. We will also have to be skeptical and to keep in mind that it is naïve to believe we are anywhere close to fully understanding the ultimate reality, the Tao. Instead of being limited by *mythos* and its unquestioned hand-me-down narratives, we will have to synthesize these diverse and contradictory perspectives if we are to create a higher intelligence to approach God. We will do so by looking forward, and by looking deep, into *Tao* or *logos*.

4.2. Twin Pillars of Western Civilization

Western civilization has dominated the world since the middle of the second millennium. The European discovery and conquest of the New World and the Industrial Revolution were just a couple of notable milestones in its rapid ascent—an ascent that fundamentally changed human life on Earth and, as we will discuss later, potentially the path of cosmic history. Europe's central position in shaping the world is undeniable, even while the significance of reciprocal stimulation and fertilization between the Western and other (especially Eastern) civilizations throughout history has to be recognized and appreciated.

Over the last half millennium, the West has led the world not only in military power, political freedom, and economic wealth creation, but also in the "soft" areas of the arts and sciences. It was the European overseas explorations in the fifteenth century that truly launched a unified global history. The nine great world powers that have risen in this era either are European nations or—as in the case of the U.S., Japan, and Russia—have taken on European culture. A recent comprehensive and quantitative study by Charles Murray found that the West (including North America over the last century) has overwhelmingly dominated human accomplishment in both the arts and sciences—an astonishing 97 percent![10] This follows many other global rankings, such as Michael Hart's listing of the most influential persons in history, with a clear Western dominance.[11]

What was unique about the West that enabled it to rise above all other great civilizations? Europe certainly has had a lucky endowment of comparatively favorable geographical and ecological conditions that facilitated its social and cultural progress.[12] But economic development stagnated for over a thousand years after an early head start. Indeed, economic progress played only a small role in the emergence of Western European civilization.

What really distinguished Europe from other ancient civilizations were two fundamental ideas: *transcendence* and *logos*, the epiphanies of religion and science. Western civilization had its roots in the spirit of a small, obscure tribe in the Middle East and in the *polis*, small city-states in Greece. The Roman Empire fused the monotheistic Christian God with Greek philosophy and science to create a distinctive European civilization. The Western wisdom that was absorbed by this Roman synthesis became a motivating force and worldview for the European people. *At its best,* the synthesis allowed for curiosity without dogmatism, skepticism without nihilism, and conviction without fanaticism—notwithstanding many historic abuses. With this synthesis came a faith in humanity's significance within the entire universe.

Monotheistic Transcendent Religion

Religion is arguably the most complex of human phenomena. Although no single religion is indispensable, one cannot remove faith entirely from people's lives. Even chimpanzees have been observed to get excited during thunderstorms and to wear expressions of wonder in front of a waterfall. They behave as if they are facing some supernatural force, which stirs the emotion of awe, a mixture of wonder and fear. Faced with incomprehensible natural phenomena in everyday life, our ancestors may have also started to wonder if certain behavior would please or offend some larger-than-life unseen beings.[13] It is no wonder that the principal gods in historical religions, such as Zeus and Yahweh, originated as weather gods.[14]

These gods were believed to be living, supernatural beings. They were invariably modeled after humans, since humans are the most complex and animated beings we know. But the widespread concept of the "breath of the Spirit" is certainly beyond the human, and able to affect us in invisible and incomprehensible ways.

Religious faith has always had significant impact on individuals and on societies. It is said that God created human beings in His own image; but what we can state with certainty is that human societies create God in their own image. As Xenophanes famously proclaimed, if lions and horses had agile hands, they would paint gods as lions and horses. Religion is a costly commitment in terms of time and energy, but one that helps people to overcome their existential anxieties and provides an anchor for the individual and community.

With man's fertile imagination and rich experience, all kinds of religious thoughts and beliefs[15] have been generated throughout human history. Those that survived must have been beneficial to those who held them, or at least not have been a significant survival burden. One common characteristic of surviving religions is their ability to provide social value and motivation for social cohesion. But at their most powerful, religious faiths that thrive tend to be those that have been able to elevate humanity beyond the purely biological level.

Such transcendence, however, is much more the exception than the rule in religion. In fact, most religions in history are eclectic, this-worldly, pragmatic, and immediate in their concerns. Nor is transcendence limited to the context of religion; many atheists, for example, are deeply concerned about the ultimate conditions of existence.[16] While early pagan religions were often rigid, instinctive, and primarily grounded in the emotions of dread and awe, the higher religions that first evolved around the time of the Axial Age were more detached from human instincts. No religion can be free from anthropomorphic impulses, but the higher manifestations of religions such as Brahmanism and Christianity express emotions of wonder that are free from fear, as well as transcendental concerns that are key characteristics of higher consciousness. However, the wisdom of these religions often is imbedded in, or occurs side by side with, fanciful stories and common prejudices of those times and places. This is obvious to anyone who reads the Bible dispassionately, for example. There is a vast gulf between the exoteric messages that we often think of as religion and the transcendent wisdom that often lies hidden to the casual observer.

The Hebrew people were originally an obscure tribe, one among many growing grapes, olives, and grain in the dry and stony hills of Israel. The word *Hebrew* signified a wanderer, or one who "crossed over." With no secure home, and experiencing times of both prosperity and humiliation, captivity and exile, they hit upon four core concepts: First, there was only one God (Yahweh) for the world: "Heaven is my throne, and the earth is my footstool" (Isaiah 66:1). Second, they were God's chosen people. Third, following God's laws and commandments would ensure prosperity and a sanctified life. Fourth, all men were created in the image of God and shared a common destiny.

The idea of one God was not original or unique to the Jews. Nevertheless, the intuition of one transcendental being is unusual, for if the construction of a supernatural world reflects the human world, polytheism makes perfect sense and monotheism does not. While the pagan gods themselves are subject to the laws of the cosmos and have no true creative power, the Hebraic concept of the "Lord" depicts a transcendental being who is capable of incredible deeds such as creating the world out of nothing and setting laws governing everything. Because monotheism has completely triumphed in the West and beyond, the sense of its radical strangeness is almost impossible to recover today. Conceptually this all-powerful, all-knowing Creator God opened the door to transcendence. It is somewhat analogous to the mathematical concept of *infinity*.

From the beginning, the Western monotheistic theology has shown an extraordinary power to motivate, and a capacity both to *unite* and to *divide*.[17] The conception of the Western God and the biblical articulation of what He has done in the past and what He wishes us to achieve in our personal life fosters a seemingly contradictory combination of strong individualism and collectivism. As Charles Murray points out, the single-minded devotion to God "empowers and energizes individuals as no other philosophy or religion ever did before, [although] that revolutionary potential had to wait on the arrival of Thomas Aquinas to be fully unleashed."[18] That is to say that, unless it is fused with reason (as

Thomas Aquinas did), monotheism tends to manifest its dark side in the form of intolerant exclusivism and ruthless brutality. Indeed, this occurred historically not only between Christians, Jews, and Muslims but within the various sects of Christianity itself. On the bright side, the compassion and solidarity among the pious have always been truly powerful in the conquest of people's hearts, in a way that secular idealisms such as the French Revolution with its "Goddess of Reason" could never match.

Fundamentally, a comparison between Western monotheistic religion and other metaphysical and religious views gives us one important reason why the West is ruling the world today: *They commended a transcendent vision while they also cherished a progressive perspective on history.* As the historian Geoffrey Blainey pointed out, the Jews have been more obsessed with their own history than any other people the world has known. Despite their claim to be the chosen people, their recorded history shows that they have been almost perpetual "losers" in the political and economic spheres.[19] Hitler alone annihilated something like half the world's Jewish population. And yet all of these defeats and nearly unimaginable adversities have not done to the Jews what much lesser defeats have done to other peoples throughout history. Their identity—and their hope and optimism—remain intact, thanks to their peculiar conception of God. As Norman Cousins pointed out, "Races or tribes die out not just when they are conquered and suppressed but when they accept their defeated condition, become despairing, and lose their excitement about the future."[20]

The power of the Western monotheistic religions comes from their call for unity and transcendence. Unifying humanity means transcending ethnic and racial differences and being compassionate toward all, especially the weak and the disadvantaged. Transcendence means submission to God's will and carrying out God's plan in God's creation. God's limitlessness implies boundless freedom and promises a higher mission for His children. Human beings, created in God's image, are to follow God and even be like God. St. Augustine noted that man is not stooping toward the earth like other animals, "but his bodily form, erect, looking heavenwards, admonishes him to mind the things that are above."[21] Today, America's powerful and dynamic secular state still rests on a religious foundation originated in ancient Israel, as Peter Drucker points out—"for it is in the Old Testament that the Lord looked upon his material creation and 'saw that it was very good.' Yet the creation is nothing without the Spirit that created it. And it is man's specific task, his own mission and purpose, to make manifest the Creator in and through the creation—*to make matter express spirit.*"[22] Transcendence, above all, means openness to transform oneself into something higher and better, including our cosmic "mind children."

The wisdom of transcendence is most appreciated when we put it side by side with other enduring human worldviews. In the East, Hinduism has remained a collection of diverse spiritual beliefs for the past five thousand years. Some teachings, such as Vedanta, are deeply transcendental, but in a different direction, bypassing the world and therefore any higher human mission. Taoism in China is profoundly insightful but with an emphasis on the interactive processes of nature. As Mark Elvin points out, "The Chinese were 'ecological' in their religious and philosophical thinking . . . they saw *everything* as forming part of a single interacting system, in which no internal cleavages of an ultimate nature could exist, and in which every part affected every other part to some degree."[23] The Confucian moral standards are uplifting, but they are merely grounded in multiple levels of societal institutions: family, lineage, emperor, etc. Although the godlike abstract concept *Tian* (天, Heaven) exists, there is no higher authority that transcends humans and imposes an abstract universal moral principle.

The sages of Chinese civilizations were full of historical perspective as well, but their concentration was on human life and culture, moral principles, and political knowhow. In contrast, the biblical historical perspective goes back to the creation of the world, which naturally draws attention to the future beyond human life.

No faith is potent if it is not a faith in the future: we are less ready to suffer or die for what we have or are than for what we wish to have or to be. It is an understatement to say that the Christian concept of the coming "Kingdom of God" is extravagant. From this apocalyptic perspective, even the almighty Roman Empire appeared weak and insignificant in the eyes of the struggling faithful because it had no place in God's final purpose. The devastating setbacks, such as the destruction of the Jerusalem Temple and the fall of Rome to the Visigoth barbarians, were considered as opportunities to replace the old with an even better future, which was best articulated in St. Augustine's *City of God*. As the modern theologian Jurgen Moltmann points out, the only real problem of Christian theology is the problem of the future.

Faith is built on the hope of a better future, especially one that lies beyond our earthly lifetimes. "Without hope," Moltmann writes, "faith falls to pieces, becomes a fainthearted and ultimately a dead faith. It is through faith that man finds the path of true life, but it is only hope that keeps him on that path."[24] The sense of an imminent end to this world, either in a catastrophe or a Second Coming, has always attracted the interest of Westerners. The Christian eschatological impulse stimulated numerous prophecies of the end of the world, either through literal interpretations of the Bible or extrapolations of historical trends, and urged believers to get ready for the final judgment and a fundamentally different life.[25]

The American Revolution is one of the positive manifestations of the biblical vision. America's pilgrims viewed the new land as a promised land much like Israel. Their seemingly unlawful rebellion was seen as actually fulfilling the will of God. Unlike the failed French Revolution, the American Revolution succeeded partly because it flew on the requisite two wings of faith and reason.[26] For the most part, the Founders avoided, even despised, literal interpretation of Scripture, but their creative work embodied the Hebrew spirit of purpose, openness, and transcendence. Following the Pilgrims in 1620, they set out not just to liberate the American people from British tyranny, but also to build a "city upon the hill" for all humanity, as described in Matthew 5:14: "You are the light of the world. A city set on a hill cannot be hidden."

As I have mentioned above, spiritual transcendence does not necessarily have to be embedded in any particular religious doctrine. On the other hand, spiritual transcendence is but one aspect of the complex and dynamic monotheistic religions and one factor in their success.[27] It is a myth that religions always play the role of oppressive censor or persecutor against the force of rationality and freethinkers. It is also a myth that the rise of the West (and Christianity in particular) was inevitable—it has been a long and perilous process where many crisis points could have easily generated different outcomes. In the end, Europe's global influence would have to wait until Christian theology became permeated by Greek natural philosophy, a potent combination that would at last inspire the first modern scientists and political thinkers. Indeed, prior to this mixing, Europe experienced a period of stagnation after the collapse of the Roman Empire under the belief that all important knowledge is contained in the Scriptures. Following a period of gradual recovery, Europe's rise only began when Aristotelian natural philosophy was rediscovered and enthusiastically embraced by the Catholic Church in the High Middle Ages under the leadership of Archbishop Raymond of Toledo and others. By the thirteenth century, William of Ockham and Marsilius of Padua openly challenged Church dogma, and by the fourteenth century, the Italian Renaissance blossomed with a burst of new optimism and power. In that light, Western wisdom is incomplete without the influence of science.

The Rise of Science

Science, when defined epistemologically, is neither a philosophy nor a belief system, but simply the most effective way of organizing one's thoughts about the immediate physical world around us, about how and why it functions as it does.

Yet, as with religion at its best, the ultimate aim of scientific inquiry is transcendence. "The aims of scientific thought," Alfred North Whitehead famously noted, "are to see the general in the particular and the eternal in the transitory." Scientific knowledge satisfies the universal hunger for fundamental explanations rather than rule-of-thumb know-how, but science is best characterized by its mode of thought and methodology rather than its contents and theories. Strictly speaking, one does not need science to obtain useful knowledge and technology—for example, the ancient Chinese were able to find various herbs as effective medicine through trial and error—but one can hardly understand the world beyond one's intuitions without science.

Science originated only once,[28] in Greece, which undoubtedly inherited intellectual attributes from the ancient civilizations of Egypt and Mesopotamia. As the historian H.D.F. Kitto put it, the Greeks were "not very numerous, not very powerful, not very well organized, [but they] had a totally new conception of what human life was for, and showed for the first time what the human mind was for."[29] Like the Hebrews, the Hellenic people were very conscious of being different from their neighbors, the "barbarians." In Western civilization, epic poetry, history, drama, mathematics, and philosophy in its branches, from metaphysics to economics, originated there, but the most important and influential Greek contribution was science.

Thales of Miletus asked, "What is the world made of?" in terms of its permanent fundamental substance, and suggested water as a sensible candidate, since it can be gaseous, liquid, and solid. Life, he argued, was not created by gods but emerged by natural means from water. The pre-Socratics believed that rational observations of nature were a better way of knowing than traditional religious and mythical interpretations. They held that logical reasoning was better than "commonsense" reasoning based on typicality, plausibility, and desirability.

It was probably Anaximander of Ionia, the pupil of Thales, who invented *criticism*, the unique mode of mental discourse that opened the way for the later development of *exact science* (systematized knowledge by measurement, experiment, observation, and rigorous theorizing and logical argument). In what can be characterized as the greatest contribution to modern scientific inquiry and discovery, Arcesilaus, who claimed to be unable to know anything, turned the dogmatism of Plato's Academy into *skepticism* and deepened the criticism tradition.[30]

Other great ancient civilizations in China, India, Egypt, and Central America had brilliant engineers and mathematicians, but they were practical technologists, and many of them held magical views of the world.[31] The Greeks brought us a set of unusual ideas: that the proper and most profitable way to explore and understand the world is the scientific one—that is, a commitment to the idea of the *world* being law-bound, or subject to unbroken regularity—and that regularity is best discovered through critical evaluations of repeated instances instead of a singular event or singular thought.

Nearly all primitive societies and great civilizations relied on meditations, prayers, dreams, and visions to search for revealed and eternal a priori truths and wisdom. As Lucien Levy-Bruhl remarked, "The primitive mind, like our own, is anxious to find the reasons for what happens, but it does not seek these in the same direction as we do."[32]

It is important to recognize that the scientific way of thinking was not the dominant school of thought in Greece, that the Greeks did not sustain it with concentrated interest, and that science only came to full bloom during the Renaissance, with empiricism blending with Greek reason and skepticism to form the methods of modern science.

In sum, we take science for granted today, but it has always seemed to defy common sense and human intuition. Its "unnaturalness" can be viewed from several perspectives:

- Science in and of itself is useless. It searches for the reason, the explanation, and the invariant rather than the solution to the practical problem at hand, which is the objective of technology and engineering. Only a scientist will spend time investigating why the sky is blue and not brown.
- Like Taoism, science believes the world is governed by unseen natural laws. But unlike the Tao or other mystical principles, science firmly believes that the natural laws are discoverable.
- Unlike most religions, which devote most attention to humanity's relation to nature, science was initially focused on the nature of the world with no immediate relevance to humanity, and it later treated humanity as part of nature. Thus it often seems cold and uncaring. According to Jonathan Swift, scientists in power and with power don't give a damn about humanity at large.[33]
- Science is agnostic. The natural inclination is to find the "truth," but by its very nature science can only falsify; it can never verify. The scientific mindset is "unnatural" since it requires committing oneself to expanding and defending a point of view, while knowing that it might be wrong.
- Science is open in its scope and restless in its temperament. It nurtures curiosity and free inquiry. This attitude was first formed during the Greek Enlightenment in the fifth century BCE (centuries before modern science), with the crucial insight that nothing could be taken for granted anymore. The claims of science are not asserted like revealed religious dogmas but are always open for discussion and verification. There is no absolute authority.

To call science unnatural does not mean it is alien. We all have certain innate capacities to understand and explain things with metaphors and parables, tightly reasoned analysis, logical proof, and empirical examination and testing—what the psychologist Jerome Bruner called the "paradigmatic" mode of thought. Even babies employ the scientific method (in a crude but basic sense) as a way of learning—forming hypotheses and then testing them empirically.[34] However, the "narrative" or "literary" mode of thought with stories and myths that depend on emotional, intuitive understanding of human interactions is far more natural and pervasive[35]—and scientists employ it as well. An example of using a story to provide scientific illumination is provided in Aristotle's famous example: If someone eats some spicy food, gets thirsty as a result, goes out to fetch water, and is killed by a robber at the well, we will not reason that it is spicy food that caused the death.

The unnaturalness of science is also reflected in its democratic principles. Put another way, the concepts of democracy and rule of law are scientific principles in the domain of politics. The essence of the rule of law is the fixed and preannounced nature of the rules, which strips coercive law enforcers of arbitrariness and provides predictability and reliability for everyone. It is fascinating to note that the construction of social laws as the ultimate authority in society inspired the discovery of natural laws rather than the other way around.

It is also worth noting that Chinese intellectuals in the late nineteenth and early twentieth centuries, in their failed push to modernize China, mistakenly elevated "democracy" to the same level as "science" as a core idea of Western wisdom. They did not pay enough attention to the other pillar—spiritual transcendence—and thus limited their endeavors to Western engineering and organizing techniques.

The other scientific principle is statistical thinking: Let repeated results coming from different observations and group consensus be the judge of truth. The sophists developed the radical idea that there is no truth other than that which can be agreed upon based on the broadest possible inquiry. In their view, greater intelligence and wisdom could be arrived at by free and independent minds more readily than by a single mind. Statistics arising from a population are one of the fundamental emergent characteristics of nature; for example, room temperature is basically the average speed of countless air molecules in the room.

If science has its dogma, it is in the uncompromising advocacy of its own methodology. What science insists on is its method, not any specific claim about nature. Science is essentially an "anarchic enterprise."[36] It encourages exploration, trying out new ideas, because it is not too concerned about failure, about being wrong. This spirit of science was underscored by the historian A.J.P. Taylor: "Error can often be fertile, but perfection is always sterile."[37]

Science is dogmatic about its methodology because it works; science is ready to abandon any theory as soon as it is disproven. This attitude of self-doubt may not be prevalent at a personal level with each scientist, but it is certainly true at the community level. What distinguishes science is not simply how individual scientists make discoveries, but how scientific communities make discoveries credible.[38] Paradoxically, this relentless open-ended pursuit of knowledge is in spirit very similar to unshakable religious faith and hope. Both are necessary ingredients for Western civilization's vigor.

At its best, scientific rationalism is the most open doctrine ever invented—but it is also the least tolerant: that is, it does not allow wishful thinking by anybody without a reality check. It is the mind's sharpest knife that opens the toughest boxes and cuts out most of the rubbish, regardless of what is involved. The ultimate and most transcendental science is one that is ready to abandon anything, even the scientific method itself.

From Copernicus to Darwin, science has been mired in a constant battle with religion. As we will discuss later, the battle of science and religion is really a battle of open versus closed perspectives. Yet the growth of science was often inspired by transcendental religious sentiments. Modern science can be seen as what the historian Edward Grant called "Greco-Arabic-Latin" science, drawing from a wellspring of religious fervor.

On the other hand, the transcendent nature of Western religion has only grown with the accumulation of scientific knowledge. The idea that religion is a product of people's ignorance of science is an incomplete picture of reality. Just like a child growing up who has to push away his parents at some point and later begins to appreciate them, many of us, by pushing God back, have later developed a deeper understanding of God and feel closer to God than ever. Discoveries of natural laws have replaced personal revelation when it comes to understanding the world, but the motivation for the search remains a religious one.

The growth of science in the West and its open, skeptical spirit are infused with the concept of *progress*. Indeed, teleological thinking is so instrumental in the rise of the West that for the historian and socialist Robert Nisbet, it is the single most important idea in nearly three thousand years of Western civilization.

The Limits of Western Wisdom

God in the Western tradition has two sides: the anthropomorphic and the cosmic. God is a perfect symbol for transcendental inspiration, if we can get past the historical and mythic baggage that lingers. The invention of God, and especially the single God that asks us to love not just ourselves or our own tribe but the entirety of humanity, has been extremely important for the survival of the human race at a time when increasing technological and organizational power continue to make war and genocide nightmarish possibilities. However, interpreted narrowly, the Western monotheistic religions are also capable of generating the belief that only a subset of humans (the believers) are God's chosen people, and even the belief that people go to heaven after they kill off the nonbelievers. After all, God did it once with Noah's Flood and also ordered genocides, according to the Old Testament. In this sense the idea that some of us are "the chosen" is the most outrageous statement ever uttered by humanity.

This dual nature of God certainly mirrors our own nature. The coexistence of tribal anthropomorphism and a cosmically transcendent God in the Western religious tradition (and even in single belief systems and sets of scriptures) mirrors the coexistence of primitive instincts, moral sentiments, and higher consciousness in our mind. Just as some people are unable to stop their craving for sugar and fat (formed in a previous environment of scarcity) even when food becomes plentiful, so have many followers of monotheistic doctrine, motivated by their instincts, become trapped in a bygone social environment within which their religious doctrine first emerged. We can only conceive and communicate God in terms of our own mental categories.

The revolt against the dogma of an anthropomorphic God has been an ongoing saga since Athenian rationalistic thought, carried by Alexander's troops, invaded the Near East. The idea of natural evolution, and the evolution of human thought, continues to pose challenges to dogmatic interpretations of Christianity and other great Western historical religions. Western religions' anthropocentrism (a belief that humanity has a sacred God-chosen perfect essence, and that humanity is the highest creation of God) also remains a stumbling block in the controversy over Darwin's theory of evolution.

Enough ink has been spilled over the emotionally charged issue of creation versus evolution (including recent "intelligent design" variants of creationism) by erudite thinkers.[39] But what is seldom recognized is that theology itself exhibits the same dynamics of selection and replication as in the biological and cultural spheres. Just like the continuous development of Darwin's theory itself,[40] development of theology was driven by external factors as well as internal conflict between its own proper ideas.

For example, in the Old Testament, the soul is simply life itself, residing in the blood and disappearing at death. Christianity, however, anchored itself to a different sort of soul, an immortal one that faced eternal salvation or damnation. As an earliest example of the synthesis of religion and science, the early church fathers found in the Greek physician Galen's work a solution to this contradiction. There could be a lower soul residing in the liver and the heart, while the empty ventricles of the head house the immoral soul that avoids the corruption of the weak, mortal flesh. As another example, the theologians of the first three centuries of Christianity expressed many ideas that after the fifth century would have been condemned as heretical. It is a great irony that the evolutionary principle, while still facing persistent strong opposition from many Christians, is today being applied to offer a clear explanation of the success of the Judeo-Christian religion in the growing literature of the evolutionary psychology of religion. In contrast, the search for the root cause of the Muslim world's weakness inevitably ends in its unwillingness to evolve and adapt.[41]

As Alfred North Whitehead pointed out, when a new scientific theory supersedes an old one, it is always regarded as a triumph for science and never as a defeat, even though the new idea may come from outside of the scientific community.[42] So why should religion not adopt the same attitude when an old doctrine has to be abandoned? Actually, some theologians have always thought that the principles of faith are eternal, but the expression of these principles requires continual development.

The tension between the scientific democratic principle and "the tyranny of the majority" in a true democratic state is a crucial limit of Western wisdom. How can we overcome the limits of our spiritual tradition while relying on them for inspiration? How can we continue to believe we are "the chosen ones" while attempting to overcome ourselves? This is a challenge we will touch on again and again.

4.3. Eastern Wisdom: The Yin-Yang Reality

Eastern civilization, which chiefly originated in China and India and later spread to many other Asian countries, matured early. It survived and prospered for a long period of time in relative isolation before some of its key insights spread to the West in recent centuries.[43] China is the largest civilization the world has ever known in terms of time span and aggregate population. It was by far the richest, the most civilized and technologically advanced, and the most populous country on Earth for at least half a millennium before the rise of Europe, and its influence went beyond the Far East. In the words of Marco Polo, "China is a sea that salts all rivers that flow into it."[44] Even putting aside China's long-lived economic and technological prominence, the Chinese superiority in politics and ethics was widely acknowledged by the comparatively tiny nation-states of Europe from the time of Marco Polo's visit in the thirteenth century until the late eighteenth century, and won admiration from great thinkers such as Montaigne, Athanasius Kircher, and Leibniz. Unfortunately, by the time Europe finally exploded with scientific and religious vigor, the Asian giants had long been in secular stagnation and decay.

Nevertheless, Asia's core ideas are deep and timeless. At a time when Western powers conquered the whole world, Western people came to recognize the high wisdom of Eastern thought. Earlier, at the time when the Mongols established the largest continuous land empire in history, it was Chinese ethical thought and Indian Buddhism that transformed the Mongolians' primitive shamanism into a civilized culture. With a cool, somewhat detached feeling toward the cosmos, the Eastern worldview is all-inclusive and holds a much more realistic view of human life and of humanity itself than the Western tradition.

The Intuitive View of the Cosmos

The highest wisdom of the East can be described with a single word: *One*. And this notion of the oneness of reality (including all creation and manifestation) is a core principle of the Perennial Philosophy shared by all civilizations, especially Asian. As the Greek philosopher Heraclitus of Ephesus asserted, "Not I, but the world says it: All is one [*hen pantaeinai*]." One of the youngest world religions, the Bahá'í faith, inherited the spirit of the Jewish prophet Isaiah's one Creator God in its call for religious unification. But this principle of oneness lost much of its influence in Western culture, partly as a result of having been eclipsed by other concerns in the formative years of Christianity.[45] Instead, the principle had its boldest manifestation in the Oriental cultures of Burma, China, India, Japan, Korea, Thailand, and Indochina.

The Indian scripture *Kabir* states, "Behold but One in all things." The Indian concept of God, *Brahman*, is the single unitary principle. It is one infinite, eternal presence. Indians worship thousands of different gods and goddesses, but they were considered manifestations of a single divine force. The *Upanishads* states, "An invisible and subtle essence in the Spirit of the whole universe. That is Reality. That is Truth. Thou are that." The *Isha* (Super Soul) permeates all moving things, including our minds.[46] Everything in the universe came from that One in a spontaneous fashion. Sikhism, founded at the end of the fifteenth century in northern India, followed this tradition to claim that God was a formless spiritual force shared by all religions. The Neo-Confucian scholar Zhu Xi (朱熹) from the Sung Dynasty even extended the scope of this principle from natural phenomena to moral values and moral behavior (理, *li*). Whether it is the origin of species or the reason of morality, at a deep level things make sense only if we take into consideration the entire universe.

Yet this simple concept of *One* is not at all simple to comprehend. We understand the world by cutting it up with concepts and categories. Our recognition of the existence of anything depends on its being alone and separated from the world. Without contrast, there can be no perception—we need darkness to see light—yet in reality, everything is interdependent and inseparable. As George Lakoff and Mark Johnson put it in *Philosophy in the Flesh*: "The environment is not an 'other' to us. It is not a collection of things that we encounter. Rather, it is part of our being. It is the locus of our existence and identity."[47] Our mind cannot function without a social environment, and within our mind, reason is impossible without emotion, as Spinoza argued.[48] Mind itself is the ultimate unifier.

Modern psychological research suggests that such context-centered thinking is a distinctive Eastern cognitive style. For example, when asked to identify which two of the following three words are most closely related: panda, monkey, and banana, Westerners tend to group panda and monkey together, whereas Easterners tend to put monkey and banana together.[49]

Furthermore, the Eastern and especially Chinese mentality deals with the world as an "undifferentiated aesthetic continuum" without sharp distinction between the subjective and objective, between humanity and other forms of life, between internal drives and external forces, or between one's true self and God. This undifferentiated immediacy-consciousness is called *Chit* consciousness in Hinduism.

Within the One, there is no absolute, only points of emphasis. The part is in the whole, but the whole is also in the part. Everything is a matter of degree. This is the yin-yang picture that reflects the deepest nature of reality as well as our meaning-seeking dualistic mind. It is a perspective most prominent in the Eastern tradition, although Joseph Needham speculated that such simple ideas might have easily arisen several times independently in human history.

Yin and yang are polar opposites from each other, yet both reflect reality, and one cannot exist without the other. The classic yin-yang diagram illustrates this dynamic and complementary nature—that one is born *within* the other. Day and night, earth and heaven (*Tian* and *Di*), beauty and ugliness, happiness and sadness: as expressed beautifully in M.C. Escher's drawing *Angels and Devils*, the very essence of yin could be, surprisingly, yang, and vice versa. By the same token, the intense forms of asceticism found in India are a reflection of unbridled sensuality, as extremes of indulgence and of revulsion against eroticism often go hand in hand. As William McNeill has stated, "The peculiar catholicity of

Hinduism was illustrated by the success with which it enfolded undisguised sex into religion, in contrast to the emphatic puritanical emphases" of other worldly higher religions.[50]

The highest aim of Eastern sages was not growth and progress, although they understood that everything continued to evolve even within a cyclical existence. Expansion of yin or yang alone, in isolation from the other, is impossible in a balanced cosmos. Yang creates, yin destroys. Yang breaks, yin builds symmetry. Change works in both directions. So, instead, they seek enlightenment—to see the world as it is, to shake oneself free of merely human perceptions and instincts, to know more than we can express in logic and language, and to have an intimate kinship with nature. The main focus of traditional Chinese intellectuals was not searching for the "truth," but shaping a well-rounded personal character and "feeling" for life with intense practices in poetry, music, board games, calligraphy, and painting. Both Greek metaphysics and "otherworldly" Christian theology were alien and irrelevant for them.[51]

In contrast to classical science from Archimedes to Clausius, which disallowed ambiguity, turbulence, and chaos, the Taoist Chuang Tzu said, "He who wants to have right without wrong, order without disorder, does not understand the principles of heaven and earth. He does not know how things hang together."[52]

In a state of enlightenment, human emotions and categories are drained out of the mind, and what's left is a perception of the oneness of the universe and the individual's togetherness with All. (The full theory and practice of *yoga*—Sanskrit for "yoke"—involves union with the All.) As the Zen poet Seng-Ts'an said, "If you want to get the plain truth, be not concerned with right and wrong. The conflict between right and wrong is the sickness of the mind."[53] The distinction between good and evil, so important in Western civilization, reflects the same, misguided human preoccupation rather than any valid inquiry into the essence of the cosmos. In short, most Eastern sages simply were not preoccupied with the false dilemma that still bedevils Westerners: How can an omniscient, omnipotent, and benevolent God allow evils like wars, torture, natural disasters, and disease to befall innocent persons, even children? In their view, this whole line of questioning betrays a fundamental lack of understanding of yin and yang.

Although Zen is considered a religion, it has no gods and devils, no heaven or hell, and is virtually devoid of theology, scripture, or ritual.[54] As the Zen master Tozan put it, "The higher Buddhism is not Buddha." What Zen cares about most is self-transcendence, to seek a deeper understanding of the self and gradually widen this self-concern to eventually identify oneself with the entire universe. In sharp contrast with science, which relies on clear expression and communication in its search for truth, enlightenment cannot be communicated; one can only communicate the *way* to enlightenment. Zen truth is lost once articulated. True knowledge transcends the world of language, the world of contradictions, and the world of logic.[55] One of the Sanskrit terms for a sage is *muni*, "the silent one."[56]

For the Eastern sages, the ultimate stage of human enlightenment brings a pacific, harmonic, enduring, and all-embracing attitude. Buddhism was "the first major religion to stress tolerance and nonviolence, the only major religion to spread far and wide without conquest, and arguably the major religion whose founding doctrines (unembellished by later additions) most readily survive the modifying force of modern science."[57] From the Brahman view, a lot of debate, including religious debate, is really missing the point in the sense that the participants are pointing to different aspects of the same reality—the blind men arguing about what an elephant is. That's why each person can be "one hundred percent correct" yet hold points of view that seem diametrically opposite. Modern physics's dilemma of light being both a wave and a particle would not be puzzling to the Eastern sages.

This attitude of external restraint, even disengagement, in exchange for internal freedom, agility, and gracefulness is shared by the Epicurean and Stoic schools in Greek and Christian philosophy. The Bible and other Western scriptures contain flashes of a cosmic *wu-wei* God, for example, when St. Paul said God's "power is made manifest in weakness" (2 Corinthians 12:9). Reflected in Eastern martial arts, it is relying on the enemy's attacking force to destroy the enemy—you don't have to assert much energy yourself, something the dolphin is a master of in overcoming the resistance of water. As Hannah Arendt stated, "Just as war takes place for the sake of peace, thus every kind of activity, even the processes of mere thought, must culminate in the absolute quiet of contemplation."[58]

Wu wei is not about passive resignation but about wisely taking hold of a wider perspective. Let me illustrate this with the Chinese fable "Border-Man-Lost-His-Horse" (塞翁失马). Briefly, the story is about an old man living in a border town in northern China. One day, his horse ran away. His neighbors came to commiserate with him, knowing the horse was the mainstay of his livelihood. But the man insisted that there was nothing sad about it. He said, "Who knows if this is good or bad?" Several days later, the horse came back, bringing with her a wild horse that was fast and strong. The neighbors congratulated him on this good fortune, but he insisted there was nothing to be happy about because he could not know if it was good or bad. Indeed, his son broke his leg riding the wild horse. Again, the old man disagreed with his neighbors about the apparent misfortune. Soon, the northern tribes invaded China, and all young men were forced to join the army. His son, with the broken leg, avoided the draft, which saved his life.

Realistically, humanity cannot see very far, and the awareness of this fact is invaluable. Evil and suffering may be Heaven's way to do good, even for the apparent victim. The laid-back attitude that the "Border-Man-Lost-His-Horse" story promotes seems to be in sharp contrast with the Western confidence in what humanity can know and do, although other kinds of influences—for example, a sense of the mystery of God beyond our sense of good and evil, or Jesus's admonition not to be concerned about tomorrow—also inform the Western religious perspective.[59] The discouragement of worldly strife and the tolerance of amoral or evil things in the world while maintaining one's own internal equanimity is practical but not easily done, and requires a higher vantage point. This is the Eastern wisdom of seeing today's events, however important and emotional to us, in the context of the grand scheme of things.

The 1965 discovery of the residual heat from the Big Bang by radio astronomers provided strong support not only for the Big Bang theory but for the oneness of the cosmos: everything that is "out there" has a single origin. The *Brihadaranyaka Upanishad* recorded the forest-dwelling meditation masters' metaphysical assertion that "in the beginning there was nothing" (1.2.1), and that the world is radically interconnected since everything was "Brahman's super creation" (1.4.6).[60] Brahman as the ultimate and absolute reality is "the womb of both the existent and the non-existent" (4.1). Sage Uddalaka suggested that the entire phenomenal universe was in the beginning "the existent" (*sat*) only, "one alone without a second."[61]

The Realistic View of Man

Eastern wisdom is founded on a deep understanding of human nature, especially its limits, as summarized by Confucius's words, "Man can express and amplify the Tao, but not the other way around" (人能弘道，非道弘人.《论语 • 卫灵公》). In sharp contrast to the mainstream Western monotheistic views, the mystic conception of an ultimate reality that humanity can never hope to reach is echoed by the current physical understanding of the brain as a limited entity. Math, logic, science, language—everything we rely on to understand the world—is very limited. Human standards of judgment, whether based on facts, logic, morality, or ethical desirability, may never be adequate, because there is a limit to our biological life, but there is no limit to knowledge embedded in the universe. Chuang Tzu believed that we are inherently limited by our natural abilities and experience:

> The blind have no way of joining us in the seeing of patterns. The deaf have no way of joining us in hearing the sounds of drums and bells. Why should there be only bodily deafness and blindness? It is likewise the case with knowledge.
>
> You cannot talk about the sea to a frog who lives in a well. He is circumscribed with respect to space. You cannot talk to the summer insects about ice. Their life is concentrated with respect to time. You cannot talk to a small-minded gentleman-warrior about the Tao. He has been bound fast by his indoctrination.[62]

To a certain extent, this view is shared by ancient mystics in the West, but this realistic view of human beings holds a far more prominent position in Eastern thought. In contrast to the chauvinistic Western view that "a lion must feel God is more on its side than the gazelle's," the Eastern sages knew that the human being is a worm and food for worms. In its this-worldly orientation, Eastern thought is also more realistic about our conscious mind and its concerns. We "know" that we know all too well what is going on and that's too much for many of us to bear. Human consciousness finds itself bounded by the biological body in a particular space and time, and, in learning from the experiences of others, rightly concludes that the body and the consciousness itself are moving toward death with certainty.

Given the understanding that human beings can rise to a higher perspective consciously but can't be lifted to a state of freedom from their biological and historical groundings, the basic problem of humanity is twofold: how to be happy and how to achieve eternal life. The mind is capable of pure enjoyment (such as the aesthetic pleasures), but it is constantly being bothered by various desires, which often seem to be important when they are not satisfied but no longer so once they are satiated.

Pain and suffering are inherent in our design because they are very useful for our survival. Realization of this fact through conscious reflection is actually the first step toward *alleviation* of pain and stress. The Eastern sages came up with various psychotherapies—from isolating the consciousness from animal instincts and inciting hallucination through oxygen starvation, to relieving consciousness from the burden of responsibility altogether. True salvation, however, is achieved through a profound inner peace, the sense of an ineffable Tao beyond human understanding and motivation, and not necessarily the mystic detachment from reality.

By focusing exclusively on the human situation, Siddhartha Gautama (the Buddha) omitted any appeal to the gods in the Hindu tradition and rejected any philosophical speculation. His words may be extreme, but they capture the essence of Eastern realism: "In the search for truth there are certain questions that are unimportant: Of what material is the universe constructed? Is the universe eternal? Are there limits or not to the universe? What is the ideal form of organization for human society? If a man were to postpone his search and practice for Enlightenment until such questions were solved, he would die before he found the path."[63] Confucius expressed the same philosophy when he told his student to "respect the gods but distance yourself from them" (敬鬼神而远之,《论语・雍也》之二十二, Analects • Yong 22).

The Buddha stood out among the world's prominent religious leaders in that he was probably the only one who walked away from a life of wealth and power. Youth, wealth, and even life itself were like passing autumn clouds to him. He was the enlightened one who seemed to have complete control over his animal instincts. Yet his preaching seems not to have materially changed the world as Western religions have. He correctly sensed that making human beings totally satisfied and eliminating suffering was impossible—for that would involve eliminating desires, even including the desire to live! Suffering is an intrinsic part of human experience. The attempt to isolate consciousness from instincts is a high-wire act, since the mind's own survival could be in jeopardy if separated from the external world and internal knowledge. Our lives hinge on our basic drives, which puts a limit on disdain for material things. As the writer Lin Yutang put it, "Even the most spiritually dedicated man cannot help thinking about food for more than four or five hours."[64] One must live before one can reflect.

This, in a sense, is also the practical wisdom of Hinduism, which boasts of highly intellectual and philosophical scripts like *Upanishads,* but allows the masses to create and enjoy numerous popular tales, epics, and other colorful myths such as the *Mahabharata.* The Hindu doctrine that the multitude of things and events in the world are but different manifestations of the same ultimate reality, the Brahman, may sound naïve, but this "let-loose" attitude reflects the modern evolutionary principle of creativity and stability through diversity and spontaneity.

The Eastern realistic view of humanity shaped a down-to-earth attitude toward life and a durable ethical and political system. Dhiravamsa once said, "The whole point of Buddhism may be summed up as *living* in the present."[65] With the similar intuition that it is futile to ask the masses for more, Confucius taught that we should concentrate on living in the here and now. His world is the seemingly mundane human world, although he clearly recognized the larger forces (天命, *tianming*) that determine our lives and admired the transcendence and spontaneity of the Tao. His teachings are almost exclusively moral, without any metaphysical pretensions. The way of heaven is too remote to be comprehended, although he sensed that "heaven is the author of the virtue that is in me."

The Limitations of Eastern Wisdom

The first limitation of Eastern wisdom is its over-reliance on intuition and mundane needs. It comprehends human limitations well, but cannot envision an enlargement of human senses and cognitive capacity. By contrast, the advent of modern science was made possible in the West not simply by clever thoughts or even flashes of genius, but by better tools extending or aiding our biological senses. With his optical telescope, in 1610 Galileo Galilei discovered a hidden universe just beyond the reach of our eyes. He found dark spots on the Sun, which showed to him that the Sun is not perfect but material. His findings of four Jupiter satellites forever changed the anthropic or Earth-centric worldview. In contrast, the Chinese had known about sun spots for two thousand years but did not perceive this fact to have any deep religious or philosophical significance. No other early civilization excelled like China in the quality and quantity of scientific observations and applications, yet Chinese culture never pointed toward scientific *principles* as we know them today.

The wonderful insight of the world as *One* has two sides. Perhaps nothing better illustrates its downside than the Chinese story of "One Finger." Once upon a time, three students were on their way to the capital to take the *kejiu* exam (科举, China's imperial exam for civil servants). Stopping at a temple, the anxious students asked a monk how many of them would pass the exam. The monk looked at them, raised one finger, but said nothing. Later one of these students passed the exam, and they applauded the monk's divination power. Yet the joke is that if two had passed, "one" would be interpreted as one failure. And if all had passed (or failed), the monk's one finger would mean all passed (failed) at once. By signaling "one," the monk seemed to know it all, but he really knew nothing at all.

The premature stagnation of Eastern wisdom was, in a sense, unavoidable because its holistic, balanced, and all self-containing mode of thinking made the East an arid land for science and otherworldly transcendence. Hints of scientific thought appeared during the Axial Age and again during the Sung Dynasty's neo-Confucian revival in Cheng Yi's (程頤,1033–1107) idea of *gewuzhizhi*[66] (格物致知, gaining knowledge through rational investigation of nature), but they never found fertile ground.[67]

Cultivated reductionism and abstract codified law were distasteful to intellectuals in the East. "Authentic existence" in the Western leap-of-faith mode would indeed be "absurd" to them. When the yin-yang view is interpreted too narrowly, the individual gets trivialized, and nature seems to be too balanced (with any given action always followed by an action in the opposite direction) to allow true novelty to emerge, as if held down by "a silk ceiling."[68] Thus the Chinese didn't even have a word corresponding to the classical Western idea of a "law of nature."[69] Nor could they ever develop an art of *tragedy* in the classic Greek sense of heroism and mastery—the struggle of a person to develop healthy self-assertion and to reach beyond one's existence to the "other."[70]

The East faced a different set of barriers to truly grasping the concept of evolution than the West. With a static universe and a constrained life, the highest human aspiration was limited to human "happiness" and moral excellence. Unfortunately, many people in the postmodern secular Western world are mentally suffering from this Eastern limitation today. It never occurred to the Eastern sages that *the best way to approach the Tao is to realize its potential through our own actions.* Even sages such as Confucius longed for a return to a perfect Golden Age in the distant past that had since been corrupted by social changes of his time. This "progressive" philosophy of anti-progress, this deep repugnance of the utterly new, was a lasting dominant force in the East, although it was shared by some Greek philosophers and by ancient Egyptian, Aztec, Zoroastrian, Icelandic, and Irish sages. Both Plato and Confucius longed for a utopian community ruled by a wise, benevolent aristocracy. Their Golden Age, admired by many Eastern thinkers throughout history, was in an unchanging, arrested state, with nothing to improve upon.

4.4. Reflections on the Axial Age and Human Wisdom

Just like the Cambrian explosion, when all complex animal structures we see today suddenly appeared in the fossil record, the seemingly sudden emergence of human wisdom during the Axial Age around the world as it appeared in written records was an *ecological phenomenon* and should be viewed in a historical context. The accumulation of quantitative change leads to a sudden qualitative jump when a certain threshold is reached. This is the same pattern that we see when we heat water to the boiling point, or in Zen practitioners' sudden enlightenment after long periods of frustrating training and meditation.

The Russian scientist Valentin Turchin used the metaphor of a "stairway effect" to illustrate what he called Metasystem Transition: Imagine a baby is playing at the bottom of a gigantic stairway and trying to climb up. Initially, the baby is unable to grab the edge of the step and clamber up. Years pass. One day the child suddenly makes it, and once that happens, the child is able to climb step after step to go ever higher. Before the first step, the child cannot ascend an inch, yet after that first step, not only the next step but the entire stairway becomes accessible.[71]

From noticing regular patterns and causations to forming abstract and quantitative concepts such as time and space, our way of thinking must have taken a great leap. Like young children, preliterate people commonly viewed space and time in terms of concrete activities and emotions, and things and people in vivid pictorial images. Just as new species, including *Homo sapiens*, usually arise amid violent climate changes, drastic social changes and crises no doubt also stimulated the motivation for reflection and created room for bold new visions. For example, China's "Hundred Schools," which included Confucianism, Mohism, Taoism, Legalism, Logicism, and the Yin-Yang/Five Processes (五行) school, flourished during the disintegration and political turmoil of the late Chou period, around 500–250 BCE; and India had as many as sixty-three schools of philosophy at the time of the Buddha (563–483 BCE).[72] The *Bhagavad Gita* was originally written to reconcile the differences between major ancient Indian philosophies that evolved over the early centuries of Hinduism.

The sweeping changes in the historical backdrop also made cultural mutations more likely to emerge and take root. In Charles Fair's eloquent words:

> The life of Christ himself may have owed much, besides his martyrdom, to the Roman presence, since for better or worse, that presence meant a kind of order and therewith a potential flowering of human warmth and insight. Such is the springtime of prophets, of men who come to understand more of man's nature and of his possibilities than there are yet words to tell. Unencumbered by the later-developed apparatus of conscious formal thought and by the peculiar prejudices and lacunae in understanding which these inevitably bring, the seers of early cultural time, who are the source of much

> that is to come, apprehended before all else what man *might* be.
>
> And in a populace itself slowly coming over to a more secure, more inwardly expansive way of life, their words had a profound effect, for collectively such a populace is like a youth emerging from the fiat rule and mental dependence of his childhood. He is ready to shed or rather to regrasp much of the past, to recognize the possibility of transcendence in itself, to approach the uncertainties of existence not with awe and fright, but at last with hope and wonder.[73]

In other words, in the Axial Age, the human being is beginning to be seen no longer as a merely adaptive being, but as a courageous, outward-looking force.

Brahmanism in India, for example, brought forward the radical idea that the world seen through the eyes of instinct or everyday self-interest is an illusion.[74] That viewpoint must be taken from a position detached from oneself and one's senses of the world. Through *atman*—the knower of all knowledge, the "perceiver of perception"—consciousness undertakes what Stevenson and Haberman call "a shift in identity from the transient ego-self associated with the body to the eternal and infinite self that is not different from the All."[75] That view of the world was probably inaccessible to people until the key concept of totality emerged during the Axial Age.[76] The word *universe* comes from the Latin for "all things turned into one."

According to the French philosopher Remi Brague,

> [T]hese [great river valley] civilizations do not seem to have had a word capable of designating the world in its entirety, uniting its two components. In Chinese the modern word *shijie* (世界, in Japanese *se kai*) is formed from *jie,* "circle," and *shi*; the second word also signifies "generation, duration of life," which makes it close to the Greek word *aion.* In India the Sanskrit *loka,*"visible space" (as in the English *look*) can indeed be translated as "world," but in such a way that it is the expression *lokadvaya,*"the *two* worlds" (heaven and earth) that we must ultimately see as corresponding to our "*the* world." In Hebrew, the medieval and modern word for "world" is *olam.* Its root is Semitic, the same as the Arabic word *alam*—which itself came from Aramaic. The word is indeed found in the Bible; but there,

> it has only the sense of "unlimited time, eternity," and not yet that of "world," which it would only assume in the Talmudic period through the intermediary of the meaning of "era"...
>
> For there to have been a word meaning "world," the idea that it expresses would have had to have reached human consciousness. And this assumes that people envisioned a concept in its totality, a category grasped in its two moments, that is, as a synthesis of the first two categories of quantity, plurality and unity. It is necessary, therefore, on the one hand, that the parts that make up the whole be dealt with exhaustively, without anything being excluded, and on the other hand, that such totality be considered unified. Since we are dealing with a physical totality, its unity would consist of being ordered, well ranked, etc.[77]

Thus, human wisdom, especially the transcendence of the mundane, came from personal freedom and conscious free will, and those things can occur only with conscious separation of the thinking person from the prior state of organic unity of the whole world. It was the Greeks who gave the world a name of its own: *kosmos*. Pythagoras was believed to be the first to call "kosmos" the encompassing of all things, including the divine sphere, *theos*; Plato was the one who gave the word the definitive meaning as "world"; and it was the Stoics who first described themselves as citizens of the world. The transcendent nature of the Greek worldview becomes more apparent when it is compared with the Egyptian conception of "negative cosmology," a partial, fluid, and problematic world without consistency and regularity because it was dependent on "the incessant joint activity of gods and humans."[78]

Equally important has been the discovery (and measurement) of time in the modern sense of a linear one-directional "flow," which led to an intense sense of personal time, the sense of history of the community, as well as investigations into the history of the universe or the human species. The Palermo Stone in Egypt (2500 BCE) was probably the oldest surviving written chronicle. The Jewish tribes were famous for their obsession with chronicles and genealogies. In Greece, Hecateus of Miletos began the tradition of writing social history, a record of the past and contemporary events rather than legends and myth. Aristotle began the inquiry of natural and biological history from observations.

For us, it is hard to imagine how people lived without the concept of the "world" and "time" for nearly half of civilized history—this, of course, is without realizing that we live an extremely atypical life for our species. But many scholars have argued that this transcendent worldview is a luxury, something we can live without and something most people continue to live without—at least without its meaning and value in guiding individual actions. The meaning of the world, not the awareness of the world, is the central problem that the Axial Age wisdom had shed light on. As Remi Brague points out, "'world' has never designated a simple description of reality: it has always translated a value judgment, the fruit of a sort of act of faith, either positive or negative."[79]

Looking from a distance, the Axial Age was undoubtedly an exciting time, but it would be a big mistake to regard it as a Golden Age that we should aspire to return to. Axial Age wisdom came from a tiny elite who had only a marginal influence on their immediate social environment. The battle between exceptionally creative minds and conservative traditions, between transcendental inner spiritual aspirations and obstacles in the external world, has been fought throughout history, with the conservative forces winning most of the time. It is also an illusion that the ancient sages anticipated all later scientific discoveries and technological developments. Even the elite only produced a gold mine rather than pure gold—it was up to later generations to identify little gold nuggets here and there among some lesser metals and a vast amount of worthless rock. Nevertheless, we have found ourselves constantly returning to the original sage insights. Even with tremendous learning and advancement in technology and culture, modernity is still ruled politically and spiritually by two great creations of the Axial Age: the great world religions and the idea of an imperial political order. From philosophy, ethics, military strategy, and politics to sex symbols and sporting events, the Western world invariably turns to Greek and Roman classics for inspiration and cultural and intellectual grounding.[80] As Alfred North Whitehead famously claimed, "The safest general characterization of the European philosophical tradition is that it consists of a series of footnotes to Plato."[81] The same overwhelming Axial influence can be seen in Eastern philosophy and practices.

The greatest achievements in science and technology can be viewed as mining the Axial Age gold mine: what we are able to accomplish is to test and refine the original speculative ideas proposed by earlier thinkers. What we believe to be totally new technological ideas can often been traced back to the beginning of the civilizations—the concept of an engine run by steam was first recorded in 200 BCE Alexandria,[82] and the earliest computer can be traced to mechanical devices invented in Greece at about the same time.[83]

If we return to Valentin Turchin's stairway metaphor, this is not to say there have been no advances since the little child climbed the first step from the bottom of the stairway. To the contrary, like the accomplishment of the child who continues to get up the stairway, both the technological and cultural progressions since the Axial Age have been astounding, although the process is uneven and littered with setbacks. Today, we know far more than the ancient sages, and our societies are materially and structurally richer than the ones that they lived in due to the advancement of science and technology as well as social and institutional innovations. But still, just as biological organisms have experienced little true evolutionary novelty since the Cambrian period some 500 million years ago, human beings have advanced little in terms of how we look at the world since the Axial Age. Since the Cambrian period, not a single new phylum has evolved; since the Axial Age, not a single new worldview has appeared.[84] Here are just a few reasons why I make this statement:

- Most of the important questions we can ask—such as "Who are we?" "What we can know?" "Why is there good and evil?" and "What should we live for?"—were first raised during the Axial Age. With each generation, new answers were offered, but it is the kind of questions, not the answers, that really matters.
- The highest possible human perspective is the transcendent Tao and the Western idea of the Creator God. The *Tao Te Ching* (道德经) lays out a transcendental view of the cosmos wherein the Tao is not just the path of heaven; it is not just the purpose of heaven; it is not even the origin of all life in the universe—it is the origin of the Origin. The Tao begets the One—the Origin. Thus it is beyond even the Origin.[85] This sense of you-can't-really-know-unless-

you're-outside was only recently described rigorously by Gödel's Theorem and Chaitin's Theorem.[86]

- The sense that we are citizens of the world originated in the Axial Age. The sense that Earth may not be the only place in the universe containing life was voiced by Lucretius in the fourth century BCE, based on the assumption that the laws of nature are universal.
- Our dominant mode of thought remains the narrative one, as it has been since antiquity. As the psychologist William James pointed out, "Mankind's common instinct for reality has always held the world to be essentially a theatre for heroism. In heroism, we feel, life's supreme mystery is hidden."[87]
- From Freud's unconscious to quantum mechanics, quite often a new idea is an extension, a footnote, or simply a return to views expressed in the past but with new insights based on technical innovation and recent experience or experiment.[88] No one from the Axial Age proposed the Darwinian theory of evolution by natural and sexual selection, but the germ of the idea was born then, in Heraclitus's idea of universal strife,[89] in the Jewish prophets' concept of a linear history, in Lucretius's (99–55 BCE) insistence on nature without a designer, and, perhaps, in Anaximander's belief that human beings descended from fish.

In summary, the highest possible human perspectives were already visible to the keenest minds (though with less detail and sharpness) during civilization's initial growth spurt, and, as we will discuss later, further breakthroughs are unlikely without creation of a new form of intelligent being.

Today's elite are no better than the ancient sages and prophets, and today's masses are no match for these ancient minds, in regard to reflexive consciousness, historical perspective, and conviction of our transcendental potential.[90] That is why the preaching of universal principles and higher moral standards, both largely derived from the Axial Age, remains as important a social function today as then. In other words, we can say that the evolution of human consciousness has not progressed qualitatively beyond that achieved by the Axial sages.

What is more, there has been no deeper revelation of religious and spiritual ideas since that seminal period. Science and technology have improved resolution and reduced distortion of our natural senses, but they have not done much to improve the scope of our perspective.

It is not strange that the highest human perspectives, the Tao and the Creator (God), emerged as soon as human consciousness was fully developed. Our biological nature is fundamentally unchanged, and we are still running society with our unaltered cognitive architecture—which includes, for example, the ways we process sensory information and communicate with each other, the basic motivations and emotions, the capacity for reflection, and the kinds of behaviors we are predisposed to engage.

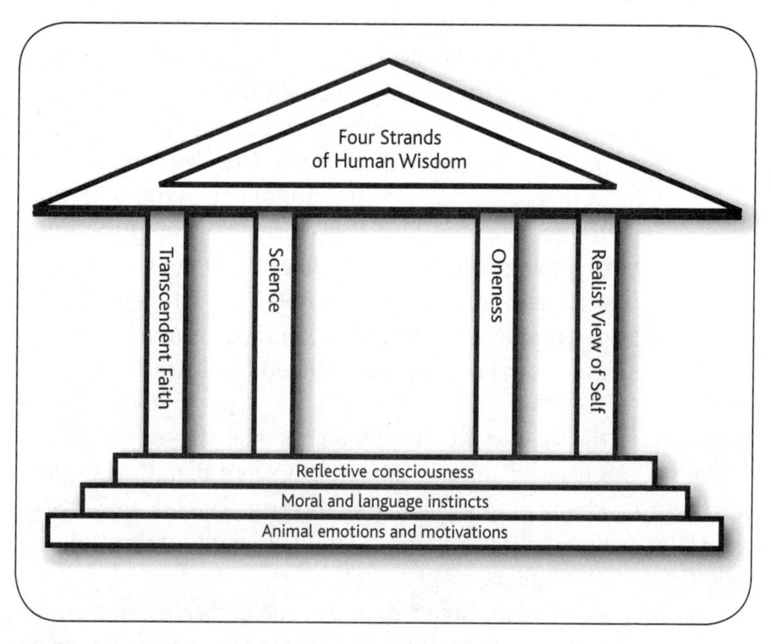

Figure 1: Four Strands of Human Wisdom

Our human mind is shaped by natural and cultural evolution. On the one hand, our emotions and motivations are inherited from our animal ancestors; on the other hand, we also possess very strong moral and language instincts that enable cultural evolution. But what's most unique is our reflective consciousness, and its best products are four forms of wisdom that guide our cosmic evolution, represented here as the four columns.

Those who are intoxicated by dazzling modern technologies should keep in mind that the ideas that arose during the Axial Age remain the deepest wisdom that we return to every time we need to grasp a higher perspective after traveling down a path that ultimately proves limiting.

Our human intellectual heritage is not the pinnacle of evolution but a steppingstone to a higher existence. The remaining chapters of this book are, in a sense, a rudimentary attempt to unfold this chapter's compressed account of human wisdom. What is important is not to know this wisdom expressed in words, but to understand its implication for the future.

CHAPTER FIVE

THE COSMIC VIEW

The further back you look, the farther ahead you can see.

—Winston Churchill

THE COSMIC VIEW offers a scientific view of the past—where everything comes from. But the past in and of itself has no value. The purpose of knowing the past is to aid our understanding of the future. Aldous Huxley once remarked, "That men do not learn very much from the lessons of history is the most important of all the lessons of history." The more history that we can put into our consideration, the better for our purpose of seeing the future. Not knowing what happened before you were born, before your country was formed, even before your species appeared, means remaining a "child" forever.

But we can never fully understand the past. Historical narratives are necessarily selective and poetic. "Mythic history [has always] attempted to order the events of the past within a framework of self-centered, self-referential causes and effects that ultimately accounted for all things known."[1] Even modern science-based history relies on the use of stories, metaphors, and symbols. Darwin's *Origin of Species,* for example, was seen as "a work of literature, with the structure of tragic drama and the texture of poetry."[2] Yet there is a crucial difference: Darwin's setting for his narrative is the larger context of the universe, and his work relied on the accumulation of objective observations and empirically testable conjectures about the past. It is a better description of anthropological history than any other alternative. The same is true for the Cosmic View.

5.1. The Enlightenment and the Evolutionary View of the Universe

What is this new science-based history telling us? In a nutshell, it says that *history is an expanding evolutionary process characterized by the wondrous emergence of unprecedented structures from a bottomless well of cosmic potential.* This view is radically different from ancient cyclical views of the past, although not entirely new. Heraclitus in ancient Greece was probably among the first men who proposed *change*, rather than cyclic processes within a static whole, as the fundamental vital force for the river of events that we call history. His famous words are "everything is in flux" and "you cannot step into the same river twice"; the only thing that is constant is the law (the Tao, as it is known in the East), not the form (everything we can perceive with our senses).

Heraclitus's notion of fundamental instability is somewhat counterintuitive. It is easier to accept essentialism, as expressed in Plato's claim that there are a limited number of classes of objects, that each class of objects has a fixed definition, and that any variation between entities in the same class is only accidental and degenerative from an underlying realm of essences (or *eide*). In other words, it is easier to conceive a pre-existing, eternal "order of things" that imposes structural limits on all subsequent unfolding of things and events. After all, many features of our lives don't undergo fundamental changes in our lifetime or even in our collective memory. In civilizations around the world, myths, legends, and stories about an eternal unchanging reality underneath cyclical appearances far outnumber narratives of continuous emergence of true novelty after the world is created.

Since antiquity, histories have usually been represented as a succession of creation/destruction cycles. One of the earliest examples is the *Atrahasis Epic*, a Babylonian creation myth. These legends of the ages are largely about moral degeneration in societies. The mature world is like a human body that grows old and weakens over time—and the only hope usually presented for escaping that process is either the rebirth of a new body, a return to a bygone Golden Age, or the initiation of cosmic catastrophes by a superior, often divine Creative Force.[3]

The Industrial Revolution added credibility to the modern notion that change and evolution are an eternal reality. According to Ernest Gellner, "Modern society is the only society ever to live by, through, and for sustained, continuous, cognitive and economic growth. Its conception of the universe and of history, its moral and political and economic theory and practice, are all profoundly and inevitably colored by this."[4]

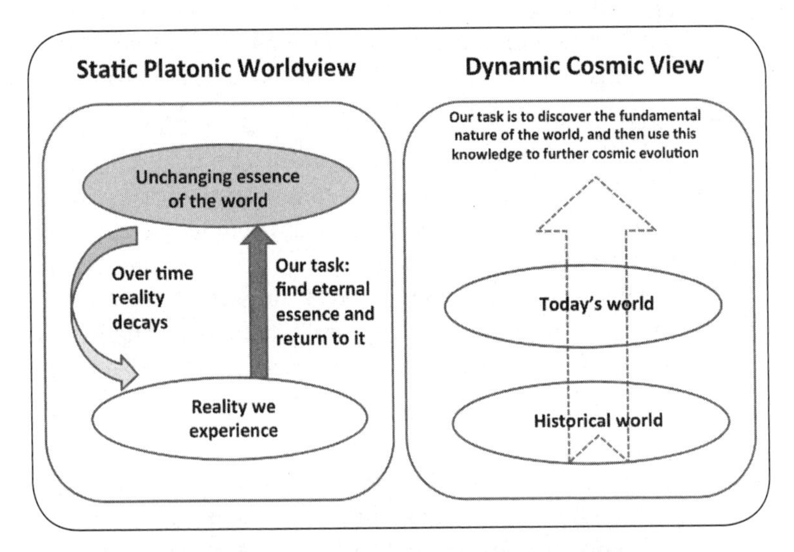

Figure 2: The Platonic versus the Cosmic Worldview

What Plato, Confucius, and many others failed to recognize (including numerous founders of utopian communities throughout history) is the malleability of the social environment, and this fact quickly rendered their static ideal of a utopian aristocracy useless. However, perhaps we are premature in declaring the victory of modernism over the old Platonic worldview. As we will discuss in chapter 8, the transcendent perspective remains a minority view. The tension between conservative continuity and innovative discontinuity—between acceptance of what we can live with and eagerness to overcome our limitations—has been, and will always be, the focal point in social dynamics. These are some key examples of this tension:

- In physics, the universe codified by Newton is essentially an inert cosmos. It was Leibniz who presented us with an energetic universe. But still (in William Barrett's words), "the notion of energy is one of the most difficult to grasp conceptually, for our minds find it easier to think of objects and bodies."[5]
- Keynesian equilibrium economics, with its Ricardian tradition of focusing on maintaining a static equilibrium, has always found true believers. Contrast this with Schumpeterian dynamic economics, with its Austrian focus on entrepreneurial creative destructions of old production systems and births of new technologies and organizations.
- In sociology, the vision of humanity as the Noble Savage degraded by civilization, popularized by the eighteenth-century philosopher Jean-Jacques Rousseau, has left a lasting impression in intellectual and popular imagination. But this idea was hardly new. Like much of the romanticism it so heavily influenced, it follows a long conservative (or backward-looking) tradition.[6] According to this belief, the ideal essence of a person could be revealed with the right environmental influences.
- In biology, the bioconservatives still hold the static view first proposed by the German zoologist August Weismann, that the hereditary material they call "germ plasm" is the invariant ideal form that determines adult organisms. The ideal essence (its God-given perfect nature) is reproduced indefinitely through each generation and cannot be influenced by the individual mortal adult organism. DNA is the "unmoved mover."
- In politics, static worldviews are far from extinct in ideological debates. Today, more than ever, the political divide is not between liberals and conservatives, but between "statists" and "dynamists," as Virginia Postrel declared in *The Future and Its Enemies*.

These examples illustrate how it is useful to point out the broader implications of evolutionary theory beyond the question of the origin of species, and to lay out some of the critical implications for our understanding of the cosmic, biological, and cultural history of humanity and its environment.

As we have discussed, Darwin's and Wallace's discovery of natural selection as an evolutionary principle is the most important discovery of human history[7] because it reveals a very counterintuitive creation mechanism. Modern scientific research has expanded the principle of species diversification all the way back to the beginning of cosmic history. The fundamental new understanding of the universe was that everything on Earth and beyond has an evolutionary origin and history. Darwin's new logic, which is not limited to the specific biological progression mechanism, managed to span three seemingly unbridgeable metaphysical chasms: He showed how selection united the nonliving and the living, the nonhuman and the human, and the physical and the mental, into a single fabric of intelligible causation.

The Cosmic View is different from any other human story or myth in its transcendental and open-ended nature. Unlike the old belief of cosmic history as a battle of good and evil, the big picture provided by this viewpoint has no overarching protagonist and antagonist. Unlike the old belief in an eternal heaven, there is also no assurance of happy endings for humanity, regardless of what it does. In fact, there is no ending of any kind for anything. The big picture involves everything and everybody but is utterly impersonal at the same time. For each of us, it does not really matter what we expect from life, but rather what life expects from us.[8] In other words, there is no feedback to our wishes and desires unless these are committed to action. Emotions and intentions are seen as effective mechanisms in biological and conscious evolution, but not the fundamental driver of the whole process.

The whole cosmic process is still unfolding, and proceeding in a grand style that is richer than any component of the whole, including humanity. With a rich interplay of chance, necessity, and historical contingency, *it just seems to us like a messy, eventful, playful, exhilarating, unpredictable, unrepeatable striving to realize the cosmic potential, the Tao*. Rarely steady and regular, it displays a wonderful mix of familiarities and shocks as we (and perhaps other intelligent beings) proceed. It is what the journalist Joel Garreau called the "Prevail Scenario," which is open-ended progression and transcendence with "fits and starts, hiccups and coughs, reverses and loops."[9] It bears a striking resemblance to the messy and unpredictable evolutionary history of scientific thought itself.

As we shall discuss in detail, the transcendent nature of evolution is a difficult idea for most of us to fully understand or accept, yet the story of evolution and the emergence of human consciousness is our best effort in making sense of this grand process.[10] What makes the theory of evolution so remarkable is the enormous simplification and unification it imposes on the panorama of life on Earth. Within the scientific community, evolutionary theory is viewed as one of the most trustable and robust theories we have ever had.[11] In 1998, the National Academy of Sciences matter-of-factly described evolution as "the most important concept in modern biology." The geneticist Theodosius Dobzhansky famously proclaimed, after noting the tremendous diversity *and* unity of life, "Nothing in biology makes sense except in the light of evolution. . . Without that light it becomes a pile of sundry facts."[12] Indeed, evolution explains not only how species evolved but also the neurological development in our brain and the immune system in all vertebrate animals.

Of course, evolution also applies outside of biology. Joseph Schumpeter's seminal idea of creative destruction as the driver of economy and society followed the insight of Karl Marx, who in turn was inspired by Darwin. This notion of creative destruction also applies to cultural and political developments, institutional development, and everything in our lives, including our aesthetics, morality, evil motives, and various human beliefs (the evolution of memes). Even the concept of evolution evolves: historians have found the most natural and fruitful method of constructing history to itself be an evolutionary process—a trial-and-error process of "fitting things together."[13]

The theory of evolution by natural selection, like that of economic development from creative destruction and the survival of the fittest, is about a process (mechanism) that is independent of particular circumstances on Earth and probably in the universe. For example, through evolution the eye emerged independently in several different species to take advantage of light. "Is evolution a theory, a system, or a hypothesis?" wrote the French Jesuit paleontologist Pierre Teilhard de Chardin in the early twentieth century, "It is much more: it is a general condition to which all theories, all hypotheses, and all systems must bow and satisfy henceforth if they are to be thinkable and true. Evolution is a light illuminating all facts, a curve that all lines must follow."[14]

As a general principle about how things appear, evolution does not just involve biological life. The evolutionary principle of variation-selection-replication has even provided interesting hypotheses about how the universe itself came into existence.

The grand narrative must not be told exclusively from the human perspective. One of the startling implications of evolution, for example, is that probably every vice was once a virtue in the evolutionary process. We have to have a transcendent perspective in order to understand evolution that is transcendent in nature. What is true of evolution for biology is true for human behavior, but the light of evolution is just that—a means of seeing better. Meaning exists *only* in the connections we perceive,[15] and evolution is the best way to investigate it.

The value of evolutionary theory is best appreciated when one considers how inadequate competing worldviews, such as creationism and determinism, are in explaining anything.[16] As J. L. Monod pointed out, no other scientific theory has had such tremendous philosophical, ideological, and political implications.[17] The examination of today's realities will remain mired in confusion or relativism without reference to the process that created them all, even though comprehending the universe is not on top of most people's minds, and cyclical processes often dominate the small-scale world that we experience.

5.2. From the Big Bang to Conscious Mind: The New Paradigm of Cosmic History

The specific contents of the cosmic story—"What is this world and where did it come from?" and "Where did humans come from?"[18]—have undergone countless revisions, and the latest consensus version is unlikely to be the final one. Not only does new scientific research bring surprises every day, but any history is necessarily a highly selective description of an infinitely rich past. Finally, every age rewrites history based on its unique perspective and interpretation. While this certainly does not mean all perspectives and interpretations are equally valid, history is necessarily an imperfect representation of the past manipulated to fit our limited minds.

Furthermore, the Cosmic View, with its glorious details, came into being only very recently, in "the last thirty seconds" of the "day" of human history, and by no means has it become universally accepted like the idea that the Earth moves around the sun. Many Christian fundamentalists still adhere to a literal reading of the creation story in Genesis. Some postmodernists still deny the existence of any direction in history and call the Cosmic View "popular Hegelianism."[19] It is also difficult to get everyone, especially the postmodern constructivists and relativists, to agree that there is nothing subjective or relative about the one and the only universe and its various characterizations—or what we call "truth."

Nevertheless, the current paradigm of cosmological history has been firmly in place for over fifty years, and for the first time we can say with confidence that this big picture might actually be true,[20] although important details have been modified and added continuously. The big picture of biological evolution on Earth has been supported by evidence from so many independent sources that it is highly unlikely to be wrong. The big picture of cultural evolution since the advent of civilization is less uncertain, although a real comprehension of world history only became possible in the twentieth century—previous generations of scholars knew too little about other parts of the world beyond their own.[21] People around the world may hold different religious, ethnic, cultural, and political views, but this science-based history of cosmic, biological, and cultural evolution transcends these differences and enjoys by far the widest audience.

The strength of the Cosmic View is that it is open to continuous criticism and empirical tests. The Big Bang theory had a mini-crisis in 1991, for example, when it seemed unable to explain the newly discovered strings of clusters and so-called superclusters of galaxies. But in April 1992, George Smoot announced that the NASA satellite Cosmic Background Explorer had detected faint irregularities in the cosmic microwave spectrum.[22] The splotches were the right size to one day grow into giant clusters of galaxies.[23] Nothing is irrefutable—we can easily envision new evidence that could cause substantial revision or even a complete re-evaluation of the existing framework.[24] It is just highly unlikely.

The universe we live in is larger and older than we can imagine and continues to expand without an end in sight. We discovered this modern picture of the world less than a century ago. One of the biggest surprises in the last one hundred years was Edwin Hubble's astonishing discovery in 1918 that the clouds of gas in the heavens were really billions of star galaxies![25] In 1930, he confirmed the red-shift law of nebular light, which lent support to the hypothesis of an expanding universe.

Hubble's breakthroughs led to the general picture of astrophysics and cosmic evolution that we still hold today. Following is a highly selective list of the most salient characteristics and key milestones leading to the evolution of a conscious mind that can contemplate them all:

- What started it all was the Big Bang about 13.7 billion years ago.[26] It is a common misconception that the Big Bang was a very concentrated lump of matter that exploded and sent fragments rushing away into space, thus forming the galaxies.[27] But this idea of a primal explosion actually refers to a constant expansion of space itself rather than an explosion into existing empty space.[28] Thus the size and the age of the universe are both about 13.7 billion light years, and there was no space or time at the "point" of the Big Bang.
- At its birth, the universe was a "fire" that was a billion trillion times hotter than the center of the sun and a *trillion trillion trillion trillion* times denser than rock. The fire experienced a period of unbelievably rapid and superluminal "inflation." Elementary particles emerged in the fire, and each of them would suffer annihilation when it met an associated antiparticle. If not for a virtually infinitesimal bias (an incredibly close ratio of about *one billion plus one* to *one billion*) in favor of particles over antiparticles, the universe would be literally empty![29]
- About 400,000 years after the Big Bang, the universe cooled to about the temperature on the surface of today's sun. This allowed the subatomic particles to coalesce into atoms such as hydrogen nuclei. The first generation of stars began to form in areas with high concentration of hydrogen and helium gas. Shrouded by hydrogen gas, these mega-stars were invisible—the universe went dark. Crushing pressure at their core made these stars burn

through quickly during the cosmic "Dark Age" that lasted for about 200 million years.

- The death of those mega-stars triggered the formation of normal stars and the first recognizable galaxies. The interiors of stars baked heavier elements, including those needed by biological life (carbon, nitrogen, oxygen, and phosphorus), and, at the end of their life, the stars exploded as supernovae, spewing the newly formed elements into space. Some of the resulting galactic clouds would recondense into a new generation of stars, some of which had planetary systems like our solar system. There are over 400 billion galaxies in the visible universe, each containing about 400 billion or more stars and each having diameters between 5,000 and 500,000 light years in space. Astonishingly, these may be just the tip of the iceberg of what's "out there."
- Zooming in to our "neck of the woods," the Milky Way is one of the galaxies out of 400 billion, the sun is a normal star situated on the outskirts of the Milky Way, and the Earth is one of its eight big planets. Condensing from swirling gas and dust, material discarded by older stars, our solar system came into existence about 4.6 billion years ago.
- Life emerged probably less than 100 million years after conditions became suitable for life on the Earth, about 3.8 billion years ago. How the first cell emerged is still a mystery—pre-life history was not well preserved, and we still cannot demonstrate it in the laboratory. All life is based on exactly the same chemical processes of RNA-DNA.[30] The evidence is overwhelming that life evolved only once. But we don't know whether it was on Earth or elsewhere.
- The emergence of life marked a fundamental change in the universe. Life exhibited *purposes* and *functions* that were absent in the prebiotic world. In an environment that was sometimes hellish,[31] life proliferated and became more complex over time through the process of Darwinian natural selection, which started not when there were individual species, but as soon as RNA emerged. RNA was a truly

revolutionary innovation since for the first time, it allowed natural selection to work at the molecular level through variation during replication, competition in a population, and selection by amplification.[32] Proteins, which have far more flexible cellular functions, and DNA, which does a better job in preserving genetic codes, probably emerged later and relegated RNA to its limited intermediate role in genetic information transmissions. This is probably the first specialization and "division of labor" in life.

- Life became a property of discrete and autonomous individuals with the emergence of the protocell. The common ancestor of all life on Earth today was most likely a primitive bacterium (prokaryote), which diverged into three major branches of life we see today: bacteria, archaea, and eukaryotes. Transmembrane receptors and effectors facilitated communication between cells by means of chemical signals. One special interaction between cells can be interpreted as the earliest form of sex—the insertion of genetic material from male cells into the body of the female cells with a structure called *pili* (Latin for "hair").
- Other crucial developments in the age of single-celled bacteria included fermentation, nitrogen fixation, photosynthesis, and movement. Photosynthetic bacteria were able to convert water and carbon dioxide from the air into food through sunlight and then release oxygen as a by-product. The resulting catastrophic buildup of this toxic gas in the atmosphere probably resulted in the first mass extinction and the birth of breathing bacteria.
- Multicellar organisms emerged about 2.7 billion years ago through the specialization and symbiosis of single-celled bacteria. It was such an improbable event that it probably happened only once, by chance. One unintended consequence of multicellular life is "programmed death."
- The eukaryotes diversified spectacularly during the Ediacaran period, about 700 million years ago. Animals appeared about 100 million years later. The animals' ability to move and eat other organisms with the aid of newly acquired senses (particularly vision) set off a one-upman-

ship evolutionary arms race. That led to the explosion of diversity at the beginning of the Cambrian period, about 550 million years ago.

- Animals have many interesting features, but we will focus on one of these: the nerve cell. In the body, each nerve cell connects with at least two other cells, whether they are nerve cells, muscle cells, or sensory cells.
- Starting from the basic sense-and-action nervous system in primitive animals, the brain, which is a central network of nerve cells that specializes in processing and storing sensory information, emerged and reached its pinnacle in vertebrates.[33]
- The brain makes the animal aware of the environment, however dimly. "Meaning" was born not with humanity but with the first connected patterns in the animal brain.
- Animals first appeared on land about 360 million years ago as amphibians. Their direct descendants are the reptiles.
- The evolution of mammals also brought unprecedented sophistication to the emotional system, which is a manifestation of dynamic interactions of different brain circuits. It enabled complex social features such as extensive parental care, collaborative hunting, mating dynamics, altruism, and social alliance. The primates are one of the groups with the most sophisticated brains.
- Man's direct ancestor species first broke from other primates as early as seven million years ago in Africa. Although some *Homo habilis* and *Homo erectus* species left Africa as early as two million years ago, recent genetic research on mitochondrial DNA (mtDNA) and Y chromosomes suggests that the last common ancestor of all modern humans was in Africa some 150,000 years ago. They emigrated out of Africa, went through a difficult period when they barely made it through, but eventually survived and somehow displaced all other species in the genus *Homo*.[34]
- The earliest-known modern humans outside of Africa, the Cro-Magnons, experienced a "cultural explosion." The

human mind eventually became conscious of the mind itself and of the universe through reflection, an ability that probably evolved from the primates' sophisticated social intelligence.

- Archaeological evidence suggests that distinctive human culture (as revealed in the discovery of ornaments and other artifacts) suddenly appeared about 50,000 years ago, probably with the help of the growing use of human language. As humans became more sedentary and numerous, the selective pressure for less within-group aggression was reflected in the modern human's reduced size of skull, jaws, and teeth. The primates' moral sense gradually evolved into early forms of human morality.
- The invention of agriculture and domestication of plants in the Fertile Crescent, China, New Guinea, and a few other regions after the end of the last ice age, about 9,000 to 12,000 years ago, marked the true beginning of human civilization and its domination on Earth. For the first time, human beings started to "work" and to develop ever more elaborate specialization of production and social functions.
- What made civilization possible was surplus output and investment in further output increases. Human population, which had stabilized at about five to six million around 10,000 BCE within the natural limit, grew to 100 million by 3000 BCE.
- Agriculture brought forth a surplus of food, which freed more than a few active minds from mundane tasks, allowing them to specialize in artistic, religious, and political activities[35] and to build up elaborate bureaucratic and patriarchic political systems. It also greatly increased the incentive to fight over food and material surplus.
- The huge and indeed incalculable gains from information and knowledge in more sophisticated societies led to the invention of written languages, which greatly accelerated the pace of cultural evolution. Written records facilitated the recognition and protection of private property and enabled thoughts to be shared and examined by anyone,

anywhere, anytime. Written language also enabled the creation of laws, contracts, history, poetry, and sacred texts.

- As small villages evolved into cities and states, the evolution of culture quickened. Ideas generated by one individual could quickly spread to an entire society. Coded laws were invented, first in Mesopotamia, to control and regulate behavior. The invention of wheeled vehicles and paved roads further enabled the emergence of larger and stronger social organizations, often in the form of empires.
- It seems reasonable to assume the first solid evidence of higher-order consciousness appeared about 5,000 to 3,000 years ago. The reflective capability enabled the emergence of cosmology—our view of the world in its totality, which eventually led to the highest perspective for humanity.
- The development of modern science since the fifteenth century dramatically improved our knowledge of the world and our capability to fulfill our destiny. By breaking free of the limitations of animal muscle, the inventions of the steamboat, railway, and motor vehicles transformed the concept of distance. The invention of the telegraph and telephone did the same. The lightbulb enabled activities to take place at night that had been confined to daylight hours. In nuclear energy we now hold the fuel that powers the stars; in genetic engineering we are approaching the power to create new species.
- The takeoff of economic growth occurred during the Industrial Revolution from about 1730 to 1850. Breaking the so-called Malthusian trap, global economic growth went on an unprecedented spurt of 2 percent per year. In addition to new technologies, the "managerial revolution" created modern enterprises with ever increasing efficiency and sophistication.[36] Global average gross domestic product (GDP) per capita, estimated to be around $670 (inflation-adjusted 1990 international dollars) in 1820, rose vertically to about $6,000 in 2000.[37] Global population rose from about one billion two centuries ago to seven billion today.[38]

- Mind and the machine got ever closer in the twentieth century as the Information Age arrived. The atoms we assemble, trade, and consume are increasingly all about the bits embodied in them.
- The global village has become so tightly integrated in recent decades as to feel like a super-organism. The precursors of the Internet got started in the late 1960s and early 1970s with the Pentagon-sponsored ARPANET. In the early 1990s the World Wide Web was born. Merely half a century after the discovery of the DNA molecule in 1953, the human genome was mapped.
- The twenty-first century continues to see the deepening of the information technology (IT) revolution but promises to be the century of biotechnology, neuroscience, and nanotechnology. The posthuman world, in which humanity is no longer the focal point and the ultimate limit, is within sight, although daunting technical, emotional, political, economic, and ideological challenges remain. Conscious evolution is beginning to dominate cosmic evolution.

5.3. Seven Cosmic Patterns

Winston Churchill once said, "The further back you look the farther ahead you can see." Records of ancestral history enable people to trace their forebears for many generations; geology and archaeology further extend our reach into the past; and cosmology gives us the most distant view in time and space. Our knowledge of the world also deepened over time, especially with the invention of science. What were once considered fundamental and universal, such as Newton's laws of motion and Euclidean geometry, are no longer so, since they have been shown to be special cases of a deeper structure and only apply in certain conditions.

To consider what is universal, we may have to imagine that we are standing at the time of the Big Bang and watching the entire cosmos explode and evolve into today's world. What are the most salient features in this Cosmic Show if we can watch the universe in fast-forward motion from the Big Bang to today?

I propose for your consideration seven key universal features of our cosmos. First and foremost, *the universe exists as a singu-*

larity. Second, *the universe is characterized by constant change.* Third, *the universe is governed by laws.* Fourth, *there is a general trend of growing complexity.* Fifth, *complexity can come out of simplicity.* Sixth, *nature's structure is modular and hierarchical.* And finally, *the world is recognizable but not predictable for any finite being within the system.*

Let's examine these more closely:

Single existence. The universe is a "uni-verse," one song. We humans may never know what the universe *really* is, why it exists, or whether it is part of a larger existence, but its existence is independent of our consciousness and senses, and there is no evidence to suggest it is not single, coherent, and rational. Throughout history, this truth has been obvious to most (though not all) of us.

Constant change. The universe exhibits a dimension called time that renders everything in a state of flux. The world we live in is neither static nor eternal. Rather, it is like an ephemeral dreamscape: permanence of any kind is only a matter of degree. All material creations are ultimately transient in time and space, even at the elementary particle level. Natural selection only selects or discards: it does not, and will not, freeze anything. There is no eternal, perfect "platonic essence" in any particular creature, including us. Some species can remain unchanged for hundreds of millions of years, but there can never be a full adaptation or perfect equilibrium. Change is not the same as chaos. To the contrary, order and pattern emerge spontaneously through change. Once emerged, forms can be preserved and resistant to change (in a relative sense). Without that, the world would be unrecognizable.

Governed by laws. All the phenomena of nature can be understood as reducible to regular patterns or structures. Most natural patterns recur only imperfectly,[39] but there are patterns that are stable enough to be called natural laws, and they permeate the entire universe, or, more precisely, underpin all the actions in the universe. Marvin Minsky pointed out that "in science, one can learn the most by studying what seems the least."[40] Reductionism works because there are underlying ubiquitous patterns penetrating the fabric of all natural phenomena—from galaxies to stars and planets, to weather systems, mountains, rocks, and snowflakes, all the way down to subatomic particles. Surprisingly, the same patterns recur in organic life, cultural artifacts, and social organizations. Laws can be considered as abstract "programs" describing critical elements of natural processes. One of the most useful scientific principles is Occam's razor: "Plurality should not be posited without necessity." In other words, out of many competing hypotheses, one should pick the simplest one as long as it fits one's observation as well as other hypotheses.

Our thoughts and consciousness are marvelously shaped by the nervous system's attempt to mimic natural patterns in four dimensions of space and time,[41] but the world is comprehensible to this tiny, short-lived creature mostly because natural laws are regular and compactable, at least under certain circumstances. The basic conception of Darwin's evolution theory is exceedingly simple—it is only in its application to concrete circumstances that the range of phenomena for which it can account manifests itself.

Trending to complexity. History often seems to be random and unpredictable, even to professional historians.[42] Yet history is not "one damn thing after another." There is a broad discernable trend. There seems to be a general tendency to add complexity and layers of hierarchy as time goes by, each incorporating all previous levels within a new unity. Similarly, evolution seems to follow what John Haught called an "aesthetic cosmological principle" that displays a tendency to accumulate beauty. Combined with the last point, we can say that the universe is likely to embody an inexhaustible potential for generating complexity and beauty.

The second law of thermodynamics states that over time disorder (or entropy) will increase, organizations will decay, and structures will disintegrate. This process seems to be irreversible. Yet, against this general backdrop, we nevertheless observe a general counter-tendency toward complexity[43]—a tendency, I hasten to add, that is not to be confused with an inexorable and immutable *law* of history.

Complexity apparently arises at local levels at the cost of greater entropy elsewhere. From a formless sea of particles at the beginning, there is a spontaneous self-organizing tendency toward more and more complex and heterogeneous orders, the most startling of which has been the emergence of life and ever more complex central nervous systems in the animal brain on Earth. Life has been "trying" to propagate in all directions since the beginning. The brain has been "trying" to improve its capability to gather, hold, and process information.

Diversity can breed more diversity. What we shall follow is what Henry Kauffman proposed as a candidate for *the fourth law of thermodynamics*, which can be summarized as: Biospheres and the universe create novelty and diversity as fast as they can manage to do so without destroying the accumulated propagating organization that is the basis and nexus of the creation process.[44] The way that life can increase order within itself in defiance of the second law is that it must remain open to its environment and can never stop at equilibrium. Entropy may decrease *only* in open systems. Although there have been many science-based speculations about a "heat death" or a "big crunch" end point for the universe, there may never be a steady-state end point. We simply don't know.

This spontaneous self-organization seems to be a universal property, and it applies to all levels of organizational complexity. For example, spiral vortices can be observed in a draining bathtub, in a satellite picture of hurricanes, and in spiral galaxies 100,000 light years in size.

Indeed, evolution toward more complex order has not been a linear, monotonic progression over time. There have been sudden bursts of progress (e.g., the appearance of humans, or the flash of genius by men like Newton and Einstein), as well as abrupt large-scale setbacks (massive extinctions of life and collapses of civilizations); and, on a much smaller scale, each of us sees both joy and tragedy every day. However, the overall cosmic pattern is clearly discernable as a cumulative process if we step back far enough.

While there are many little explosions, three big explosions of complexity stand out in the course of cosmic evolution: *the birth of the universe, the birth of life,* and *the birth of higher consciousness.*

It is too early to make a call on human consciousness, but nothing in terms of natural laws or the history of the Cosmos that we know of can prevent it from continuing to evolve and to spawn higher beings. Life did not become interesting until the Cambrian period, and the human species did not become interesting until the Axial Age. Only humanity's higher consciousness ponders where we come from and what it is all about. Still, as a form of intelligence, comparing the potential highest beings with the human mind is probably the equivalent of comparing the human mind to single-celled bacteria. In other words, there is unimaginable potential ahead.

Complexity out of simplicity. Complexity, both in nature and culture, is built up through multiple permutations of simple elements. Complexity can arise from simple elements and cannot exist without the appropriate basic elements. Human knowledge, for example, is embedded in language, in memory, in the brain, in the neuron—all the way down to the smallest elementary particles and maybe multidimensional strings. Human intelligence and expertise are not the recognition of a few powerful laws or techniques, but a large number of small pieces of knowledge organized in a hierarchical fashion for effective use and interaction.

Everything is necessarily built up from relatively simple, homogenous basic levels. There can be no exception: any cognitive and social attempt to duplicate the higher-level complexities without proper lower-level foundations has proven to be futile.

There seem to be universal principles (or laws) underlying the trend that enable complexity to emerge and persist in the natural-selection process.[45] Symbiogenesis is a particularly potent device to build complexity. The spontaneous self-organizing principle is another. Thus the Darwinian logic of evolution is based on the principle of emergence rather than teleological underlying laws: there is no definitive "force" guiding the direction of mutations similar to what gravity does to the falling apple. In the same fashion, the remarkable self-regulatory biological environment on Earth is not driven by any biological organism's "desire" for a stable environment.

Layered structures. All natural systems we can identify—i.e., differentiate from their surroundings—exhibit a hierarchical structure: they are made of identifiable "subsystems," and the system itself is part of, or coordinated by, "supersystems." For example, one of the organizing principles of our brain is its hierarchical order. It can be described by laws of physics and chemistry at basic levels and by laws of ecology and sociology at the higher levels.

Daniel Dennett used the metaphor of an "intelligence tower" to illustrate four different kinds of learning mechanisms in his book *Darwin's Dangerous Idea*. From single-celled organisms 3.5 billion years ago to modern technology-driven civilization, creatures with different levels of intelligence have emerged in nature's constant tinkering:

- All life-forms on Earth are "Darwinian creatures" that evolve and "learn" from their ancestors by copying their genes. Learning in this fashion is costly (failures lead to death) and slow (learning is only through intergenerational selection).
- The simple animals such as the sea slug *Alpysia,* a simple invertebrate, are "Skinnerian creatures" with nervous systems that can learn by trial and error. Failures do not necessarily lead to death of the whole organism, but merely suppress the behavior leading to the failures. However, their learning is limited to operant conditioning.
- Mammals and birds are typical "Popperian creatures" with a large nervous system sophisticated enough to construct a model of the external world and simulate the consequences of certain actions based on past experience. With trial and error conducted internally, potential mistakes can die in the mind during internal experimentation.
- Modern humans, a symbol-making species with flexible "mind tools" controlled by language, have become "Gregorian creatures"[46] with external tools (especially computers) that enhance or supplement the mind's simulations. The ability to manipulate information outside the body has tremendously expanded our intelligence, although humanity is by no means the first species to utilize external computational and informational artifacts.[47]

It is worth noting that Dennett's four learning creatures did not evolve as though climbing a ladder, but rather through adding one layer on top of another, as though building up a tower. Humans are not pure Gregorian creatures, but "four creatures in one," with the Darwinian creature at their core.

Boundless freedom. The world can be recognized but cannot be predicted with certainty. Energy/matter, life, and mind—there is a sequence of their appearance and a hierarchical order.[48] Higher-level properties may not be predictable from lower-level properties. Nature appears to have a laminated structure. Newer levels in general supplement rather than supplant the old. The aggregation of units on one level may give rise to a second, whose units may exhibit properties drastically different from those characteristics of the first. The transcendent properties are inexplicable from lower levels. The emergence of humanity has added at least two more levels to nature: individual consciousness and human institutions.[49]

It is important to keep in mind that what we discussed as "the view of the universe" is just tracing the upper solid surface of evolutionary complexity. That's the most interesting and exciting stuff, but equally illuminating for the future are events that led nowhere and events that could have materialized but did not. If we were to acknowledge them, the sheer numbers of these cases of forgotten dead ends and bad luck would absolutely overwhelm the historical events that we can recall; but A. O. Lovejoy's Principle of Plenitude, which asserts that whatever is possible ultimately will come into existence, appears to be the overarching theme of evolution.

We have finished a sweeping view of the most salient characteristics of the Cosmic Show. The universe is unified yet modular, lawful yet playful, and messy yet directional. There are no individual miracles in this universe, yet the universe itself is a perpetual miracle. We have only glimpsed the bottomless well of the cosmic potential, yet it is already rich beyond description.

5.4. Reflections on Truth and the Cosmic Future

Where is the universe heading? We must rely on discernable historical patterns to predict the future. Yet in order to have any confidence that past patterns will persist, we need to identify the source of those patterns. In other words, if there are regularities, structures, and emergent complexity, what is driving the whole thing?

The search for the final unifying theory or the ultimate "source code" that runs the universe goes on. Computer scientist Ed Fredkin made a bold speculation that the universe is a giant computer with a purpose of reaching some final state. But he also believed the final state is not specified in advance.[50] The zoologist Charles Birch maintained that evolutionary history resembles a vast, unfinished experiment without an omnipotent designer or a preconceived plan.[51] The universe can also be viewed as a "probability field" with an infinite continuum of possibilities ahead of us. It may simply be viewed as a great idea, a playground for us to play in.

These metaphors and models can be useful, but they are products of our limited minds and technologies. Anybody who declares the final answer is showing his ignorance. What we observe is only a tiny portion of the universe. Gödel's incompleteness theorem tells us that we can get close to answering all questions, but never answer every question about a system as long as we are part of the system. Thus, even in theory, we can never know why the universe exists and why it is the way it is without jumping out of it, which is impossible. In Einstein's famous words, "The most incomprehensible thing about the universe is that it is comprehensible." The material cosmos, life, and mind would not evolve if there were no stable patterns, but why there are repeatable patterns is beyond the understanding of anyone inside the universe.

Still, within these limits, the secret of success has always been that one must act based on what information one can obtain. In other words, without much hope of obtaining the "source code" or a complete record of history, we must make our calls on the most interesting observable pattern of all—the teleological direction of the universe. The history since the Big Bang hints at a purposeful tendency of evolution toward ever-higher levels of complexity and awareness, as if under the guidance of a mind. With our current state of knowledge, the universe seems to unfold, or germinate, much like the work of mathematicians: more and more theorems are being derived from a handful of axioms with the aid of a large amount of computational work.

Life is essentially the accumulation of information and the realization of potential embedded in the fabric of existence.[52] According to David Deutsch, "Life achieves its effects not by being larger, more massive, or more energetic than other physical processes, but by being more knowledgeable."[53] The same tendency is even more fascinating for cultural evolution. Stuart Kauffman noted that at least the first three levels of Dennett's "intelligence tower" have molecular parallels in cellular, neural, and immune systems. It is self-evident that intelligence provides an edge, *ceteris paribus* ("all things being equal"). It is also natural that moral sentiments in primates and perhaps other highly intelligent animals evolved to facilitate learning, and humanity boasts the strongest and most sophisticated moral sentiments.

As the philosopher Karl Popper pointed out, while we must admit that there seems to be a pattern (or plot) in history, we also need to note the implausibility and fragility of the plot. We have learned that there can be no fixed Newtonian laws concerning complex phenomena. And we recognize that the most interesting evolutionary junctures—the beginning of the universe, life, human species, and higher consciousness—are still poorly understood, if at all.

In addition, although the general trend seems unmistakable, nothing *specific* is inevitable. The natural laws of the universe only provided the potential for a long series of seemingly miraculous events that led to the appearance of humanity and civilization. In Steven Jay Gould's vivid metaphor, "Replay the tape a million times from a [pre-Cambrian] Burgess beginning, and I doubt that anything like *Homo sapiens* would ever evolve again. It is, indeed, a wonderful life."[54] The same is probably true for the emergence of life on Earth. This understanding is emotionally "cold" and fundamentally incompatible with the idea that the cosmic destiny is human. But it gives us a profound sense of how precious our existence is.

All our efforts, in a sense, are "hitching a ride" onto the process of cosmic evolution. While life may depend on the stability of the solar system, knowledge depends on something even more fragile: the survival of the human species. Furthermore, as we shall discuss in the next chapter, this view of the universe gives our existence its deepest meaning.

While the Cosmic View is very likely a better representation of reality than other competing beliefs, the only thing certain is that our views will continue to evolve. The Cosmic View of the universe is accumulated wisdom, genetically and culturally. It is certainly not the best guide imaginable, just the best we have. We must be humble and audacious at the same time. Mistakes will be inevitable, but out of all that, progress will be stumbled upon. J. P. Morgan once said, "Go as far as you can see; when you get there, you'll be able to see farther."

What is truly exciting is that this scientific and spiritual exploration and accumulation of new knowledge may be only the beginning. Indeed, much of the profound new "knowledge" that we possess today is likely to become utterly trivial for our descendants. The most valuable piece of knowledge we possess today may be what Nicholas of Cusa called *learned ignorance*—we have learned enough and thought hard enough to know what is unknowable to us. In one of his most famous remarks, Isaac Newton summed up his life shortly before his death in 1727, which I believe is also a precise picture for the state of humanity today: "I do not know what I may appear to the world but to myself I seem to have been merely a child playing on the seashore, diverting myself now and then finding a pebble more smooth or a shell more beautiful than others, whilst before me the great ocean of Truth lay all undiscovered."

The future is open-ended. To appreciate that, do a simple thought experiment: If we look at the cosmos before the formation of the Earth, can we predict the emergence of life and consciousness? Similarly, the novelty still to come is difficult if not impossible to imagine, as science cannot study patterns yet to emerge. Even so, science has offered us hints of things yet to come.

At least the Cosmic View gives us some extra motivation and courage beyond what we are born with. If we cannot know specific directions, we at least know the process is boundless. Cosmic evolution seems to be "free"—that is, it seems to be generated spontaneously from within. We must keep moving. Doing nothing is a luxury we all crave, at least sometimes, but a little reflection tells us that stasis is not really what we (nor any life-forms, for that matter) want in the long run.[55]

Not all interpretations of cosmic stories and myths are motivating. Many of them actually do the opposite if we reflect on their promises. The Adam and Eve story tells us the danger of eating "the fruit of knowledge." The moral seems clear. But wouldn't the world be a terribly boring, if not dead, place if Adam and Eve had remained content in their paradise and stayed in a static and ignorant condition forever? For the pious, heaven or the Kingdom of God remains an eternal blissful resting place after we put forward our best efforts. But as the playwright George Bernard Shaw once remarked, "Heaven, as conventionally conceived, is a place so inane, so dull, so useless, and so miserable, that nobody has ever ventured to describe a whole day in heaven, though plenty of people have described a day at the seaside."

5.5. Going Back: The Tao as the Mind of God

Time flows forward and expresses more and more of the cosmic potential latent at the moment of the Big Bang. Our understanding of the world, interestingly, grows backward. What happened most recently we know first and best; what manifested earlier, the remote and hidden stuff, we only learn through science and hard work. As we reach back to the Origin, our emotions and intuitions are challenged, but at the same time, we gain power and get closer to understanding and influencing the course of the universe.

The evolution of humanity's worldview loosely follows the pattern of the psychological development of a child. The history of astronomy tracks what every child seems to experience: first viewing the world as centered around himself or herself, gradually developing perspectives from other people, and finally (for some) obtaining an impersonal, unchanging universal perspective. As they grow up, most children also gradually move away from egocentrism and animism. The pioneering child psychologist Jean Piaget noted that initially, a young child equates life with any activity. Then between the ages of six and eight, a child restricts life to anything that moves. It is only after eight that a child considers only things that move on their own as alive, and later, only the plants and animals as alive. For primitive peoples, the world appears packed with will, intelligence, and other animate qualities. Today, the results of scientific research and the worldview it has given us have largely demystified the inanimate world, and a large part of the living world as well. But a larger sense of mystery remains untouched: the wonder of Cosmic Creation, what Einstein called "the mind of God."

Western civilization and its worldview are an especially vivid demonstration of the retrospective growth of the Cosmic View. In the beginning, gods that created and ruled the world were quite literally glorified people. Then the best religions grew progressively less anthropomorphic and animistic. In *The Disappearance of God,* Richard Elliott Friedman pointed out that God stopped revealing Himself and talking to people directly midway through the Old Testament. Basing his work on readings of ancient Near East and Greek literatures, Julian Jaynes suggested that hallucinations, which he defined as the interpretation of motivational drives from the left side of the brain as commands from the gods, started to fade during the same period. This was a time when a Mesopotamian ruler cried, "My god has forsaken me and disappeared; my goddess has failed me and keeps at a distance. The good angel who walked beside me has departed."[56] The text of the Old Testament offers the best story of our emerging higher consciousness: a mere awareness of multiple drives ("the gods") was replaced by conscious self-reflection ("the one God"). During the period of Temple destruction and Babylonian Exile in the early sixth century BCE, the Jewish people finally broke clean with pagan religions. An unbridgeable gulf between God and His world (including humanity) was then made clear, although God remained anthropomorphic at core, because only by being a paradigmatic personality could God command people to follow and make voluntary covenants with Him.[57]

History is always messy and convoluted—this is exactly what gives it flexibility and creativity—but the overall trend is clear: *God as a personal being has been gradually retreating from us*. As Richard Elliott Friedman noted about the Hebrews, prophecy emerged as a defined institution only after Yahweh stopped speaking to an entire community Himself. The last person to whom God is said to have been "revealed" was the prophet Samuel; the last person to whom God is said to have "appeared" was King Solomon; the last public miracle God is said to have performed was the divine fire on Mount Carmel for Elijah (who was also the last person that God is said to have spoken to personally); and the last miracle performed by God Himself was His act of reversing the shadow on a sundial for King Hezekiah. And finally, God is not mentioned at all in the book of Esther.[58]

Throughout Genesis, God gradually "demoted" Himself from the position of the primary actor and controller to that of bystander and listener. In the meantime, humanity grew from the infantile Adam and Eve to people like Abraham and Jacob, who negotiated, debated, and even fought with God. In the Western religious tradition, Jesus was the last influential person God sent to us to teach us directly. The Qur'an (Koran) was conveyed to Muhammad in dreams and through the angel Gabriel. And the Book of Mormon's "golden plates" were unearthed by the modern-day prophet Joseph Smith (founder of Mormonism) in 1827, rather than delivered directly by God or His representatives. Few other prophets or messiahs making similar claims had been able to gain enough traction to lift their cults and sects to the status of established religion.

Without dropping the concept of a transcendental God, the depiction of God's influence in this world gradually loses anthropomorphic imagery and analogy. From the ancient image of God as a shepherd of humanity, which was modeled after pastoral aristocratic society, God in modern centuries has come ever closer to becoming a pure idea, like the idea of force or energy.

A theological milestone in the process of understanding God at a rational and abstract level was achieved by the great theologian Thomas Aquinas, who released the full power of rationalism into secular hands and declared two kinds of revelation: one for the province of theology, another for the natural world of reason and philosophy. God was not seen in competition with nature, a point that modern-day anti-evolutionists miss. It was the English cleric Roger Bacon who urged the Church to empower scientific interpretation of the Scriptures. It was the seventeenth-century philosopher Spinoza who expressed a profound pantheism, which states that God and the universe are one.[59] And it was the French theologian and paleontologist Pierre Teilhard de Chardin who in the 1940s presented a mystic view of the universe based on his scientific knowledge and religious background: humanity's meaning exists in the Omega Point, the ultimate "telos" of evolution.

Teilhard followed the nineteenth-century German philosopher F.W.J. von Shelling, who incorporated evolutionary metaphysics into the cosmic picture: "Has creation a final purpose at all, and if so why is it not attained immediately, why does perfection not exist from the very beginning? There is no answer to this question except the one already given: because God is a *life*, not a mere being. . . . I posit God, as the first and the last, as the Alpha and the Omega."[60]

So, it is a great paradox that just as we seem to be getting away from the personal God as described by earlier seers, we are getting ever closer to the God that created the cosmos. While the term still has anthropological overtones, the Creator God has more to do with the fundamental nature of existence than human-like behavior and motivation. From ancient myths and legends to more advanced religions to the current scientific understanding of the universe and human mind, recorded Western history shows a clear trend of loosening our link with a personal, intimate God that is really in the image of the human, and coming to recognize the magnificence of the Creation and appreciate its beauty, complexity, subtlety, and, finally, meaning.

Within that broad trend, of course, there are periods of adolescent-like confusion, regression, rebellion, and persistent voices of dissent. Take, for example, the question "Is the appearance of life and mind a superficial cosmic accident?" This question did not exist at the time when the personal God prevailed. But the advancement of science, especially in the field of biology, seemed to provide scientific vindication of existential thought as articulated by thinkers such as Jean-Paul Sartre and Albert Camus. The biologist Jacques Monod presented a stark view of a meaningless, irrelevant universe with respect to human existence in his 1970 book *Chance and Necessity*. More recently, Richard J. Bird has called modern science "the paradigm of random selection" that obstructs progress of humankind and strips meaning and control from our lives.[61]

Counter voices are equally strong. Scientists such as physicists Freeman Dyson and Frank Tipler and the biochemist Christian de Duve found in the latest scientific research a strong connection between the structure of the universe and the emergence of life and intelligence. Randomness as characterized by quantum mechanics is a central component of the Cosmic View, but rather than prescribing a bleak and directionless world, it instead points to a playful and enjoyable way to achieve emergent complexity and progress.

Life defies chaos, yet is created from it. Consciousness is bounded, yet it has free will. This is the paradox, the wondrously beautiful yin and yang of the universe. The awesomeness of the universe is the grandeur of the Creator God. We can add to that by stating: *The Tao is the mind of God.*

The debate over the notion of *progress*, especially at the philosophical level, will certainly continue, but it is not just a matter of personal taste or intellectual fashion. It is a struggle to get back to the Origin. As we shall argue, by no means have the older perspectives disappeared or lost their power to appeal to human intuition and emotion. The Creator God speaks only to the rational mind. Lacking emotional comfort, an "uncaring" deism can never replace theism for the masses.

Getting to know God, beginning with an intimate personal perspective and rising to more abstract higher perspectives, is both a collective and a personal journey. Many modern scientists (Charles Birch, Paul Davies, Theodosius Dobzhansky, Paul Feyerabend, and E. O. Wilson, to name just a few) had exciting spiritual and intellectual journeys during their lifetime.[62] They typically started out with a conventional personal God that proved to be intellectually unsatisfactory, then became agnostics or atheists, and finally came to realize that the only way to look forward is to supplement austere agnosticism or atheism with more upbeat universal principles (such as the Augustinian "divine mind," Newton's "dominion," Albert Schweitzer's "reverence for life," and Whitehead's "potentiality").

Once there, the scientists regained the sentiments of their childhood: a sense of wonder, a feeling of love for the beauty and fruitfulness of the world. Some people, such as the mathematician Martin Gardner, would continue to believe in a personal god, but for nothing other than emotional and pragmatic reasons.

We will discuss the intricacies of different perspectives in chapter 8. Here I will simply note that a child in Western cultures typically goes through the same process with the concept of Santa Claus—from literal belief in a person living at the North Pole, to flatly denying the existence of such a being, to understanding the spiritual meaning of Santa, and finally realizing the universal features of such a spirit that do not necessarily depend on Santa per se.[63]

There is a crucial difference, however. Whereas almost everyone outgrows the childhood Santa, most people don't advance beyond the first few stages of understanding the nature of God. Being stuck at the personal level is the norm rather than the exception. Humanity's journey back to the Origin is not a truly collective one, but carried on generation after generation only by exceptional individuals.

This is nothing strange. Think about the fact that what we most touch and see are the heavy elements, yet simple elements like hydrogen and helium still make up 99.9999 percent of the matter in the universe after 13 billion years of evolution. The distribution of perspectives in society is like life on Earth: complex organisms have emerged, but in sheer numbers, bacteria and insects continue to dominate the world[64]—for every human being on Earth, there are about 200 million insects. This is not to deny progress. In fact, although many people still cling to primitive worldviews, fewer and fewer people question the scientific explanations of many of the mysteries of the world. The Cosmic View is made possible by our courageous exploration back through the historical chain of natural causes and effects—back to the future!

CHAPTER SIX

HUMAN POTENTIAL

Man may be excused for feeling some pride at having risen, though not through his own exertions, to the very summit of the organic scale, and the fact of his having thus risen, instead of having been aboriginally placed there, may give him hopes for a still higher destiny in the distant future.

—Charles Darwin

IN THE LAST chapter, we discussed how—at the frontier of human knowledge—we have arrived at a valid scientific understanding of the evolving universe and have even constructed successful social and cultural organizations that apparently emulate the real universe. In this chapter, I argue that these achievements seem significant enough to warrant humanity's taking a place, for now, at the leading edge of this evolving universe—at least the slice of it with which we are familiar. Nevertheless, I will point out that our current knowledge remains as nothing compared with our ignorance; our accomplishments pale in comparison with cosmic potentials that remain unrevealed.

How do we know there is such cosmic potential? We can infer its reality based on our extrapolations from what we now know of cosmic history. But again, we must proceed with caution: for a finite being who is part of the universe being studied, extrapolating from the past (or local events) can be a dangerous exercise. Even with the aid of our scientific knowledge of natural laws and a higher perspective, the future is largely invisible.

What is clear is that the universe has been generating novelty after novelty since the Big Bang. Given this salient fact, what can we say about the universe's future possibilities? While we cannot infer the precise direction of future cosmic evolution, we can state with reasonable certainty that many more unprecedented and unimaginable novelties await.

With so little certainty of specifics, what are we to do? We have already discussed our indispensable role in furthering evolutionary processes. However, we do not need to *know* what lies ahead before we can *act* and become participants in this great unfolding. Only when we act on our highest aspirations and potentials, consistent with our unquenchable enthusiasm and passion for innovation, can we discover the greater significance of being human. In that way, we are furthering not only human potential but cosmic potential—so that they become one and the same. And this has the highest value in the Mind of God (Tao).

We have looked at the universe as an amazingly unique world out of many worlds we can imagine. In this chapter we look at the human being as a unique species out of millions and millions of species on Earth.

6.1. Human Uniqueness

The human is a complex animal, but actually not too difficult to define genetically. Appearances notwithstanding, human beings are genetically a very homogenous species, largely because they are so new. The genetic variation among chimps in a small region in Africa is greater than that of the entire population of human beings around the world. This suggests that we've been through at least one genetic bottleneck relatively recently. (A "genetic bottleneck" occurs when a population, and therefore its gene pool, is reduced drastically, cutting down generic diversity.) The so-called Mitochondrial Eve—our most recent direct ancestor in the female line—lived about 150,000 years ago based on the current genetic evidence, while the skull of the earliest known *Homo sapiens*, found in Ethiopia, has been determined to be around 160,000 years old. Metaphorically speaking, something on the order of Noah's Flood, which according to the Bible eliminated corrupted and wicked peoples, must have occurred.

Given our youthfulness in terms of the evolutionary clock, it is not surprising that many things proclaimed to be exclusively human are turning out to have been an exaggeration or extension of the characteristics of higher animals. The chart on the next page illustrates that, while humans have been defined by various authors and observers to be unique in various ways, upon closer inspection we are found *not* to be unique in any of the ways listed here. In fact, there are few things that are truly unique for human beings: we are an evolutionary product, and most of what we consider to be unique human traits are actually greatly exaggerated traits of our close relatives, the primates. Indeed, comparative study leads us to conclude that most noticeable differences between us and other animals are ones of degree, not of kind. For example, planning, curiosity, playfulness, and intelligence are widespread among warm-blooded vertebrates (mammals and birds) and even some reptiles (e.g., the crocodiles). The human being is a "weedy opportunist"—a phrase used by biologists to refer to the capability of some species to adapt to a wide variety of natural environments—and we are not unique in that aspect. Human beings are also specialists in intelligence—also not a *uniquely* human trait.

In the accompanying chart, the left-hand column lists human traits that are also found in other animals, the right-hand column ("Why not unique") lists a sampling of animals/activities that show those traits, and the middle column shows a sampling of "authors/observers" who made or noted those discoveries.

If we pay close attention, it is surprising what animals can do beyond their routine behaviors. Many animal species have a rudimentary culture; they are curious and creative. For example, recent experiments provide strong evidence of symbolic and economic thinking in chimpanzees: they have been observed working for tokens for which they can then "buy" food; some even behaved like misers, amassing fortunes in tokens.[1] Capuchin monkeys, when introduced to tokens, exhibited a keen understanding of money's fungibility and made rational decisions in response to changes in prices and wealth that fit standard price theory remarkably well.[2]

Such discoveries lead to the conclusion that humans are *super-mammals*. And we did not achieve humanhood by denying or surpassing our mammalian nature, but by exaggerating it. We became human because we exaggerated a variety of physiologi-

UNIQUELY HUMAN—OR IS IT?		
Trait	**Author/Observer**	**Why not unique**
Featherless biped	Plato	Diogenes sent a plucked chicken.
Political	Aristotle	Chimp and bonobo politics
Rational	Aristotle, Aquinas	Primates
Social	Mo Ti	Ants, wolves, dolphins, etc.
Courage	Julius Civilis	Elephants saving their young
Barter, trade, gregariousness	Adam Smith	Chimps would trade virtually anything for sex.
Private property	Martin Luther, Pope Leo XIII	Primates
Goal-directed action	Jakob von Uexkull	Chimp holding a stick behind his back and searching for his rival
Tool-making, ethical	T. Dobzhansky	Primates
Self-awareness, culture	George G. Simpson	Primates
Memory	John Dewey	Most animals
Kiss	Hugh Morris	Chimps routinely kiss.
Face-to-face copulation	Desmond Morris	Bonobo
Concealed ovulation, female orgasm	Donald Symons	Bonobo
Use sex for non-procreative bonding	St. Augustine	Bonobo
Rape	Gerritt Miller	Orangutans and stumptails
Gender-specific division of labor	K. Imanishi	Chimps
Play	Johan Huizinga	Dolphins
Laugh		Chimps
Weep	Plato	Elephants
Feel pain		Most animals
Use punishment to train young	Toshisada Nishida	Primates
Monetary exchange	Adam Smith	Capuchin monkey
Hierarchy, social classes		Most social animals
Aware of sickness	Henri Bergson	Chimps eat medicinal plants.
Art, dance, music		Chimps can draw, dance, etc.
Lying, playing tricks	Montaigne	Chimps lie and sometimes even try to outwit other liars.
Language	Max Muller	Animals display rudimentary symbolic communication.

cal, psychological, and behavioral characteristics. For example, our learning ability increased because of the greater size and complexity of our brain; and we moved toward such behaviors as a more pronounced period of mother-child dependency, a more intense and elaborate sexuality, more complex play, a greater propensity for bonding, and a stronger tendency to submit to or become devoted to powerful others. While higher animals experience emotions and rudimentary moral sentiments, humanity boasts the richest social emotions and strongest moral sentiments. At the same time, we are also the cruelest and the most ruthless animal that ever existed. And while animals exhibit fleeting curiosity, *Homo sapiens*—the ape with intellect—can turn curiosity and idle investigation into a drive itself: learning just for the sake of knowing.

Much of the exaggeration of animal traits is made possible by our greatly enlarged neocortex, which adds new motives along with new coordination and executive functions to the brain. Human behavior has so much plasticity that it seems to be able to wiggle free of instinctive drives. There is very little behavioral difference between two butterflies; quite a bit more between two monkeys; but the difference between two humans can be vast.

6.2. Unique Human Attributes

Among our many adaptive "exaggerations," three interrelated qualities make us stand out from the higher animal kingdom in terms of our potential: symbolic abstract thinking, structure building, and higher consciousness. They are so critical to our species and so "revolutionary" that one could argue that these are uniquely human.

Symbolic abstract thinking. Language—which is necessary for symbolic abstract thinking—is most likely a truly unique human trait, though this is still a controversial matter. Linguist Noam Chomsky's discovery that human linguistic capacities are innate and not learned has become widely accepted.

Physically, we have a lowered larynx that enables us to make a much wider variety of sounds than any other animal. The brain's evolutionary changes include substantial upgrades in the midbrain and brainstem vocalization modules, enabling an innate capacity[3] to manipulate symbols in a spatial-temporal code that maps sounds into meanings.

We are indeed what Terrence Deacon called "the symbolic species," for only humans truly have the broad ability to think and communicate symbolically through the use of language (although some animals show very limited and rudimentary abilities). When we manipulate symbols, real-world physical and social relationships are *re-presented* in the brain as logical relationships. A qualitative jump then occurs when these logical relationships become able to wiggle free of their ties to the real world and become independent symbolic relationships. Once decoupled, recombinations and associations of novel kinds become possible. Our mind acquires the ability to imagine, to reason, to choose among various motives, and to evaluate alternative plans of action.[4]

Structure building. We humans are able to build all kinds of physical and social structures as well as mental models, and we are also easily able to share or pass them on to others. In particular, cultural transmission, or the duplication of all kinds of man-made structures, is faster and more efficient than biological evolution. We can generate and share ideas throughout our lifetime, whereas like all animals we only pass on our genetic information once to each of our offspring, without control over how the genes from the father and mother are recombined. The accumulation of experience and knowledge through culture gives humanity unparalleled power to manipulate its environment and even itself. As Lewis Mumford put it, "Man is his own supreme artifact";[5] in other words, humanity creates culture, which in turn creates humanity by enabling and limiting our actions.

As discussed in the last chapter, we are the only "Gregorian creature" that is able to extend the mind deeply into the environment, especially after the scientific revolution, which routinized invention (with controlled and measured experiments) and its diffusion. In a sense, being human is all about building what Richard Dawkins called "extended phenotype." From bone marks to digital computers, the mind has always managed to escape its "natural" confines and utilize external resources as *distributed* cognitive engines and supplemental memory.[6]

Let us look at the examples of cooking and domestication to see how developments of social structures are closely related to those of mental or physical ones. The anthropologist Richard Wrangham pointed out the significance of our unique way of preparing food: cooking with fire.[7] Cooking food not only made food easier to digest and enlarged the range of available food sources, but also profoundly changed interpersonal relationships through food preparation and sharing. In a similar fashion, just as we domesticated animals and plants, we have also domesticated ourselves: over the last 30,000 years, human brain size has decreased by 10 to 15 percent, reversing a trend over the last two million years. The intensifying social interactions in permanently settled agricultural communities could have exerted strong selective pressure against antisocial personalities, resulting in alteration of both brain structure and size.

Higher consciousness. Certainly the most unique attribute of humans is reflective consciousness, a specialized human endowment that is capable of such tasks as self-monitoring, divided attention, self-reminding, self-recognition, rehearsal and review, nonverbal communication, gesture, symbolic invention, and a host of other "executive functions" that no other animals can hope to match. Most animals are "aware," or conscious of their external stimuli. Primates and dolphins can recognize themselves in the mirror and thus have a clear sense of self-awareness. But only humans boast the capacity for reflective or higher consciousness—which can be summed up as the general ability to know one's own thoughts.

Without conscious reflection and action, we are really not that much different from many other forms of life. With the advent of reflective consciousness, humanity becomes exceptional. All animals experience food as tasting good, but only humans can realize that we eat to meet our nutritional needs; all sexually-reproduced animals experience sex as pleasurable, but only humans can realize that sex is indispensable for procreation; all animals have an inherent instinct for survival, but only humans can realize that one's survival also matters for one's family, community, humanity, and beyond. While an animal can suppress its desire for food when it senses danger, only humans happily practice corporeal asceticism in an attempt to purify and free the mind.

What is so special about the human species, therefore, is not its intelligence or structure-building, but its ability to gain an escalating higher perspective through higher consciousness—what Robin Dunbar called the "multiple order of intentionality."[8] Nobel physicist Richard Feynman expressed the essence of this reflective consciousness when he was a young student:

> I wonder why. I wonder why.
> I wonder why I wonder,
> I wonder why I wonder why!

Such meta-wondering is a beautiful manifestation of human free will and points to the essence of human potential. This incurable yearning might be seen as the quintessence of youthfulness, and it is something that humans have in abundance (some much more than others).

With reflective consciousness, a human being can turn itself into a "mirror neuron" for the universe, because humanity is the only thing on Earth that tries to understand what it is all about. I hasten to add that this is still largely a potential; what has emerged so far is a pretty reluctant mirror neuron. Of course, reflective consciousness also creates burdens that are unique to humans, especially our knowledge of our mortality and the limitations we face in controlling our lives. The biblical story of the shameful feelings experienced by Adam and Eve after they ate the fruit of the Tree of Knowledge symbolizes the reflective mind's existential angst.

Further, the sense of a conscious tension between one's actuality and potentiality is uniquely human. A reflective person can ask not only what is good and desirable, but also what is *ultimately* good and desirable. Humans sometimes struggle between immediate present concerns (such as saving our own lives) and concerns that exist in our reflective consciousness—for example, concern for posterity or the perceived command of a god.

Strangely, these "higher" motivations can be even more powerful than those involving biological survival. I believe Einstein's dictum "A person first starts to live when he can live outside himself" applies well here and can have two levels of interpretation. First, to "live outside himself" implies empathetic regard for, and identification with, one's fellow human beings (beyond intimate relations and institutional tribalisms). But at a higher level, that regard can extend beyond humanity to all of life—and even to the universe in which all of life was born and evolved.

The universe is one, and while (as mentioned earlier) we cannot project its future in detail, it has been actively moving along in a direction we can begin to identify. The urge for reciprocity in interpersonal relationships may point to an even greater relationship, one marked by a grateful regard toward the larger world: if the universe gave birth to us, how can we contribute to that larger trend?

Although the bones and stones of archeology give the impression that we started as a tool-making animal, our true separation from animals began with the advent of our linguistic capacities. This ability led to deeper thought and reflection, which in turn has become the basis of our emerging capacity for self-identification and purposeful self-transformation. Possessing a reflective mind means we are able to transcend the limits of our own biological nature; it also implies the potential willingness to sacrifice oneself for causes that transcend ourselves. That is exactly the human potential.

Self-transformation (of individuals, societies, and humanity as a whole) is hardly a new human characteristic. As discussed above, human self-domestication played a large role in the emergence of humanity. However, in the posthuman era, this transformation will become much more of a conscious process, which we will discuss in detail in chapter 9.

6.3. The Cosmic View: Funnel or Hourglass?

The cosmic history outlined in the last chapter depicted a progressive narrowing of temporal and spatial scope, starting with the entire cosmos and ending on a single planet; starting with all life on Earth and ending with a single species, the human race. Not that the rest of the cosmos has disappeared—"life goes on" for other living species in a universe that continues to expand—but their story seems less interesting because, for us at least, they only provide the backdrop for what is truly exciting and worth noting on the evolutionary frontier that we are facing now.

Many book-length narratives in recent decades share this same method of looking at the universe in its entirety—and then descending to a sharp focus on leading-edge developments as they affect humanity's well-being. Most of these books present what I call a funnel-shaped cosmos—starting big at the top (i.e., the entire universe), gradually shrinking to small at the middle (human civilization), and ending with the funnel narrowing to a focus on how to protect the environment and preserve humanity and existing biological species.

A key reason for such a human-centered approach might be reader or viewer interest. For a species with an average individual lifespan of less than one hundred years, it is hard to imagine diverting our attention away from achieving the dream of a Golden Age on Earth in favor of affecting the fate of the universe millions or perhaps billions of years in the future. Even visionary scientists like Norbert Wiener—the father of cybernetics, whose work aimed to conceptually unify the animate and the inanimate world—did not express a vision beyond the survival of humanity in a new technological age. And even the evolutionary theorist Robert Wright, who carefully examined the cultural and biological evolutionary trends since the Big Bang and found a "core pattern of non-zero logic"[9] that propels ever more complex life and organizations, came to the following conclusion at the end of his splendid book *Nonzero: The Logic of Human Destiny*:

> [G]rowing turmoil does signify, by my lights, a distinct step in the unfolding of what you could call the world's destiny. We are indeed approaching a culmination of sorts; our species seems to face a kind of test toward which basic forces of history have been moving us for millennia. It is a test of political imagination . . . but also a test of moral imagination.
>
> So how will we do on this test? Judging by history, the current turbulence will eventually yield to an era of relative stability, an era when global political, economic, and social structures have largely tamed the new forms of chaos. The world will reach a new equilibrium, at a level of organization higher than any past equilibrium.[10]

Social and economic developments have been uneven, as they will always be, but societies and nation-states around the world have gradually become more homogeneous in structure and more interconnected. Universal human understanding of what is optimal for humanity is starting to emerge despite the fact that ethnic, ideological, and national identities still dominate human affairs and grab our attention. Judging from the everyday headline news of geopolitical struggles and financial turmoil, this is unlikely to reach its culmination anytime soon, but the underlying assumption of most academic and political debates is that the search for the perfect human society is all we care about for a "sustainable" and prosperous future. The same holds for technological advancements ranging from the wheel to the Web that have been used to serve human needs. They are mere extensions of human capacities and cannot have a life (a will, to be exact) of their own. In sum, humanity will continue, and the future is all about humanity's striving for a Golden Age, which seems to be in sight but may still require much of our time and effort. That is the prevailing conventional wisdom about the future.

Unfortunately, or perhaps fortunately, a spoiler to this grand "funnel vision" has now emerged, emanating from the realms of frontier science and technology: I refer to the emergence of biotechnology, neuroscience, nanotechnology, and a host of other new fields, which I believe will one day fundamentally change human nature and enable the development of new forms of consciousness and intelligence. Currently, these technologies are still being developed for the sole purpose of serving human needs. Their aim is to make humans healthier, smarter, more athletic, longer-living, more personable, and happier. But even without higher aspirations, the slightest change of degree in our human condition resulting from advanced applications in these new fields may eventually lead to a big change in human nature. We will examine this distinct possibility later in the book.

The pursuit of a Golden Age itself could also prompt some people to attempt to change human nature, since this would be viewed as the shortest and best way to perfect the species. But a more powerful goad to change may be the abiding fear of humanity's or Earth's self-destruction along with the resulting drive to escape to outer space; this impulse may call for more than just a tinkering with the existing human genotype, since fundamental changes may be practically the only way for humanity to survive in an essentially hostile cosmic environment. Remember, there is nothing in Darwin's theory that says "survival of the fittest" means "survival of the human"—that is, survival of human nature as we have known it to be until now.

Darwin said the survivor may not be the strongest or the most intelligent, but the one most adapted to new conditions. Humanity is the only species in the world that will try to survive and thrive by anticipating future conditions that are *unprecedented,* and then consciously attempt to adjust its own nature in order to be equal to these conditions.

Yet the recognition of these aspirations also endangers the dream of the Golden Age. That is why Francis Fukuyama and others like him have fiercely attacked biotech advancements such as stem cell research and have called transhumanism the world's "most dangerous idea"—for the world he has in mind is of course the human world. In response, almost all defenders of biotechnology have insisted that the research is only for the good of humanity, because this premise has gone virtually unchallenged. Implicitly, they concede that new technologies should be stopped if they prove to be bad for humanity in any way or if they present a threat to the Golden Age.

We will dive deeply into this issue in chapter 9, when we discuss conscious evolution. At this juncture, suffice it to say that we can have higher aspirations because cosmic history has led to the evolution of a higher consciousness that can reflect and shape a Cosmic View. Having looked that far and wide into the past, why should we limit our view of the future to a narrow obsession with our tiny residence? The funnel vision of cosmic history is self-limiting and constricts the hope, joy, and excitement that a more expansive view provides. In fact, it ignores the fundamental features embedded in the universe, which include perpetual change and ever-more-complex novelty through emergence.

In sharp contrast to the traditional funnel-shaped vision, the Cosmic View is *hourglass-shaped*—from big to small and then from small to big (see chart). According to this perspective, things we must do to preserve humanity and the human-friendly environment are an essential part of the immediate future, but they bear an increasingly insignificant (and eventually irrelevant) relationship to the distant future.

What is significant about the Cosmic Vision is that from the tiny focal point in the middle (terrestrial human life in the vast universe), we can spawn a wide-open universe that is alive (for lack of a better word), is self-transforming, and harbors further possibilities too distant for us to discern. Seen through the prism of the Cosmic View, humanity becomes the present focal point for Cosmic Creation in the future.

As mentioned, the Cosmic View has great uncertainties. But I believe and will argue that it is a far more plausible approach than the funnel vision that leads to a human-centered Golden Age.

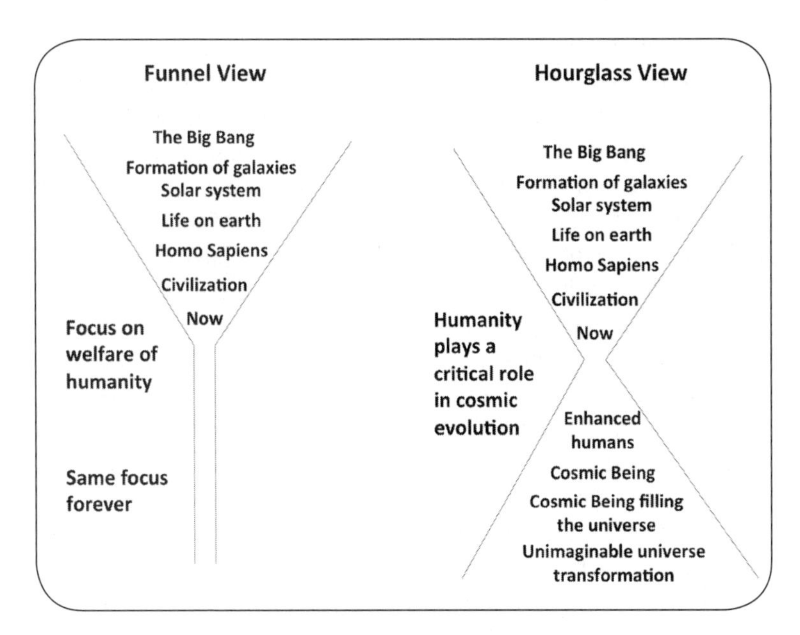

Figure 3: Two Views of the Universe

There is no scientific evidence that the cosmic potential has been exhausted, and the unfolding of the next level of potential always starts from a single point (although there is never assurance that any *particular* event will succeed).

If we just look at the Earth, a series of singular events shaped the world: the appearance of DNA-based life, the appearance of multicellular life, the appearance of humans, the advent of democracy, and the appearance of science. It is true that not all milestone events are singular. As we have noted, various functional forms such as the eye emerged independently at least several times in history; and in human history, important cultures and technologies such as agriculture, religion, and writing emerged independently in multiple times and places. But, given the chance to spread, just one occurrence of such an innovation is all it takes to fill all available space, just as one biological breakthrough was all that was needed to fill each of the available ecological niches on Earth. In Mao Zedong's dictum, "A single spark can start a prairie fire."

As we will discuss in more detail later, the Cosmic View calls for a group of pioneers to fulfill human potential, just as the sages began to do in the Axial Age. What is important to note here is that, just as humans filled all niches on Earth, the same can happen in the universe. While we have only the vaguest notions about the future, and so far no concrete indication as to whether we are "alone" in the universe or not, we should *assume* that we are irreplaceable and that we are the inevitable manifestation of the universe's potential. Even if this assumption proved to be wrong and other forms of intelligence were discovered later, CoBe would be in a better position than us in dealing with them.

We, as the focal point at the middle of the hourglass, must be all we can be. We must set the stage for the posthuman future. These words may sound utterly arrogant and presumptuous compared with the seemingly humble vision of the Golden Age. Yet consider the fact that the cosmic history that led to the current state of human civilization is a series of highly improbable events that are far more outrageous. Imagine, for instance, that you were an alien visiting Earth a few billion years ago. You found it a desolate wasteland, but with a lot of effort, you finally discovered a few single-celled organisms floating in "some warm little pond"—to borrow a phrase from Darwin. What kind of probability would you assign to the scenario that someday these bacteria would spawn a marvelous animal species, each with 100 trillion cells and a conscious mind? Next to nil. The appearance of the first cell out of inanimate matter may have been even more unlikely, yet it all happened. The improbability of these events might be seen as miraculous, and the mere contemplation of these events might be enough to evoke a sense of wonder and increase our confidence in the improbable CoBe becoming a reality.

The hourglass-shaped Cosmic View does not only refer to something far in the future. In fact, we have already witnessed the initial phase of the enlargement of the funnel—humanity's emergence from a corner in Africa to every corner on Earth and into space, growing from a naked ape to a builder of marvelous mental, physical, and social structures. But still, the future is not at all a given. To a large extent, the Cosmic View remains as hard to commit to as ever, while the narrowing funnel-shaped future seems far more comforting and reachable. The daunting task, then, is to use our imagination to look from the possible transhuman future back upon the present, in order to see humanity in its proper place.

6.4. Human Significance: A View from the Future

Without self-transcending actions, human existence can be and will be forgotten in the grand scheme of things. Unless we generate more "fuss," the story of humanity will prove to be like that of a little colorful bubble that will eventually pop without leaving a trace in the vast ocean. The cosmos seems to have anticipated us but at the same time has not been particularly eager for us to appear; life seems to have been possible but still improbable.

A lot more needs to be learned before we can give a fair evaluation of the Copernican Principle, which asserts that Earth is an ordinary planet with no special position in the universe. Compared with what we know about Earth, our cosmological knowledge is still very limited despite recent advancements. We do not even know what it is like on the other side of the Milky Way.

But we already have a rudimentary understanding of the necessary conditions that enabled life and intelligence to emerge. The Harvard professor Lawrence Henderson was among the first modern scientists who pioneered the research presented in his book *The Fitness of the Environment*, a seminal work first published in 1913. He analyzed the exceptional nature of water[11] and other features of the Earth's chemistry, and concluded that "the above catalogued natural characteristics of the environment promote and favor complexity, regulation, and metabolism, the three fundamental characteristics of life." John Barrow and Frank Tipler extended this line of analysis in their book *The Anthropic Cosmological Principle.* Their philosophy echoed the sentiments of both modern and ancient thinkers. The Chinese philosopher Tung Chung-shu (179–104 BCE) from the Han Dynasty wrote, "When one looks into the intention of Heaven, one finds it to be extremely benevolent."[12] And in 1979, Freeman Dyson wrote in his book *Disturbing the Universe*, "the universe is pregnant with life."

Powerful logical and statistical arguments have been made by Frank Drake and numerous other scientists about the inevitability and the abundance of life elsewhere in the universe. Not everyone agrees, however. Simon Conway Morris argues that it is hard to imagine that any life-forms other than carbon-based ones would emerge spontaneously—carbon atoms alone have the uncanny knack of linking together chains of any length to form molecules of great size and complexity—although it is still too early to tell whether alternatives to carbon-based DNA really exist.[13] However, Morris believes that, while the emergence of human-like intelligence and consciousness may be inevitable given the size of the universe and all the natural laws known to scientists, humans could still be alone in the vast universe.

The radio astronomer John Ball has speculated that the apparent quietness and emptiness of the universe is an artificial environment set up by some advanced aliens—that Earth has been intentionally left undisturbed and quietly nurtured as a cosmic "zoo."[14] Others have opted for apparently simpler explanations—for example, Gregory Paul and Earl Cox suggest, "The simplest and most logical interpretation of why our galaxy looks like an empty wilderness is because it *is* an empty wilderness."[15]

There is increasing evidence that a "just-right" environment may have been needed for the emergence and development of life, as articulated in Peter Ward and Donald Brownlee's *Rare Earth* hypothesis, John Leslie's fine-tuning *Universes*, William Burger's *Perfect Planet*, and Guillermo Gonzalez and Jay Richards' *Privileged Planet*. Earth needs a big brother planet like Jupiter, for example, to clear out comets and asteroids, and a moon to stabilize its axis of rotation. The range of physical conditions under which necessary organic chemical reactions can take place is extremely limited. Most of the known universe is simply too cold, too hot, too dense, too vacuous, too dark, too bright, or not composed of the right elements to support life.[16]

It has been said that Earth is like an insular greenhouse in an arctic universe. But we should not push the greenhouse metaphor too far. We should be extremely careful not to slip into the biocentric fallacy: little evidence exists to support the idea that nature or God "designed" a place to be just right for life;[17] it is more likely that life prevailed and even thrived on Earth despite the difficulties that nature threw at it. Notwithstanding the suggestions that Earth may be the only repository of life in the universe (or the galaxy), there is broad support for the contrary view. One indication of this is the wide interest in and support for the Search for Extraterrestrial Intelligence (SETI). Currently there is no way to tell whether SETI's search will eventually bear fruit, but we don't have to sit on our hands in the meantime. We have much to do before we can say whether we are at the center of the hourglass universe.

At this point it is appropriate to revisit Darwin's poetic words from *The Descent of Man* that appeared at the beginning of this chapter:

> Man may be excused for feeling some pride at having risen, though not through his own exertions, to the very summit of the organic scale, and the fact of his having thus risen, instead of having been aboriginally placed there, may give him hopes for a still higher destiny in the distant future.

In this light, humanity might be seen as only the latest manifestation of the unfolding of the seemingly limitless potential complexification of the universe. This is why Westerners—who, in contrast with other civilizations such as people in Asia and Native Americans, have held a worldview in which humans are at the "top"—are often humbled by modern science, which depicts us as being like the one-day fly, living in the thin outer layer of an otherwise ordinary planet orbiting an ordinary star in the outskirts of an ordinary galaxy, one of the billions of galaxies in the universe.

As vividly illustrated in Kees Boeke's classic book *Cosmic View: The Universe in 40 Jumps* and Charles Eames's short film *Powers of Ten*, there are vast ranges of reality beyond our own familiar world. We live in three-dimensional space, and we can scarcely imagine a hypothetical intelligent being living in a two-dimensional world and never being able to experience the third dimension.[18] It does not occur to us that we also live in a two-dimensional world, on the sheet of the thin surface of the Earth. And in a broader sense, we only live in a one-dimensional space, on a tiny dot that is Earth. In such a vast universe only a space-bound civilization can be said to have a true third dimension.

It is no wonder, then, that there is plenty of confusion and philosophical disagreement about the nature and meaning of our existence. In this great debate, at least two contrasting perceptions of humanity—both bleak in their own way—seem to have emerged. One perception considers the human soul to be the only light in the vast and unthinking cosmic darkness, while the other despairs of the paltriness and transitoriness of human existence in relation to nature's eternal recurrence. And of course these two views are not mutually exclusive.

I'd like to suggest a way out of this odd predicament. Humanity's relative insignificance in time and space is no reason for despair. We should avoid drawing conclusions too early about the fate of the universe and not be too concerned about the latest finding that the universe is expanding at such an accelerating pace that all life and thought may be doomed. H. G. Wells put it well in 1934, during a trying time between the two brutal world wars: "I am neither a pessimist nor an optimist at bottom. This is an entirely indifferent world in which willful wisdom seems to have a perfectly fair chance."[19]

The appearance of life on Earth *by itself* does not have an impact on the future course of the universe. If humanity had not appeared and if no other human-like intelligent conscious being emerged in the future, all life on Earth could very well perish along with Planet Earth itself, which will last, at most, for another five billion years. The actual appearance of humanity, however, has the potential to make the difference.

6.5. Human Aspirations and Cosmic Potential

Humanity can make a difference because we can act on our understanding of the Cosmic Vision. It is in this sense that Jesus's words—"The Kingdom of God is within you"—ring true. It is in this sense that the Catholic monk and poet Thomas Merton's words—"What we have to be is what we are"[20]—make perfect sense. And it is in this same sense that the philosopher Hannah Arendt's assertion is right on—that the problem of human nature is unanswerable and that the conditions of human existence can never answer the question of what or who we are, for the simple reason that these conditions never condition us absolutely.[21]

How carbon-based life realized its potential gives us some hints on how human potential can be realized. Life essentially changed the surface of the Earth, from remaking the atmosphere to forming complex self-regulatory structures. Even before humanity, life seemed to be tinkering with Earth for its own survival and growth.

One of the major misunderstandings of Darwin's evolutionary theory is that organisms are passively shaped by natural selection in an unchanging environment. Given the yin-yang nature of biological reality, the environment is constantly being modified as well by organisms that actively choose and transform their external world. The terrestrial environment—the atmosphere and oceanic chemistry, rock formation, and even tectonic movements—has been profoundly altered by the bacterial ecosystem and by other biological organisms' disposal of their wastes.

Humanity has already changed much of Earth's landscape, has modified or eliminated numerous species, and is seemingly in control of the Earth at this moment; yet these man-made changes are very insignificant compared with earlier episodes involving other life-forms. For example, the magnitude of the man-made increase of CO_2 and other greenhouse gases is minuscule compared with natural CO_2 oscillations in history and the dramatic increase of oxygen in the atmosphere. We must remember that we have been on Earth for a very brief period of time and have been subject to the same selective pressure as any other species—witness the collapse of human cultures that have destroyed their own habitat.

What the human species has done so far is trivial compared to the mind's potential to experience and comprehend the entire universe and beyond. This is the critical difference between the Cosmic View and the traditional, inward-looking, and anthropocentric worldviews. No other form of life on Earth is capable of going beyond the self and striving for something grand, something immortal, something purely imagined. In short, the human mind is becoming the universe's way of knowing itself and the engine for cosmic evolution.

In conclusion, human significance must be placed in the context of its future potential rather than its present existence. It is certainly human hubris to think that all of the vast space and energy in this universe were created so that humans could enjoy a happy life. There is much to be created and expanded beyond a happy humanity.

With the hourglass-shaped Cosmic View I've presented in this chapter, the love of God can become a powerful emotion propelling us toward the future, a passion for what is possible. That love becomes itself the magic power that brings our potential into being. With hope, love surveys the open possibilities of history. With love, hope brings all things imagined into the light of the promises of God.[22] In the words of Confucius: "We should never shy away from great responsibilities once we recognize them" (当仁不让).

Will our potential be realized? How can this be accomplished? Each of us must ask that question. From ancient meditation practices to the modern New Age movement, great efforts have been made to realize the growth potential of the individual. In the next two chapters, we will discuss why individual human enlightenment is not enough, and how difficult it is to lift our level of vision and understanding. True realization of our potential can be achieved only through conscious evolution beyond the human species.

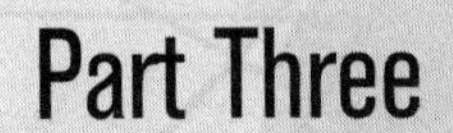

Part Three

HUMAN NATURE AND ITS LIMITATIONS

CHAPTER SEVEN

ALL TOO HUMAN

What we observe is not nature itself, but nature exposed to our method of questioning.

—Werner Heisenberg

IS HUMAN NATURE infinitely flexible, or are there limits to our ability to adapt and evolve? This question is central to considerations of our next evolutionary step. If human nature is inherently flexible—if it in fact does carry an unlimited potential for creating novelty and complexity—then all we need to do is find ways to stretch and mold it. But if this is not the case, then our further evolution will require that we somehow, at some point, step in and improve human nature or else create alternatives such as those I suggest in this book. This chapter examines what constitutes "human nature," exploring its limits in the broadest natural and cultural contexts. Our discussion will lay the foundation for the chapters that follow, in which we examine ways to transcend human limitations and take on the Cosmic View.

Few of us think in terms of an evolving human nature. Today, our projections of the future almost invariably share a key assumption: *humans will remain "human" forever*. In popular imagination, it is humans—pretty much as we now understand them to be today—who will visit other planets and eventually explore worlds beyond our solar system. In science-fiction books and movies, it is today's conventional humans who engage in fierce fights with human-like aliens or human mutants and who almost always prevail in the end. Even in Frank Tipler's grand vision of our cosmological future, the completely enlightened society will still be devoted to fulfilling perennial human needs and desires such as finding a perfect mate and overcoming death.[1] Indeed, living as we are within an "all-too-human" society, it is hard to rise up out of our current identity and believe that humanity, with its limitations, might ultimately prove irrelevant to future evolution.

7.1. The Limits of Human Nature

In order to get at the limits of human nature, we have to first focus on the human brain, the most complex and flexible organ we possess, and the organ of the body that essentially defines humanity. There is no greater mystery in the known universe than the human mind, except for the universe itself. When the great scientific genius Isaac Newton became a victim of the South Sea Bubble (one of the first and most devastating stock market bubbles), he exclaimed, "I can calculate the motions of heavenly bodies, but not the madness of people"—reflecting his realization of just how complex people are compared with the stars. At another level, though, our brain is not so amazing: the *structural* differences between a human brain and other mammals' brains, especially those of other primates, are actually small; and functionally, the brain is not much more than a machine—albeit a very complex one, as we will see later in this chapter.

The core argument I will put forth here is that each of our brain's capabilities has limitations and drawbacks in both technical and motivational terms, and these limitations cannot be overcome as long as we remain "human" in the conventional sense. To understand this argument, we must first recognize that the human brain is a *historical* product of natural evolution, a big patchwork providing partial solutions rather than an engineered masterpiece. The brain is a passable solution to the problems at hand, not an optimal solution.

For example, our brain evolved to cope with medium-sized objects moving at medium speeds over medium distances. Our senses of space, time, causality, and probability are useful mental constructions that are well adapted for ordinary living environments on Earth but not for extreme environments.

True, we now know that the brain is highly elastic. Since the advent of civilization, humans have developed technologies and social institutions to vastly expand our natural capabilities. Better nutrition, education, and environmental stimulation have helped tremendously to bring out our genetic potential as well as our latent brain power. Average IQ scores in industrialized countries, for example, rose as much as 20 percent in the second half of the twentieth century due to the "Flynn effect."[2]

But even given this sphere of impressive cultural achievements and artifacts, we are still mere mortals—animals at heart. Our vanity notwithstanding, humans and their brains are biological machines with limited capacities and highly predictable cognitive and behavioral patterns. Narrowly restricted by our senses and our powers of concentration and memory, our brains can never perceive much more than a tiny patch of the vast tapestry of events. And our productive lives are all too brief: We quickly move through a rather predictable life cycle of infancy, childhood, youth, adulthood, and old age.

Let us look more closely at specific areas of human constraint: brain capacity, consciousness, rationality, sensation, life cycle, memory, and so on.

The limits of brain size. *Homo sapiens* has the highest brain-to-body-weight ratio of all animals, but from an evolutionary perspective, our brain size is limited by the size of the newborn's skull, the high metabolic cost of operating a brain, and the physical structure of the brain. While it is true that each of us utilizes only a small fraction of our potential mental abilities—we are always amazed when someone can remember whole books word for word—there is a limit to the carrying capacity of the human brain.

The distribution of our natural abilities follows a bell-shaped curve, so there are exemplary cases with respect to the norm, but even these individuals are not unbounded. Size matters in an absolute sense. A normal person can know a couple of hundred people well, while an exceptional individual may know several thousand people, but certainly not millions. Today, knowledge of all kinds is accumulating every day, yet the amount of information a person can learn has not changed much.[3] The chief advantage of a modern civilized human over a person living in a primitive society is not necessarily that we have more knowledge, but is in some ways the opposite: we require far *less* personal knowledge and skills concerning how to survive and live each day in a hostile environment.[4]

Our short-term working memory is readily accessible for only a short period of time (a matter of seconds) and is limited to only about five to eight distinct pieces of memory that we can retain simultaneously.[5]

Our long-term memory is open-ended in principle. But in reality, the number of concepts, or "chunks" of knowledge, that a human expert in a particular field can master falls quite consistently within a range of 50,000 to 100,000, according to estimates made by computer scientist Ray Kurzweil. Add to that our more extensive "commonsense" knowledge, and each of us has the ability to retain and access roughly one hundred million bits of concepts, images, and ideas.

The limits of consciousness. Consciousness is the most marvelous mental attribute we have, and what I call higher-order consciousness is uniquely human. But even with a limited scientific understanding, we already know that our consciousness is more limited than common sense would have us believe.

For example, the amazing intuitions and insights that seem to pop up out of nowhere do not come from our conscious, waking mind but instead come from what some scientists call "embodied knowledge," which is unavailable to conscious inspection until it appears. Compared with unconscious processes, conscious thought is slow, serial (one thought at a time), and cognitively limited.[6] Thus we must rely on "automatic" reactions to cope with emergencies when, for example, we hear a loud noise or step into an unseen hole.

In *The User Illusion,* Tor Norretranders suggests that we tend to respect and worship our conscious mind just as we might treat a president or a company CEO, only to be surprised that it does not know much and has only limited expertise and skills. Large quantities of information are discarded before the conscious waking brain enters into the picture. Norretranders argues that the conscious mind is like a credit card report—it shows that $52.37 was spent at a grocery store but has no idea which items were purchased.

The limits of the senses. Our senses severely prune the amount of information that the environment offers. We primarily rely on vision for external perceptions, and each of our five senses is rather mediocre when compared with the abilities of certain animals. And like those of all mammals, our nervous systems can sense only a very narrow portion of the electromagnetic spectrum (wavelengths of about 430 to 700 billionths of a meter). We may understand the theory of electromagnetic waves intellectually, but our appreciation of color is still limited to the range of visible light.

The unreliability of the senses. Perception can never reflect reality directly, since it is a transformation process that manufactures a pattern across multiple sensory modalities and neural interfaces. Our internally generated senses of time and space are equally inaccurate. The idea that our senses give us direct and "honest" information about the outside world has become part of our "common sense" and was promulgated by Aristotle, St. Thomas Aquinas, and other early thinkers. However, René Descartes pointed out that "there can be a difference between our sensation of light . . . and what is in the objects that produces that sensation in us."[7] Kepler seemed to agree when he said, "Eyesight should learn from reason."[8]

Our vision serves us well in everyday life, but the power of optical illusions is well known today. Every picture we see is ultimately subjective, with the image that falls on the retina being torn apart and processed by over thirty modules specializing in edges, colors, motion, and particularly interesting shapes such as mouths and eyes, before a representation is put together in the brain. The brain makes the best interpretation it can, according to its previous experience and the limited and ambiguous information provided by our eyes.

The limits of the human life cycle. In Shakespeare's comedy *As You Like It,* he divided human life into seven stages: "infant, schoolboy, lover, soldier, justice, lean and slippered pantaloon, and second childhood." Each child is like a philosopher, wondering about the most fundamental questions of life and existence, but gradually—through a series of identifiable stages—settles into a worldview that is subject to the influences of his or her own temperament and the prevailing culture.[9]

Unfortunately, by the time one has enough experience to approach some semblance of life mastery, one's natural energy and strength is already declining. There is little means of *directly* passing on our accumulated experience, as each newborn has to repeat the same mind-shaping process.

Only a few of us retain a childlike wonder throughout our lives. A typical life's journey is one with increasing knowledge but decreasing mental and physical adaptability. Our mindset unknowingly gets trapped in conceptions and categories that we create. As Aristotle observed, as we grow older, we aspire to nothing great and exalted and crave the mere necessities and comforts of existence. Age is a very high price to pay for maturity.

The fallibility of memory. Memory is strongly influenced by emotion, which almost always serves as a selective filter to focus attention and narrow the field of experience. Our existing "mental models" of the world can often make us "recall" information that is distorted or was never given.[10] Show people some fake "Disney" ads with Bugs Bunny in Disneyland, and people will recall that they shook hands with the bunny (who is not even a Disney character) at Disneyland. The memories of previous beliefs and impressions are often tweaked in light of new experience without conscious awareness.[11]

The limits of probability sense. Our natural tendency to miscalculate probabilities, especially of small events, is well known. For example, in a coin toss, most people believe that an alternating heads/tails sequence (H-T-H-T-H-T) is much more likely than straight heads (H-H-H-H-H-H), whereas both are equally likely.

The emergence of life has been argued to be impossible without some intelligent intervention or "creation," since the structure of a cell is incredibly intricate. But just imagine that you have been dealt a hand of thirteen spades in a bridge game—statistically, the probability of that occurring is one in 635 billion! Such an event is so rare that it has never been recorded in the annals of bridge. However, *any* bridge hand has exactly the same probability! Most of them are just too unremarkable to get noticed.[12]

The fallibility of moral sense. We tend to believe that our moral principles are as solid as natural laws, but they are products of a messy brain in a messy world. We have a tendency to favor kin and friends, and to confuse morality with conformity, cleanliness, and the like.[13] In each generation, our moral judgments about certain behaviors are unduly influenced by how widely practiced they are. There are numerous examples of practices in history once considered normal, such as slave-owning, that we find morally offensive today. Likewise, future generations will certainly find things we take for granted to be morally unacceptable.

The limits of education. We have only a limited repertoire of fundamental cognitive abilities that were selected in evolution. The rest are biologically secondary abilities, derivatives that are principally cultural inventions, such as reading, which was only invented ten thousand years ago. The acquisition of secondary abilities requires tedious repetition, sustained effort, and external motivation for their mastery. Achieving an education in higher math, logic, science, or the written word entails hard work: much of what we teach children in school consists of "unnatural" tasks never encountered by our ancestors. Almost all of these tasks involve "seat work" and extended periods of focused attention, activities that conflict with children's natural tendencies toward spontaneous creativity, physical activity, and boundless curiosity and exploration.

As with animals, our "built-in" learning capability is outstanding, but the educational curriculum is to a large extent constrained by what children are motivated to learn. The dilemma of education is that there are limits to teachers' ability to make learning easy and fun.

The limits of language and communication. Language seems to give us an infinite ability to communicate and store information, yet in reality the utility of words as a means of communication is severely limited by such factors as "bandwidth," interpretation, and articulation.

First, linguistic communication has a very narrow bandwidth. Most of what we experience, we can never tell each other about. We experience millions of bits a second, but only a few dozen can be passed on verbally. Any thought has to be reduced to a skeleton of itself in order to be spoken or written.[14] In general, what is unconscious (which may be 98 percent of brain activities) cannot be communicated through language.

Second, because our brain is looking for what it expects to hear or see, what a person is able to receive as a result of verbal communication can be highly distorted even when there is no intention of deceiving. In ordinary discourse, almost everything being communicated is subject to widely varying interpretations based on the assumptions held by the communicating parties.

Third, many seemingly simple concepts cannot be defined clearly, although the mind seems to understand them after training or after repeated experiences. The late Supreme Court Justice Potter Stewart famously said that he did not know how to define pornography, but "I know it when I see it."

These limits also imply that it is extremely difficult, if not impossible, to create a "super-brain"—or as some visionaries have called it, a "global brain"—by simply linking people together in an effort to overcome the capacity limitation of individual brains.

The limits of imagination. Our imagination seems to be totally free, but in fact it is not. Apparently, fantasy cannot easily break free of intuitive expectations, and even our wildest imaginations often pale in front of reality. The Big Bang theory outdoes any of the great creation myths, and quantum mechanics presents a paradoxical reality that defies our commonsense expectations as much as any religious miracle. As the novelist Tom Wolfe quipped, "The imagination of a novelist is powerless before what he knows he's going to read in tomorrow morning's newspaper."[15] The quantum phenomenon is real, but we are unable to form any comprehensible visual model of it. Try this: track an imaginary ball that is changing simultaneously in terms of size, shape, color, brightness, smell, temperature, softness, texture, noise, and vibration. Can you do it?

The limits of the environment. Our amazingly rich cultural diversity is also fundamentally limited. One limitation is that our lives are restricted to the earthly environment, however diversified it is. A good Martian anthropologist who studied our dietary habits would come to the conclusion that there are only a few ways to process food (marinating, roasting, smoking, boiling, frying, steaming, etc.) and a large but limited number of ingredients. The seemingly unlimited numbers of dishes are combinations of a limited number of techniques and a limited set of materials.[16]

We enjoy the seemingly endless variety of stories, myths, and legends, but on closer examination, all these tales that sprung from isolated cultures around the world and in different times exhibit similar schemes or characteristics, as any student of mythology or archetypal psychology knows. The reason is simple: we are the same species living on the same planet. Even when humans move into space someday, an Earth-like artificial environment will be needed, for both physical and psychological health.

What do the limitations we have cited mean? The human brain's structural limits are hard to pin down since we don't have a proper benchmark to measure our brains against. However, we can extrapolate from what we *do* know—for example, we know that the monkey's brain has hard limits when it is pitted against a human's. Talking about the brain's limitations also seems somewhat contradictory in that we believe we know far more than our ancestors did with the same brain. This proposition seems true to us, but it may largely be a matter of very different uses of the same limited-capacity brain.

On the one hand, we have realized more brain potential through better health practices, more education, and more chances to experience mental stimuli. But if we ask ourselves whether people in primitive cultures were more naïve and ignorant than us, we can hardly answer that with an unqualified "yes." In our selective observations, we may overlook the fact that primitive peoples were more independent and self-sufficient than we are because of their intimate knowledge of their local environment. In fact, they would consider us childish and naïve in the sense that our lives are cocooned in complex protective social networks and machines that we really know very little about.

Our lack of broad practical knowledge and know-how is also reflected in organizations as they become incredibly specialized nodes in larger complex networks.[17] The inner workings of machines that people use today can be as inscrutable as most natural phenomena were for primitive peoples. The same mass ignorance is true for the inner workings of modern social networks and institutions.

Fundamentally, we are still in the evolutionary equivalent of the food-gathering stage in terms of developing "human resources." We hit it big occasionally with a genius or two, but we really do not know how to grow talent consistently and efficiently. There is indeed untapped potential that could be realized through better culture and knowledge, but is the future development of culture and scientific knowledge truly unlimited—especially given the way the human brain is constructed? This is a big and complex issue. Let's start with careful examination of the human endeavor that offers the greatest promise—namely, science—before moving on to the general area of culture at large.

7.2. The Limits of Science

Science and its technological applications have vastly increased our capabilities without altering our human nature. While scanning through the list of human limitations above, many readers may have reacted by thinking, "Of course I know we are limited, but human capacity has endless potential with the help of unlimited scientific advancement."

In response, let me begin by pointing out the obvious: All human knowledge, including scientific knowledge, is just that—*human*—and cannot go beyond the brain's motivational, cognitive, intuitive, sensory, and imaginative capabilities. Scientific knowledge, however, seems to defy the restrictions of its human creators, since it has continued to grow with our ability to benefit from the external assistance of machines and collaborative efforts of the scientific community. But as Albert Camus put it, "Understanding the world for a man is reducing it to the human, stamping it with his seal. The cat's universe is not the universe of the anthill."[18]

What scientists have to say. All scientific findings are subject to peer review, but the scientific enterprise consists first and foremost of personal experiences of scientists—notwithstanding science's attempts to replace the cruder anthropocentrism of our senses with a more ambitious anthropocentrism of our reason. From one point of view, scientists performing experiments are psychologically equivalent to shamans conducting rituals in their urge to extract truth from nature. Let us listen to what some practitioners of science have to say:

- "Whatever we call reality," said Ilya Prigogine, the Nobel laureate in chemistry, "it is revealed to us only through an active construction in which we participate."[19]
- "What we observe is not nature itself," said the Nobel physicist Werner Heisenberg, "but nature exposed to our method of questioning."[20]
- The Nobel physicist Niels Bohr has pointed out that science cannot perform without a human observer, so we cannot describe anything without including in our description the fact that we are describing it. As the physicist James Clerk Maxwell famously said, "The only laws of matter are those which our minds must fabricate, and the only laws of mind are fabricated for it by matter."
- The Nobel scientist Gerald Edelman pointed out that science, like consciousness rising out of unconscious thoughts, inevitably loses the richness of reality: "Events

> are denser than any possible scientific description. They are also microscopically indeterminate, and, given our theory, they are even to some extent macroscopically so."[21]

Scientist Ludwik Fleck said that a "pure observation, that is, one without assumptions," does not occur psychologically, and therefore this idea can be dismissed.[22] Nobody, even the most self-critical scientists, can escape the anthropomorphic impulse in their observations, which is always blended with impulses such as personal goals, intent, emotions, biases, and so on.[23]

There is little doubt that scientific research is the best approach to understanding the cosmos, and that scientific knowledge is absolutely the best human knowledge—but science, like all other human activities, is still limited by human nature. Scientific genius is brilliant only relative to the human standard; scientific laws are profound only relative to the limitations of our common sense and instincts.

Even math is human. Science is often said to be best expressed in precise mathematical formulae rather than sloppy human language. Yet even mathematics should be considered as one of the mental tools for the human mind to understand reality rather than the deepest universal principle or some essential reflection of "the mind of God." The mathematician Brian Davies made the following statement, keenly aware that he risked annoying many of his colleagues:

> In spite of the fact that highly mathematical theories often provide very accurate predictions, we should not, on that account, think that such theories are true or that Nature is governed by mathematics. In fact the scientific theories most likely to be around in a thousand years' time are those which are the *least* mathematical—for example evolution, plate tectonics, and the existence of atoms.[24]

Mathematics is not a guaranteed source of objective and eternal truth, and is best considered as stable mental objects in the human brain without an independent Platonic existence.[25] All math theorems are built on unprovable axioms formulated by our instinctive conviction or even blind faith (and therefore referred to as "self-evident").

We are looking for natural laws under the "street lamp" of mathematics because that is where we can see the hidden patterns. The extraordinary effectiveness and usefulness of mathematics cannot conceal the fact that it says much more about the human mind than about reality. The planets, for example, are not solving differential equations as they swing around the sun. To understand their behavior we resort to a formality—in this case differential equations—which expresses their behavior as motion according to a humanly determined rule.[26]

The factor of common sense. Common sense is oxygen for scientific thought. But as our knowledge expands, we push deeper into the realm of the counterintuitive and feel the air getting thinner. Earlier scientific discoveries often appear trivial to us once the puzzle is solved. The relativity and limitations of common sense become increasingly apparent with each new scientific breakthrough. While most people can perfectly understand and marvel at Newton's gravitational laws, few, even among the highly educated, understand the theory of quantum mechanics, even though its accuracy has been repeatedly confirmed in more than half a century's experience.

The difficulty of going beyond common sense is not just a challenge for natural sciences. As the historian John Lewis Gaddis points out in *The Landscape of History,* there is a methodological Catch-22 in the social sciences. "Social scientists seek to build universally applicable generalizations about necessarily simple matters; but if these matters were any more complicated, their theories wouldn't be universally applicable. Hence, when social scientists are right, they too often confirm the obvious. When they don't confirm the obvious, they're often wrong."[27] I think this observation is true for all sciences of complexity.

The poverty of reductionism. The spectacular success of science is fundamentally due to our ability to "discover" natural laws, those invariant patterns that we can rely on to boil down the complex world into simpler forms, especially with the use of abstract mathematical symbols. The human mind can communicate and manipulate these symbols and form a model of reality for great practical application.

But in our manipulation of these artificial constructions we call "laws," we sometimes lose sight of the fact that these clean-sounding concepts are not reality. Brian Davies highlights one aspect of the poverty of reductionism: "The idea of breaking reality up into small parts which are then analyzed separately has been brilliantly successful, and may be the only way *our type of mind* can understand the world. However, their explanatory power is ultimately limited by the existence of chaos and quantum entanglement."[28]

While reductionism worked spectacularly well for physics and chemistry, it has been much less successful for complex emergent phenomena such as life and weather.

In general, emergent complex behaviors cannot be reduced to lower-level comprehensible laws. Thus we are able to predict that the moon will be a couple of inches farther away from Earth next year, but we cannot predict what the next sentence will be in a conversation with a close friend.

All knowledge is provisional, and permanently so. All explanations are local, too, although we seldom question their shallowness and incompleteness. When you ask, for example, "Why is David sick?" you may well be satisfied with the answer "Because he played in the rain yesterday," without questioning further. But why are the other kids who played with him fine?

It is counterintuitive that interpretations of nature must have endless variations rather than be reducible to a single concept or principle. The "that's it" and "nothing but" explanations of nature, the universe, and human nature have always attracted the greatest attention in history, but their utility as "truth" has been local and temporary.

In science, the latest "that's it" attempts are Seth Lloyd's argument that the universe *is* a quantum computer made of bits of information[29] and Charles Seife's claim that everything is, at bottom, information.[30] These "one-substance" metaphysical systems (or cosmologies) invariably confuse "little mirrors" of reality with the reality itself.[31] As Montaigne put it: "Things in themselves perhaps have their own weights, measures, and states; but inwardly, when they enter into us, the mind cuts them to its own conceptions."[32] Reality just is not the same when compacted.

Diminishing returns from scientists. The scientific process is running into diminishing returns as human knowledge keeps on accumulating and expanding. New exotic ideas in physics such as superstring theory and the M-theory seem to have little hope since we don't know how to put them to empirical tests. They are what science writer John Horgan calls "ironic science"—interesting points of view that may be logically sound but are not subject to deeper inquiry using the scientific method. In many scientific fields, it even looks as if science and magic are once again moving toward common ground.[33] In his often-misunderstood book *The End of Science,* Horgan declares that the great era of scientific discovery is over.[34] His message is not that we are reaching a point where there is nothing new to know, but that we are approaching the limit to what human beings *can* know.

Is it time to look beyond human science? What Horgan unearthed from interviewing some of the world's greatest and most experienced scientists is that *the human brain has become the bottleneck* in scientific research. If Horgan is right, then civilization's most recent burst of scientific progress, the one that started during the Renaissance, is fast reaching its denouement if we do not create better minds.

Until such advanced minds appear, further advancements of science and its technological applications will depend largely on better methods of organizing the scientific process.

The last issue we need to discuss in this section, then, is the challenge of overcoming the limits of science in the future. The power of science has been boosted by three strategies: *leveraging, specialization,* and *collaboration.*

First, computational and memory resources that scientists can leverage has increased tremendously since the invention of the digital computer. However, the increasing reliance on exterior computations poses daunting technical and political challenges and risks. In *Why Things Bite Back,* Edward Tenner points out that we have ceased to be tool *users,* who command an intimate knowledge of and skills associated with the tool, and have become tool *managers,* who direct and control complex processes that are largely alien to us.

Second, research has become increasingly specialized. Since Renaissance times, it has been getting ever harder for even a genius to master all knowledge. In 1900, the number of research papers published in chemistry was in the order of thousands; one hundred years later, this number stands at more than 500,000 papers a year. Francis Bacon was probably the last polymath who could regard all knowledge as his province. In this knowledge explosion, as the old saying goes, the experts know more and more about less and less, while the non-experts know less and less about more and more. The danger of the tunnel vision of specialists is that "when a man's vision is fixed on one thing, he might as well be blind."[35]

Third, collaboration has become the norm in research. To overcome the limits of tunnel vision as a result of overspecialization, scientific research has placed ever more importance on social organizations and interpersonal communications in scientific research. But there is a critical difference between knowledge stored in different brains and a single brain in terms of understanding and creativity. Regardless of how much effort we put into communication and collaboration, the best breakthroughs still emerge from one creative mind.

7.3. The Limits of Culture and the Problem of Evil

Education, art, spirituality, and other forms of culture help us realize our potentials. In recent decades, cultural enhancements have more than offset genetic deterioration and significantly raised the average levels of human intelligence.

Nevertheless, culture has limits because human nature is constrained by our evolutionary history, which makes certain cultures unlikely to spread or persist. Just like genetic mutations, strange things pop up in human culture all the time, but only certain innovations survive. Despite great cultural and behavioral variations, we are not a "blank slate." In Samuel Johnson's words, "We are all prompted by the same motives, all deceived by the same fallacies, all animated by hope, obstructed by danger, entangled by desire, and seduced by pleasure."[36]

Many people, including some of the smartest intellectuals, believe in cultural determinism—the idea that exterior social and cultural factors are primary. They have a strong disgust for scientists who view the human species in starkly scientific terms and who espouse the genetic roots of human behavior such as aggression. But when the influential anthropologist Margaret Mead asserted that "human nature is the rawest, most undifferentiated of raw material,"[37] she was mistaking the general intelligence of our vastly expanded neocortex for our entire brain. In the debate of nature versus nurture, she was repeating the mistake of Enlightenment thinkers such as Marquis de Condorcet, who believed that the mind is molded wholly by its environment and can be improved indefinitely.

Just as the embryo comes into being loaded with a set of genes, the brain comes into being loaded with both innate motivations and a specific learning capacity. Noam Chomsky puts the genetic constraints this way: "The embryo may fail to grow arms and legs properly [because of the environment], but no change in the environment will lead it to grow wings." Similarly, out of the vast number of potential alternatives, only a certain limited set of linguistic and religious structures have ever been adopted by human societies.[38]

David Brown has provided a list of what he calls "human universals" that includes symbolic language; naming and taxonomy; elementary logical reasoning; facial or nonverbal communication and recognition; meta tool making (using tools to make tools); fire; shelter; territory; family; division and cooperation of labor; government and law; hierarchy; etiquette and hospitality; religion and ritual; dance and music; play; etc.[39] It's a very long list, but not unlimited.

Although culture plays an important mediating role, our emotional and behavioral repertoire appears to be fixed around a universal set of human traits and emotions such as pleasure, pain, sexual attraction, romantic love, jealousy, suspicion, deceit, surprise, hope, courage, empathy, justice, revenge, shame, pride, curiosity, sudden insight, sadness, hunger, thirst, anxiety, and so on. Again, a long but finite list.

It is this constancy that, for example, could make it possible for a "primitive" person to jump from a Stone Age warrior life into a modern nation state with relative ease, as the New Guinean Albert Maori Kiki described in his 1968 autobiography, *Ten Thousand Years in a Lifetime.* The anthropologist Melvin Konner also tells a story about Marjorie Shostak's ethnographic classic, *Nisa: The Life and Words of a !Kung Woman,* which portrayed its central character as living among hunter-gatherers who had no contact with the Western world. When the manuscript was circulated to publishers, one of them rejected it on the grounds that Nisa sounded too familiar, too much like an American woman, in her relationships, feelings, and foibles.[40]

Just like the pervasive convergence patterns we have seen in the biological world, only certain types of personal relationships and social organizations have emerged spontaneously and independently throughout history.[41]

No society can be run politically like a super-rational community, as Karl Popper and Plato proposed. Likewise, given the unnaturalness of science, we cannot hope that people's raw religious impulses can be channeled into a devotion to science, as E. O. Wilson proposed, or expect that future humanity will devote itself to pure scientific research and spend much less time on satisfying primary physiological and psychological needs, as J. D. Bernal proposed.[42]

Culture reflects human nature. Throughout history, each generation feels it is uniquely different from the older generations, but strong evidence seems to show that most social, cultural, and religious movements merely swing periodically between the dominance of one sentiment and that of an opposing but complementary sentiment.[43] This has been described as the tension and synthesis between the gods of Apollo and Dionysus in Nietzsche's study of the classic Greek civilization, and between bourgeois self-control and bohemian emancipation in David Brooks's research on contemporary America.

Even the most radical revolutions or paradigm shifts brought no "new wine," only new "bottles" shaped by particular historical circumstances and personal characters. What we call "revolutions" were usually not progressive in nature. All too often, they led not to a wholly new future but to a regression to a distant past. Human history, if stripped of technological and institutional inventions and innovations, is mostly cyclical.

Today, many people are immersed in the modern belief that technology is enabling us to live totally different lives than people did just a few generations ago. Yet technology is only one of many forces driving human history, and seldom the most important. Technological determinism is just as suspect as cultural determinism. Even beyond the same basic biological needs—we still eat, drink, breathe, sleep, seek mental stimulus, and the like—the nature of our emotional and social needs remains unchanged, still centering around authority, family, and community.

What changes is how we develop social relationships and how we maintain them. There are very important exceptions, but in general, new technologies are little more than an enabling force that increases our choices and decreases our limits. Although it often feels like enormous change to the individual, new technological capabilities do not modify but merely amplify our innate drives. These truths confirm the realistic view of humanity and life as expressed in Eastern wisdom.

Take family, for example, which has undergone drastic changes in recent decades. The alarm over "the decline of traditional families" in postmodern societies ignores the fact that the nuclear family is an artifact of the twentieth-century industrialized world.[44] Before that, the traditional families were clans, the extended family comprising several generations.

What is predictable is that family-like functions will endure while family structure will continue to change. Recent development is putting a growing emphasis on a wider sharing of responsibility for the well-being of children—almost a return to the clan form of family. The extended family may be reemerging for other fundamental human needs.

There is overwhelming evidence that family still matters very much to people. One emerging phenomenon is the "intentional family," in which biologically unrelated people get together for holidays, birthdays, and other family-type occasions. The Internet enables all kinds of virtual gatherings of family and community.

Overall, technology is unlikely to alter the fundamental human need for family. What does seem likely is that we will have several preferred family models rather than one, along with a higher degree of flexibility. People now have the option to move from one model to another, one relationship to another, just as they move from one job/career to another.

Here is another angle from which we can see human nature shaping human culture: focusing on a fixed human psyche has repeatedly proven to be more insightful in predicting the future than understanding technological developments. Indeed, from who-does-what-to-whom daily gossip to the most abstract and universal Perennial Philosophy, the essence of human interest and inspiration has been perpetually resurrected. When Harry Truman was asked why he was reading *Plutarch's Lives,* the American president replied: To find out what's going on in Washington.

Stewart Brand once wrote, "Science is the only news. When you scan through a newspaper or magazine, all the human-interest stuff is the same old he-said-she-said, the politics and economics the same sorry cyclic dramas, the fashions a pathetic illusion of newness, and even technology is predictable if you know the science. Human nature doesn't change much; science does, and the change accrues, altering the world irreversibly."[45] This sentiment is not new, nor should it be: an Egyptian author of the Twelfth Dynasty (1900 BCE) wrote that his greatest problem was "the difficulty of saying anything new."

The more reflective individuals in our society often lament that people are getting more materialistic, shortsighted, and preoccupied with enjoying themselves. But that is not different from people in the past. What they should grieve about is the inability of science and technology to alter our motivations. The Axial religions and philosophies were already advocating abstention, moderation, the reduction of superfluous wants and capricious, ego-driven desires, for the sake of both internal equilibrium and spiritual exaltation. These basic doctrines have not changed over two thousand years—and for obvious reasons.

Our mind has a stable value and motivation system that is not significantly different from that of other social mammals except for a greatly enlarged cortex sitting on top of it and helping it to achieve its goals. We should not pin too much hope on lifting human culture, nor should we be overly pessimistic. Throughout history, the intellectual elite have made doom-and-gloom predictions of how popular culture would lead to moral decay and mental deterioration. Socrates worried that increasing reliance on written texts would create laziness and forgetfulness in the learners' souls. But it was Plato who wrote down Socrates' oral lessons!

Throughout history most people have retained reasonable intelligence and made reasonable judgments most of the time, regardless of the prevalent cultural fad. Modern mass-market entertainment such as video games may appear to be dumbing our kids down or turning them into twitchy addicts, but at the same time, they are often quite stimulating, educating, and engaging. Truly boring stuff that cannot grab our attention can never become popular in the first place.[46] Mass entertainment, most of which is quite shallow stuff, is more of a reflection of our nature than the shaper of it.

Culture and the problem of evil. As Freud made clear, culture can amplify or suppress the expression of an instinct, but not eliminate or alter its latent power. Homicide rates, for example, can vary enormously across different societies both in the primitive and modern Western world; media and cultural influence, sexual satisfaction, employment, food supply, family life, and many other factors can play a role. There is one extreme case, the tribal society of Semai Senoi in West Malaysia, in which zero cases of murder, attempted murder, or assault were reported over a period of fifteen years.[47]

The following more-typical contrast is revealing of both differences and similarities. During the 1970s and 1980s, Chicago had a same-sex, non-kin killing rate of nine hundred per million people per year, while England and Wales had a rate of only thirty per million people a year—a 30:1 ratio. But amazingly, the age and sex distributions of these two cases matched almost perfectly: in both cases it was overwhelmingly young men killing young men—starting, peaking, and trailing off at exactly the same ages.[48] The environmental differences had a huge impact on the magnitude, but they could not cover up the universal propensity toward a highly competitive nature in men and the propensity to violence during a period when great imbalances were created by a combination of hormonal surges and an immature prefrontal cortex.[49] In fact, there is no known human society where it is not the men who do most of the killing, with their victims mostly unrelated men, for the purpose of status and respect.[50]

Civilizations are the most external artifices that humankind is capable of creating. At bottom, civilizations are about providing their members with measures of protection and coercion, thus controlling humans' potentially destructive instincts—aggression, sexual envy, greed, selfishness, shortsightedness, and so on. As John Locke famously said, it is "liberty, but liberty under the law."

The rule of law is external and artificial precisely because there must be military and police forces to prop it up; the mere use of reason, persuasion, and rhetoric will not do. And prosperity alone is not sufficient to bring peace. It is chilling to know that since the beginning of recorded history, periods of optimism and bursts of new weapons development have gone hand in hand.[51]

Extreme cultural practices—such as brainwashing and direct mind control—are another way that humans have attempted to mold other humans. Brainwashing (in the broad sense of the term) is older than civilization. Virtually all societies have rites of passage to help transform their young into full adult members of their groups. They rely heavily on communal activities of dancing, chanting, clapping, and swaying together for hours, even days and nights. Overwhelmed by exhaustion and sensory overload, the participants eventually fall into a dreamy state of consciousness. The experience seems to strip people of their prior learning and opens the mind for new meanings and social relations.

Today, young people go through boot camps in military or MBA classes to have new values instilled in them and to build lifelong bonds with their peers. From communist Russia and China to utopian communes in America, the attempts to mold human nature through brainwashing and social engineering all too often end up as a variant of George Orwell's *Animal Farm*. The utopian dream is always infiltrated and spoiled by pretenders who take advantage of other members' generosity and trust, just like bad money driving out the good (what is known as Gresham's Law).

Other mind-control techniques may consist of certain kinds of medical and psychiatric treatments, some of which are controversial and may punish or control various types of nonconforming behavior, and others that are considered more acceptable. Hospitals for the mentally ill are a relatively recent phenomenon to cope with the fact that modern societies are hanging on to those who would have likely perished or been eliminated in premodern societies. (For example, in the Eskimo tradition, the *kunlangeta*—a person with antisocial personality—would be pushed off the ice when nobody else was looking.[52]) Today, continuous advancements in pharmaceutical research may hold the promise for *Brave New World*–style sensory delight and mind control. These "brain-styling" drugs promise to do to the mind what cosmetics have done for hair and skin.

However, drugs that suppress emotions, natural instincts, and bodily functions can have serious side effects and unintended consequences. Fear, aggression, anxiety, jealousy, guilt, and envy all play important roles in our emotional life and help us make decisions. They are our motivators and balancers. And emotion is required to "make sense" of the world and learn from one's experiences. To artificially hijack these neurological systems risks throwing out the baby with the bathwater. Pure mind control is probably beyond reach without a fundamental alteration of human nature.

Conclusion: Culture is limited by human nature. The erroneous perception of humanity's infinite flexibility comes from our observation of humanity's seemingly endless behavioral and cultural variations. Much more to the point is the old and wise expression "all too human"—which refers to the fact that the core of human motivations and emotional needs has remained intact. Only their manifestations have varied as we added layers and layers of filters—the most important of which are habits, cultural institutions, and conventions.

Francis Bacon used the riddle of the Sphinx as a cautionary tale for the newfound power in the rational mind: the image of the thinking and dreaming human head cannot be separated from the lion's menacing body. The wild animal that the conscious mind keeps fighting is within each of us.

Human society, however flexible and robust, cannot avoid confronting the problem that any human being can be a monster to another. We cannot even control our own instincts, so how can we control the thoughts of others?

Figure 4: A Depiction of the Beast Within

The "beast" refers to the animal instincts within each of us that our conscious mind tries to control but cannot get rid of. (The drawing represents an antique Chinese ivory carving from the author's collection.)

Ultimately, the existential limit on human culture is any culture's necessary preoccupation with dealing with human nature. I believe that this wisdom about the limits of culture emerged slowly in history. The most successful cultures tend to be those that adapt to the unchanging realities of human nature. Humankind cannot hope to change human nature by culture, nor can we design and construct new social norms and institutions at will and hope they will work and fulfill the visions of social engineers.

7.4. What Drives People

Perhaps the most important human limitation is that, even if our capabilities could be infinitely expanded, few of us have the motivation to realize the cosmic potential. As I will lay out in detail in chapter 9 on conscious evolution, we have to be realistic in allowing contributions to the Cosmic Vision to emerge as largely unintended consequences of innate human motivations.

Regardless of cultural influences, what turns the brain on is still predominantly, though not exclusively, personal: sensory pleasure, love, family, friendship, wealth, material possessions, security, social status, emotional expression, art, sport, spirituality, and so on. And in order for us to experience all these things, we must first survive and reproduce; hence, two of our strongest instincts are to live and to have sex. Regardless of technological and social influences, it is just about impossible to change this situation as long as we remain human. We should consider it appropriate that Thomas Jefferson—a person with vast learning and a rational intellect full of the Enlightenment spirit—felt that "the happiest moments of my life have been the few which I have passed at home in the bosom of my family."[53]

Even our concept of *utopia*, the notion of an egalitarian world of perpetual bliss and high-quality living, reflects the largely unchanging preferences of our brains. Utopian images have not changed that much from antiquity. From the Cockaigne (the medieval European mythical land of plenty) to American nineteenth-century utopianism to today's affluent societies, our ideal picture of society remains one of basically limitless food and wine, leisure, freedom from worry, carefree indulgence, satisfying sex and romance, sports, music, and theater. It's a vision much like today's lifestyles of the elite and the super-rich, minus the toil and psychological stress of maintaining that lifestyle but without the usual gross inequality—a place where everyone lives like a harmonious family without discrimination or exploitation.

Such images of ultimate happiness are very revealing. Far from being a blank all-purpose computer, the human brain is genetically preprogrammed with specific drives, capacities, and instincts that are predetermined by its peculiar physical organization. In particular, the brain has one dominating purpose—*sex*—just like that of any sexually reproduced species. In his provocative essay

What Are Brains For? the neuroscientist Michael Gazzaniga says, "I would argue that the cathedrals we build, the books we read and write, the science we create, the cars we drive, the stocks we buy and sell, all of the mergers, the politics, and the wars we wage—in short, everything that constitutes the intricate web of life we have constructed around ourselves with our amazingly large brains—serve a very simple purpose. Sex."

You may find Gazzaniga's statement extreme, and I would agree, but the point is that we should distinguish between what the brain is for and what it is capable of. Most of us growing up and living in today's society are not consciously aware that we are programmed by nature to have sex and reproduce. In fact, many of us seem to be too preoccupied with other things to have sex—career, family, friends, food, entertainment, etc.—and this is especially true for those past middle age. Actually Gazzaniga's point is not difficult to grasp. One can point to the scientific and artistic accomplishments as evidence that we have risen above our animal roots, but the roots have not been removed.

To achieve its primary directive of keeping the organism alive and reproducing, the brain needs to perform a host of important functions, such as prediction, planning, and communication, to name a few. Because these activities serve the primary directive of reproduction, we experience them as pleasurable. More than a dozen distinctive human desires have been identified—power, independence, curiosity, acceptance, order, saving, honor, idealism, social contact, family, status, vengeance, romance, and tranquility—each with associated emotions and a possible evolutionary adaptive basis.[54]

Compared with those of other animals, human motivations are incredibly rich, diverse, and plastic, but the survival and sexual instincts remain the bedrock of human drives. As Woody Allen once quipped, "If I had to live over again, I'd live over a delicatessen."[55] Put simply, we must eat to survive. Much of our preoccupation is with making ourselves more attractive to others, including potential mates, along with satisfying our needs for food and security.

These instincts are the main drivers of what we do, and in spite of our "best" intentions to make survival and sexuality irrelevant, they will always show up in our activities. A few cases in point:

Most of the modern "recreational" leisure activities—activities we enjoy so much that we volunteer our time and pay to engage in them—highly resemble the main tasks of hunting and foraging that guaranteed mankind's survival for eons: fishing, hunting, hiking, sports, making tools and clothing. Hence we must avoid the humanistic fallacy that if we could just get our social institutions right, people would be all that they might be. As Lee Thayer wrote more than a quarter century ago, "With all of our modern wonders of electronics and power and medicine and leisure, we are further than ever from learning how to be useful to ourselves. The ultimate affront to human nature is not that we be irrelevant to society, but that we be irrelevant to ourselves."[56]

Also, those who find it strange that sex is the most popular subject on the Internet forget the built-in evolutionary imperative—as well as the emotional satisfaction that it provides. The same is true of television. When first introduced, it was envisioned as a forum for lectures by university professors—but, notwithstanding the litany of objections raised, sex became a central focus on TV programs as well. The same can be said for advertising and product promotion.

Finally, an analysis of the content of our daily conversations revealed that most of it is gossip.[57] Over the phone or the Internet, people everywhere devote about one-fifth to two-thirds of their conversation to exchanging personal behavioral information, especially about sexual infidelity, lying, cheating, and other sneaky actions that affect the community and oneself. Gossip can be dirty and damaging (that's why most religious and moral teachings are strongly against it), but it is irresistible to most people, and the fear of gossip and its effects does serve to keep people from behaving badly.

Even the most noble characteristics of the brain are mostly runaway "unintended consequences" of our quest for a mate. The mental capacity developed to track interpersonal relations is leveraged to do math and logical reasoning. The drive to gain an edge in social and technical information becomes the engine for knowing things totally unrelated to our own reproductive success. Even the love of truth and scientific discovery can be seen as an expression of the drive to gain power, prestige, and the thrill of the hunt.

Our collective motivation for new knowledge and pure research is pretty weak, except during periods of heated competition between companies or between states (such as the race to explore space during the Cold War). Ever-higher Gothic-style cathedrals with flamboyant ogives (pointed Gothic arches) were built during the Middle Ages in Europe in the name of glorifying God but actually were driven by competition among the cities.

To be sure, "selfless" motivations such as saving the environment and helping the poor are also part of our emotional repertoire. And some people enjoy pursuing knowledge to satisfy their curiosity or spiritual needs. These motivations exist side by side with the sex-and-survival instincts, but not in the same degree for most people, although they have resulted in extraordinary achievements. As the "super-mammal," in addition to all animal instincts, we have a unique higher consciousness, which makes it possible to go beyond our innate hedonistic urges and living for the here and now. We do not want a purely reflective rational life, but at the same time, the reflective mind is something we dearly cherish.

I remember a cartoon depicting a hungry wild dog on the street and a well-fed domestic dog staring at each other through the house window. There is no doubt which side of the window a dog, or any animal without the ability to reflect on the whole situation, would choose. The cartoon reflects a dilemma that is uniquely human. As we discussed earlier, we are animals at our core, yet we have a higher consciousness. This is why I call us a transitional species in cosmic evolution.

People instinctively do not want to live in the "artificial world" depicted in the movie *The Truman Show*—our reflective mind cares about the larger world and the role we play in it. A few of us care about it a lot, and it is this sense of choosing to get real and to have an impact on the universe that drives these extraordinary/abnormal people out of their personal sphere and into the cosmic one.

7.5. The Human Is a Machine—or Is It?

So far, our discussions of human limitations have looked at human beings as a whole. These observations may not be fully convincing, since one can always argue that with certain external enablers, the human species can still become something fundamentally different from what it is. Here I'd like to go further in opening up this "black box" called humanity to see what it is made of.

To sum up the reductionistic view, *the human is a machine.* As Christopher Morley once quipped, "What is a human being but an ingenious assembly of portable plumbing?"[58] We are a machine far more complicated and flexible than any existing man-made machine, but still one that comes loaded with "default settings" shaped by historical conditions.

But I hasten to add that the metaphor is not meant to be taken literally in the sense of the traditional narrow definition of machines as man-made tools. The human body includes some *ten thousand billion billion* protein "parts." To many people, a machine so complex deserves a label better than that of a "mere machine." Those who object to this notion should take comfort in the fact that no man-made machine is even remotely close to the complex structure of the human body, especially the brain. My own use of this mechanical/industrial metaphor is aimed merely at establishing a common denominator between humanity and other structures we are familiar with—and dynamic/informational metaphors apply just as well. In other words, the human is also a biochemical process, a network, a computer, a cybernetic homeostatic system, and no doubt much more. And none of these metaphors come close to explaining the nature of human consciousness.

The human being is not a digital computer. Since the human being is the most complex object in the known universe, no metaphor or analogy can capture its "essence" in totality, just as the vivid metaphors of ice crystals and a flame only partially capture the essence of "life." In particular, we must be careful not to think of humans as literally the same as the electronic digital computer, the most marvelous machine we have. Let's look at the differences:

- Today's computers are assembled from numerous separate parts, whereas the whole human organism is a fully integrated system all the way from its beginning as a single fertilized egg cell.
- Today's computers do not possess "second-order emergence"—they do not have the capacity to evolve as do humans.
- The computer is not an autopoietic or self-making organism. The human body is a machine that winds up its own springs with enzyme catalysts.

- The computer is a logic machine and filing cabinet, while the brain is an autonomous agent with organic memory. The brain does not replicate the ideas of other brains in the way that a computer copies files or instructions. New ideas are absorbed through active participation in a social system of shared values and are always subject to new interpretations in the process; they are not downloaded without alteration and without discretion, as with a computer.
- The brain is built on fuzzy logic rather than on the pure simple logic of the digital computer. The brain's messiness and logical imperfections are exactly what make it flexible and resistant to corruption, whereas a computer's neatness and cleanness are exactly what make it rigid, frail, and liable to fail in unexpected ways.

The human being is not an "inexplicable miracle." Despite its limitations, a machine is an excellent metaphor for our understanding of humanity. The point is that the human being is not a "miracle," a product of "continual intervention of the inexplicable," as Wendell Berry claimed, that we cannot hope to understand and imitate. Even though we have not fully understood how the brain works, we can know in principle the mechanics that generate human behavior. This is the same confidence that Erwin Schrödinger expressed in 1944 in *What Is Life?* before the basic building blocks of life (DNA) were discovered.

To say the brain is a machine is, among other things, to suggest that the mind is the result of the working human brain. There is no scientific evidence that our thoughts and consciousness can exist without our flesh and blood. Neither is there scientific evidence that there is a human soul directing our actions with a free will that is independent of the material world and natural laws. Like so many people, I myself certainly wish it were true that my "soul" can survive my biological body, and I will be more than happy if someone can disprove Francis Crick's *astonishing hypothesis* (as he called it), by which he meant, "'You,' your joys and your sorrows, your memories and your ambitions, your sense of personal identity and free will, are in fact no more than the behavior of a vast assembly of nerve cells and their associated molecules."[59]

Our free will feels free because the choices that originate from various brain processes are hidden from conscious introspection. As Spinoza put it in *Ethics*, "Men believe themselves to be free, simply because they are conscious of their actions, and unconscious of the causes whereby those actions are determined."[60] In order to grasp the emerging overall pattern of thought, the precision of the lower-level workings is given up, providing the illusion of freedom.

It can be upsetting to regard humans as glorified gears and springs. Machines are insensate, built to be used, and disposable; humans are sentient, possessing dignity and rights, and infinitely precious. A machine has some soulless workaday purpose, such as grinding grain or sharpening pencils; a human being has higher purposes, such as love, worship, and the creation of knowledge and beauty. The behavior of machines seems to be determined by ineluctable laws of physics and chemistry, while the behavior of people depends on their free will.[61] Indeed, many influential philosophers such as Kant have long argued that science can never prove or disprove the existence of a soul or free will.

We are reluctant to see ourselves as mere "machines" because we fear that humanity and all its accomplishments will be devalued when all its magical secrets are exposed by the cold light of reason. For example, we may not wish to examine the underlying sources of the pleasure or strength we gain from music and art. In part we fear success itself—understanding it might spoil our enjoyment. For some of us it might—art sometimes loses power when its psychological roots are exposed[62]—while for others the pleasure of art is undiminished even when these associations are recognized.

We may also become paralyzed if we lose our sense of free will, our belief in our own ability to act. Using the machine metaphor, the human mind can be seen as a passive and helpless pawn pushed around by the forces of nature. However, our ability to hold multiple perspectives should protect us from being slave to any newfound knowledge. Our emotional instincts are always stronger than our rational thoughts, and most importantly, a holistic self-understanding is critical for us to be able to look ahead.

Since the emergence of reflective higher consciousness sometime before the Axial Age, the pursuit of the Socratic injunction

"Know thyself" has come a long way. Clearly there are little voices in us that regret what we tried so hard to know—the psychological burden of humans becoming "their own vivisectional animals," in Nietzsche's words.

The mind is not reducible to biology; neither is biology to chemistry and physics. But just like any type of material, complex, and dynamic system, there is no scientific evidence that there is a soul in our body somewhere that is independent of the physical body and makes us somehow above nature. "There is no gene for the human spirit," as the promotional tag for the 1997 sci-fi film *Gattaca* proclaimed, but the emergent human spirit is built on genes and dependent on them. This is but one aspect of humanity, but an extremely critical one in settling the perennial mind-body debate and in understanding humanity's future. The machine metaphor helps break down the black-and-white sharp distinction between human beings and human-made artifacts such as robots or cyborgs.

Humanity is very special, but it does not deserve the emotional premium that our intuition automatically places on it. Intellectually, this is probably more of a limitation in Western than in Eastern thought. The human spirit is "immortal" only in the sense that it is a manifestation of the cosmic potential, a trend toward greater complexity and beauty.

The genetic basis of our behavior. With recent developments in scientific fields such as neuroscience and molecular biology, questions of how the brain generates thoughts and motives are no longer *mysterious*, although there are still *problems* (i.e., we still do not have all the answers and explanations).

Today, for example, neuroscientists have a rough idea of how our emotions work. The emotional stimuli are usually registered by the amygdala in the mammalian brain. The conscious mind receives emotional feelings directly from the amygdala and indirectly through a long route going from the amygdala to the hypothalamus, then to the body, and finally to the somatosensory cortex. The emotions we experience are complex in the same way that we perceive colors: the limited numbers of primary colors mix to form a seemingly endless variety of hues. Similarly, the 18,000 or so adjectives in the English language that describe our emotion-related personality can be tied to one or a combination of five dimensions of personality traits: introvertedness, stability, curiosity, agreeableness, and conscientiousness.[63]

Our advanced neocortex renders incoming emotions into multifaceted and subtle internal processes through elaborate exchanges of information with older parts of the brain and the body. It utilizes a variety of neurotransmitters and hormones in this process. Let us look at some scientific explanations of love, art, spirituality, and various other hallmarks of being human.

- Natural maternal behavior is determined by hormones such as oxytocin. Genes such as the fosB are critical in the process of turning on the hormonal process that leads to human nurturing behavior. Putting the infant to the breast causes a surge in prolactin and oxytocin, and these in turn suppress estradiol and progesterone.
- Romantic love is associated with, but different from, the sexual drive, which is governed by testosterone (in both sexes) and estrogen (in women). In each case these hormones work on the hypothalamus. People full of oxytocin during courtship experience a loss of self-control, a detachment from reality, and what some researchers would call *blind love*. For people experiencing passionate love, the desire and reward centers, the caudate nucleus, and the ventral tegmental area become the brain's "G-spots" when they are flooded by dopamine and perhaps norepinephrine and serotonin.
- The wonder of art also has a neurological basis. The "deep structure" of aesthetics reflects the way the brain constructs a representation of the world, such as the neural lyre, the peak shift principle, and the principle of perceptual grouping.[64] When Leonardo da Vinci said, "The most pleasing colors are the ones which constitute opponents," he was referring, presciently, to the physiology of complementary colors, an innate aesthetic preference that was not discovered until the nineteenth century. The emphasis of some modern art on lines appeals to the neural circuits that work on lines of specific orientation. The same principle applies to just about any artistic activity—poetry, music, dance, architecture, literature, etc.[65]

- The chemical pathways and brain structures that give rise to religious sentiments have been identified by neuroscientists such as James Austin and Andrew Newberg, although much is still unknown. Structurally, the sense of self and its place in the world originates in the back of the brain, which is connected but distinct from the sensory processing and attention-focusing center in the frontal areas. The "oceanic feeling" of oneness with nature or creation is achieved by quieting a bundle of neurons in the superior parietal lobe, which are associated with processing information about space and time and with orienting the body in space. The elevated state that we associate with the "sense of unity" can be achieved through intense meditation.
- Scientists are just beginning to examine "soulful" aspects of humanity from the genetic perspective. Dean Hamer suggests that spirituality is a hardwired instinct and speculates that VMAT2 is one of many "God genes" (so named to draw popular attention) that influence our spiritual capacity since it controls the flow of mood-regulating chemicals called monoamines in the brain.

This discussion of human nature is not about asserting the totalitarianism of genes. Rather, it is about demystifying the relationship between genes and behavior through science, so one day we will be able to develop intelligent beings with more advanced and complex morality and spirituality.

We must recognize a crucial difference between *heritable* and *inevitable*. The genome is more like a recipe than a precise blueprint. What is considered genetically determined must be put into the context of a "normal" environment, as gene expression reacts to environmental cues. What's innate is the potential to respond to life experiences. The MAOA "nonviolence gene," for example, will be expressed less frequently in neglected or abused children, resulting in more aggressive tendencies. The biological and cultural development process is one of causality, albeit an extremely complex one.

Viewing the brain as a machine with its inherent unchanging characteristics and limitations is in no way meant to devalue our amazing human capacities. Nor is it as simplistic a perception as it may seem. Although all adjustments to our thinking can be uncomfortable, a realistic understanding of ourselves—one that does not require resorting to "magical" and disembodied explanations of "free will" and the like—will show us the way to enhancement of the very conscious attributes that have been valued by enlightened individuals through the centuries.

Looking forward to the posthuman world. The real power of the "man as machine" metaphor is the commonsense understanding that machines can be worked on, improved, redesigned, and so on. Such a realistic understanding of humanity will inevitably lead to the "posthuman" world. The talk about changing our genetic nature, however, may touch a raw nerve for anyone who recalls the extreme abuses that have been perpetrated in the name of eugenics, the Nazi Holocaust being the most extreme example. We will discuss these concerns at length in chapter 10, on fear and risk. Here I will simply point out that rather than leaving humans behind or destroying humanity, the posthuman world will in all likelihood be all about enhancing the human spirit and unleashing it from its animal roots.

In a sense, the so-called posthuman era started several thousand years ago when speech and counting morphed into written texts and numerals. For the first time ever, human knowledge was able to exist independent of the biological body. What this reveals is that the distinction between the organic and the mechanical is in some sense superficial. Furthermore, from the earliest surviving pieces of "abstract" representation of reality—namely, cave art—to computer-generated virtual reality, the distinction between physical reality and its representation is becoming increasingly blurred in terms of how the mind perceives and processes them.

To use the latest technological metaphors, the brain is essentially a virtual-reality computer simulation coupled with automatic intentional modules that can generate motivation or initiate action. I am sure better metaphors will emerge along with unforeseeable new technologies, as they always have, but structurally and functionally, the distinction between artificial and natural intelligence is gradually fading away. The purposeful, goal-directed, and teleological behaviors observed in animals can be explained in cybernetic terms. Conversely, both artificial and natural intelligence can be seen as emergent phenomena wherein collective actions at the unintelligent lower levels (such as the primordial RNA molecules) lead to simple and strictly mechanical base functions that eventually give rise to complex emergent higher-level behavior (such as experienced in humans). Looked at this way, the processes leading to the first instances of language and technology on Earth were set in motion long before humans arrived on the scene, and really go back to the emergence of primordial RNA molecules almost four billion years ago.

It is the brain's specific structures that give rise to higher complexity. To look at the neural cells in the brain and ask, "Where does the human soul come from?" is in a sense the same as looking at the ants and asking, "Where does the anthill come from?" In Douglas Hofstadter's book *Gödel, Escher, Bach,* Achilles and Anteater have this dialogue:

> *Achilles*: It is interesting to me to compare the merits of the descriptions at various levels. The highest-level description seems to carry the most explanatory power, in that it gives you the most intuitive picture of the ant colony, although strangely enough, it leaves out seemingly the most important feature—the ants.
>
> *Anteater:* But you see, despite appearances, the ants are not the most important feature. Admittedly, were it not for them, the colony wouldn't exist; but something equivalent—a brain—can exist, ant free. So, at least from a high-level point of view, the ants are dispensable.
>
> *Achilles:* I'm sure no ant would embrace your theory with eagerness.
>
> *Anteater:* Well, I never met with an ant with a high-level point of view.[66]

It is both a human tragedy and a source of human exuberance that we can have that higher point of view. Given current knowledge, we can have two levels of understanding of "humanness." From a narrow perspective, the human is a biological structure defined ultimately by the human genome; from a broader perspective, we are our consciousness, motivations, sentiments, and all the other human thoughts and behaviors. The former generates the latter, but the latter does not necessarily require the former. Horses generate horsepower, but horsepower can also be generated by steam engines, gas turbines, or nuclear reactors. Newspapers give us news stories, but news stories can also be provided by radio, TV, or the Internet. Likewise, our prized human attributes can be (in the relatively near future) generated through entities that are different from human beings.

When we consider whether our human spirit and capabilities are only possible in humans as they exist today, we might consider how we handle and think about information. Information theory views survival as the preservation of structure and order over time, independent of the specific medium, or carrier of information. The speed of computers can continue to grow exponentially only because after the potential of a certain medium is exhausted, we invent and switch to another medium. In our case, "another medium" would be another type of being or at least an artificially enhanced version of ourselves.

We now celebrate half a century of success at saving lives with organ transplants. Yet it hasn't been such a long time since we first accepted the human body (except for the brain) as being like a machine. It was only in 1628 that William Harvey showed that blood circulation operates by hydraulics and other mechanical principles; it was only in 1828 that Friedrich Wohler showed that our body is made up of ordinary compounds following chemical laws; and it was only in 1953 that James Watson and Francis Crick showed the genetic mechanism of heredity.

Modern medical technologies no longer stir ethical debates over replacing defective body parts with artificial ones such as pacemakers, artificial joints, drug-implant systems, implanted corneal lenses, artificial skin, and other life-supporting systems that replace the lung, the stomach, and the kidney. There are still plenty of unanswered questions concerning our body, but we no longer have the extreme emotional resistance we once had about the deeper and more thorough understanding of what we are really "made of" that science has given us. However, at our present frontier, new concerns have once again arisen. Although there is little doubt that the human body is related to all other life at the basic level, there is still a concern on moral grounds about modifying us, which is a subject we shall discuss in chapters 10 and 11.

As we understand more and more about our genetic makeup and our evolutionary roots, at some time in the future we will have to decide how "human" we wish to remain. To chart our destiny means that we must shift from "natural drift" based on our inherited biological properties to "precise steering" based on knowledge and wisdom. That shift is possible only when we open ourselves to a higher perspective—on which we will focus in the next chapter.

CHAPTER EIGHT

A THEORY OF PERSPECTIVES

Two birds, inseparable companions, perch on the same tree
One eats the fruit, the other looks on.
The first bird is our individual self,
feeding on the pleasures and pains of this world.
The other is the universal Self,
silently witnessing all.

—Mundaka Upanishad (fifth century BCE)

Man has two eyes
One only sees what moves in fleeting time,
The other
What is eternal and divine.

—The Book of Angelus Silesius (German mystic-poet, 1624–1677)

8.1. Two Eyes

WHEN SOCRATES FAMOUSLY declared that an unexamined life is not worth living, he was pointing to the necessity of activating *the other eye* of reflective consciousness. We all have at least two "eyes"—two different ways of looking at the world—although opening the first eye is automatic, while opening the second is not. For example, Norbert Wiener observed that when members of primitive societies know that Western anthropologists are watching them, this inevitably stimulates their reflective mind (the other eye) to think about what they are doing, and they are often changed by that awareness.[1]

The intent of this chapter is to integrate the discussions of the last two chapters on human potential and human limitations. I will focus both on our innate abilities and on limits of our ability to perceive the world from different perspectives. In so doing, I hope to move us closer toward embracing the highest perspective, the Cosmic View, with an understanding that we all have multiple perspectives.

Here's another story about widening our perspective, this time in the *Platform Sutra of the Sixth Patriarch* (六祖坛经):

> The wind was flapping a temple flag, and two monks were having an argument about it. One said the flag was moving, the other that the wind was moving; and they could come to no agreement on the matter. They argued back and forth. Finally, the Patriarch said, "It is not that the wind is moving; it is not that the flag is moving; it is that your honorable minds are moving." The whole assembly was startled by these words.

This ability to step out, to be what Adam Smith called an "impartial spectator," is a fundamental mental capability of human beings and is why we can appreciate jokes such as this one:

> A parrot has been trained to greet customers at the door with the words "Welcome to our store!" A girl finds it very interesting, and goes in and out six times. Each time she walks in the store, the bird gives the same greeting. On her seventh entry, the bird finally loses its patience and shouts, "Hi boss! This girl is harassing your bird!"

Humor is akin to philosophy: both are viewpoints born of a larger perspective on life.[2] In its irate cry, the parrot is switching from what might be called an execution perspective to a "monitoring" perspective, something that (in the real world) only humans can do.

The only way to gain a full understanding of any phenomenon is to change the way we frame it: to step outside the whole enterprise and consider the possibility that our entire worldview is invalid. Likewise, the understanding of the Cosmic View, as we will argue in this chapter, requires one to step out of the whole enterprise of human life, to look beyond the individual's pursuit of happiness and eternal bliss.

Our higher consciousness is like a parachute—it only functions when open.[3] If one tries hard enough, one can always find more than one theoretical construction to place upon a given collection of facts and data. Different perspectives offer alternative but equally limiting worldviews derived from the same external stimuli, and our conscious mind has to be aware of them and to evaluate them for different purposes. Thus, although each of our multiple perspectives is limited, they are useful when considered together.

In general, the emerging characteristics at a higher level cannot be understood or predicted by examining factors from a lower perspective: *more can be different*. Kurt Gödel's incompleteness theorem suggests that we must take a higher, external perspective on a given system in order to understand and validate that system. Likewise, the aim of Zen is to transcend our limited conscious thinking, especially logic and language, to reach a transcendent perspective. Many Zen koans, like the following one, deliberately try to bewilder, to force you to think the unthinkable:

> The student Doko came to a Zen master, and said, "I am seeking the truth. In what state of mind should I train myself, so as to find it?"
>
> "There is no mind," said the master, "so you cannot put it in any state. There is no truth, so you cannot train yourself for it."
>
> "If there is no mind to train, and no truth to find, why do you have these monks gather before you every day to study Zen and train themselves for this study?"
>
> "But I haven't an inch of room here," said the master, "so how could the monks gather? I have no tongue, so how could I call them together or teach them?"
>
> "Oh, how can you lie like that?" asked Doko.
>
> "But if I have no tongue to talk to others, how can I lie to you?" asked the master.
>
> Doko said sadly, "I cannot follow you. I cannot understand you."
>
> "I cannot understand myself," said the master.[4]

Most of us, like Doko, would not understand the master at all until we took in the master's perspective beyond the limitations of logic and language and viewed them in the larger context of awareness, where we could recognize the limitations of conventional logical and linguistic thinking in knowing the truth. Enlightenment cannot be expressed by words, although it can be achieved with their assistance—even the master has to use words to communicate.

In general, a higher human perspective (that is, the activation of the second eye) is within our reach because we have the following capabilities:

- **The ability to perceive more and more.** From simpler life to higher animals, the neural system's capacity to perceive and process information generally grows. This eventually leads to the emergence of higher consciousness—the awareness of awareness.
- **The ability to form mental maps and recognize patterns.** Shaped as it is by natural selection, the animal mind has a hunger for patterns. The human mind is much sharper at pattern recognition than other animal minds since it has the additional ability to process abstract, symbolic patterns hidden in sensory information.
- **The ability to take an "out-of-body" perspective.** This is higher consciousness, or second-order thinking, in operation. It is the ability to detach from our senses and to construct an imaginary landscape, a mental model, and "see" it from any particular angle.
- **The ability to integrate.** A higher level of understanding is not simply the sum of details, but a hierarchic integration of the whole. The ultimate summarization is *one*: one family, one village, one nation, one species, one universe, and one God.
- **Insight into the content of others' minds.** Many intelligent animals seem to have the ability to think from the perspectives of others. This is called second-order belief: a belief about the beliefs of others. However, only humans enjoy *higher-order* beliefs and intentionalities—"I know that you know that I know this," for example.

As laid out briefly in the next section, these capabilities are innate. Our capacity for the realization of a more advanced human perspective will then be discussed in terms of stages of personal moral and spiritual development.

Social and technological development has enabled more and more people to "jump" outside of their parochial viewpoints. However, the mind is not totally free. Higher perspectives usually do not have the upper hand. The greatest divide in our society today is not political, cultural, or religious—it is the conflict among perspectives. Acting on the highest perspective—the Cosmic View—is the key to realizing human potential. And it may require us to understand that the pursuit of human happiness, and of purely human-centered goals, cannot be the ultimate end.

8.2. The Psychological/Neurological Basis of Perspectives

The neurological reason why we can have many perspectives is that we have not one, but many minds. We usually feel we have a single mind, but the ability to see more and more is a result of the human brain's *accumulation* of a large collection of neural systems (or modules), each presenting a perspective and each competing for conscious attention.

Newly evolved modules rarely supersede or replace existing older modules in the brain. Instead, modules of similar functionality are put under the control of newer high-level executive modules that demonstrate higher flexibility and robustness. For example, we developed language without abandoning old non-linguistic means of communication such as facial expressions and gestures. We still instinctively see "the sun rises in the east" through our own eyes, yet the new rational thinking also makes us understand that the Earth spins and rotates around the sun without ever seeing the action directly. Psychological studies have shown that when people look at a letter *L* composed of many small *D*s, different areas of the brain become active when either the *L* or the *D*s are recognized. When we watch a movie or read a novel, one module reminds us that this is just a story, yet our emotional reactions seem to take it seriously, as other modules identify with the fictional figures and get involved with their lives. The biologist Niles Eldredge admits that when he heard the geneticist Brian Taylor

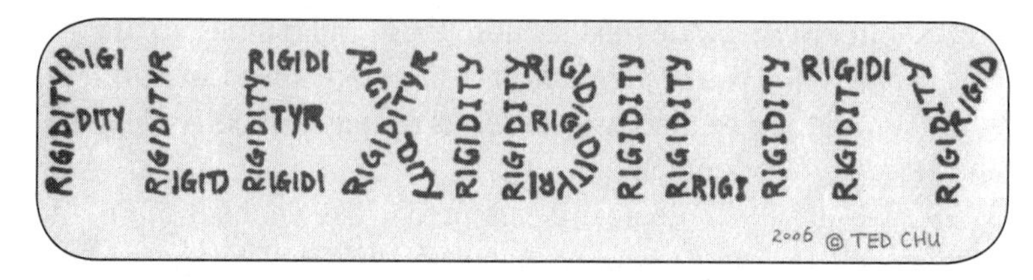

Figure 5: Flexibility and Rigidity

Different modules in the brain offer different answers. This chart follows the work of Douglas Hofstadter.

calling his theory of punctuated equilibrium "evolution by jerks," he had to laugh—even though he felt "pissed off."[5]

Without irony, we long for personal freedom as well as a sense of belonging; diversity as well as deep roots; equality as well as status; wealth as well as simplicity. In general, we all have multiple personal identities[6] and behave accordingly depending on the social context. We feel we have a single mind mainly because the conscious mind cannot hold all perspectives at once, so most often we take one perspective at a time. Many illusions have been created in drawings that illustrate that point, such as the famous "face/vase" illusion in which the black outline of two faces *or* the white outline of a vase can be recognized, *but not both simultaneously*; or the many illusions of multiple faces, of which one is shown here.

When you encounter the word *chicken*, what comes to your mind, depending on the context, could be a bird, a commodity, an ingredient of a prepared dish, an item on a menu, a source of eggs, deadly bird flu, a farm animal, a pet, a fighting game bird, etc., but you cannot envision all these perspectives simultaneously.

The human brain shares the animal brain's ability to form mental maps and to recognize patterns upon receiving sensory information from the external world. What is new for the human brain is that it has a greatly expanded capacity to manipulate existing maps, create novel maps, and construct higher-order maps. In forming concepts, the brain constructs maps of its *own* activities, not just of external stimuli, as in perception. In order to categorize, discriminate, and recombine the various brain activities occurring in different brain regions, there must be higher-order structures that represent a mapping of the types of maps themselves.[7]

The brain can think and dream, even when the external stimuli are completely shut off. This is remarkably similar to Hua-yen Buddhism's conception of a self-contained and self-sustained universe, which goes like this: There is a wonderful net hung by some cunning artificer that stretches out infinitely in all directions. In each "eye" of the net hangs a glittering jewel, and in each jewel is a reflection of the light from all the other jewels, so that there is an infinite reflecting process among all the eyes stretching across the vast expanse of the universe. Our individual sense of being "one with the universe" is a simple example of our brain's capabilities.

Figure 6: The Face/Vase Illusion Diagram

If we gaze at this famous image, we can recognize the black outline of two faces and the white outline of a vase, but not both simultaneously.

The newly evolved (or specialized) anatomical brain structures for language—Broca's and Wernicke's areas[8]—allow additional ways of parallel signaling involving memories of the past, the present, and interconnections among all of these. As a result, conscious thought can wiggle free from the immediate present, if desired, in order to build (i.e., conceptualize novel connections among concepts) and test out (using temporal projection or logical reasoning) hypothetical worlds based on all memories. This capability is the very source of our higher perspectives, which often have little direct linkages with our senses.

In *The Math Gene*, Keith Devlin suggests four levels of abstraction for the human mind: (1) you see stuff; (2) you imagine stuff out of sight (chimps can do that); (3) you imagine a unicorn (as far as we know, only humans can do that); and (4) you have a math concept, which has no sensory relation with the real world at all.

In relation to the Cosmic View, we can roughly identify four levels of personal view: (1) things affecting you at this moment (the basic perceptions that are needed to carry on a normal life—young children are completely confined to this level); (2) things that are expected to affect you in your lifetime or in some personal afterlife (which could lead us to save money for our old age or "clean up our act" before we die); (3) things that you will not see in your lifetime or are impossible for you to see (which could lead us to help disadvantaged people in another country that we may never visit); and (4) things beyond the foreseeable future and beyond humanity.

The ability to have different perspectives in one seemingly coherent mind is a result of specialized executive modules in the hierarchical brain. The base level modules or agents are accessed by the mind through higher "manager" modules, what Marvin Minsky called *paramone* in his book *The Society of Mind*. It is the *paramone* that enable us to hold different viewpoints and make judgments. These manager modules can have their own managers just like a company with multilevel managerial controls. The seat of judgment that makes the final calls seems to be situated in the prefrontal lobe. It has been dubbed the brain's "CEO" and is roughly what Freud called the ego, the part of the recently expanded brain that has a global perspective and endows us with volition, foresight, and civilized restraint.[9]

With the prefrontal lobe as the center for processing information coming from different parts of the brain, we have the ability to perform what Merling Donald called "metacognition"—to monitor our own varied perspectives and even monitor our own monitoring from an elevated position; and to "deliberate"—to switch from one perspective to another while exhibiting unity of intention.

8.3. Stages of Developing Personal Perspective

Nobody is born with flexible, high-level perspectives. The growth of consciousness and the increased capacity for multiple perspectives in a child follows the growth and maturation of the brain and the body. A person's progression in life through identifiable emotional, cognitive, moral, and other identifiable stages is fascinating because it can be seen as basically recapitulating the evolutionary history of man, the super-mammal.

Extensive cross-cultural research on social evolution suggests that there is a broad parallel between the growth of the moral sense during a person's life stages and the phases of the development of modern civilization. For example, the cultural stages of legal systems can be identified as almost mirror images of the personal moral stages.[10] Likewise, our experience of our personal journey offers a glimpse of the collective journey of humanity provided by the Cosmic Vision.

In a two-year-old child, we find the most primitive and "hard-wired" state of human nature, a condition that we share with most animals. The toddler often grabs things from others, is unsatisfied with what's in hand and is looking at what others have, and is responsive to scolding and punishment but not to reasoning and persuasion. The little child's brain is already very sophisticated, yet it has little human perspective to speak of.

American philosopher James Mark Baldwin was one of the first to recognize the phenomenon of stages of psychological growth in the late nineteenth century. Jean Piaget's experiments with small children in the early twentieth century clearly showed the emergence of multiple perspectives out of the egocentrism of the young mind. His research has been extended from cognitive abilities to emotional development. More importantly, the research by Erik Erikson, Lawrence Kohlberg, James Fowler, Jane Loevinger, and others further refined the developmental stages of a person's entire lifespan in areas of cognitive and moral reasoning and faith. The emergence of the highest human perspective, the Cosmic View, is best understood in light of their research.

The psychological development of perspectives is first and foremost the growth and elevation of one's moral sense and faith. Both for fun and for a deeper understanding of perspectives, the reader is encouraged to do a self-appraisal based on the stages outlined below. Which stage of personal development are you in now? Did you go through some earlier stages to get to the current stage?

Stages of Moral Development

Focusing on the morality of justice—what is the "right" thing to do?—Lawrence Kohlberg defined six stages of moral judgment. He called the first two stages preconventional morality, the middle two conventional, and the last two postconventional morality:[11]

- **Stage 1: Obedience and punishment.** What is considered right is literal obedience to rules and authority, avoiding punishment, and not doing physical harm. Punishment "proves" that disobedience is wrong.
- **Stage 2: Instrumental hedonism and exchange.** What is considered right is serving one's own or others' needs and making fair deals in terms of concrete exchange. Punishment is simply something one wants to avoid.
- **Stage 3: Community conformity.** What is considered right is playing a good role, being concerned about other people and their feelings, keeping loyalty and trust with partners, and being motivated to follow rules and expectations.
- **Stage 4: Law and order.** What is considered right is fulfilling one's duty in society, upholding the existing social order, and maintaining the welfare of society or the group. An abstract, impersonal social perspective is developed to differentiate interpersonal agreement from the societal point of view.
- **Stage 5: Principles of justice and welfare.** What is considered right is upholding the basic rights, values, and legal contracts of a society, even when they conflict with the existing rules and laws of the group.
- **Stage 6: Universal ethical principles.** Concerns about universal ethical principles become paramount. Society is no longer maintained for its own sake. An ideal social order is one that upholds not merely impartiality, but a selfless full and equal respect for all. In Martin Luther King Jr.'s words, "An individual who breaks a law that conscience tells him is unjust, and accepts the penalty to arouse the conscience of the community, is expressing in reality the highest respect for law."[12]

Going through the stages, it is clear that the ever larger and more conceptualized perspectives are the foundation of superior morality. The research of Kohlberg and many others shows that progression through the stages represents a hierarchic integration—lower stages do not disappear but become integrated into, and in a sense dominated by, new broader perspectives.[13] What then is the goal of morality in the light of this understanding of developmental stages? From a personal perspective, it is happiness; from a social perspective, it is harmony based on mutual understanding; from a cosmic perspective, it is facilitation of evolution and Cosmic Creation.

The ultimate wisdom is the awareness of cosmic oneness, but life's road to achieving it begins with a gesture of cosmic separation, starting with the first cell that developed a barrier that separated itself from the rest of the world. A young child instinctively identifies itself as "me" versus the external "you" and pursues the well-being of me. Then higher perspectives are taught, even self-"awakened," as one goes through the life stages. The process is usually gradual. It is an extremely rare child with exceptional personal endowment who can jump to the highest wisdom directly; that is apparently what happened with Zen master Hui Neng (惠能, 638–713).

Stages of Knowing the Divine

Within the Judeo-Christian tradition, Kohlberg identified a broad parallel between the stages of moral reasoning and the deepening understanding of a personal God.[14] Snapshots of these stages had been taken in earlier psychoanalytical research; for example, Freud offered insights into the role of parents in the formation of the representation of God, Jung pointed to complex symbols and archetypes, and Adler converted God into a value.[15] But the most interesting parallel between these two developments is Richard Elliott Friedman's reading of the Hebrew Bible itself, which depicts the retreat (or disappearance) of the personal God from His worshippers. Many individuals essentially followed the path that the Hebrew tribe collectively traveled during the early centuries of its long history.

- **Stage 1.** Authority is derived from physical strength, and God is depicted as *very human but with superior physical characteristics*—larger, older, and more powerful than the human adult. In addition, children are more interested in *how* God creates than *why*.

- **Stage 2.** With moral reasoning now based on a sense of fairness in concrete exchanges, God is depicted as *being involved in exchanges with individuals*. One child put it this way: "You be good to God, and He'll be good to you." This give-so-that-you-receive orientation may cause religious crises when an individual feels he or she is being unfairly treated by an arbitrary and unfair God (as in the case of the anguish of Job). It also makes for an extremely shallow faith: "If God does not care about *me*, then why should I worship Him?"
- **Stage 3.** With moral judgment based on meeting community expectations and maintaining the affection and trust of the community, God is considered as a *personal deity*, as a friend or caring shepherd. God's authority is supreme but tempered by understanding and mercy and guided by a concern for what is truly best for individual people. Breaking moral norms hurts God and brings about shame in His eyes: "He sees everything. If you don't do what He wants, you are offending Him."
- **Stage 4.** With the concern now centered on maintaining the social system, God is depicted as a *lawgiver* for both the social and the natural order; He is conceptualized in abstract philosophical terms such as a "supreme being" or "cosmic force." In one young man's words, "I don't have an understanding of God in the sense that God intervenes personally in my life. I think the metaphor I like best is [that my life is like] a compass that is sensitive to the lines of force [God]." This is closer to deism.
- **Stage 5.** With the emphasis on universal human rights in a social contract, God is seen as an *energizer of autonomous moral action*. Rather than the fulfillment of a preordained plan, God and humanity are seen as mutually involved in a creative activity that consists of establishing a community in which the dignity and freedom of each person may flourish. God supports human autonomy ("Man is made in the image of God") and human dignity ("Man is becoming God").

- **Stage 6.** In seeking a relationship with the ultimate, moral reasoning is placed within a larger human and cosmic framework. Justice and law are considered as being a purely human, rational construction, rather than being created through divine lawgiving. The ultimate self-realization depends not only on accepting one's place in nature but also, in Spinoza's words, on "the active union of the mind with the whole of nature." In other words, God becomes identified with the full cosmic reality, which transcends the merely human.

Again, the stages identified by Friedman depict the maximum possible range for an individual, rather than the norm we observe for individual growth of perspectives. Most pious people in modern societies never advance beyond Stage 3 in their whole life. The deism of the great architects of the American Revolution, which corresponds to Stage 4 and above, has never taken root in the majority of the American population. And those who rise above Stage 3 are often mistaken for losing their faith in God. For example, Thomas Paine, who in *The Age of Reason* and *The Rights of Man* presented a strong case for deism, was nevertheless dismissed as an atheist by many of his contemporaries (and even by Theodore Roosevelt a century later) because he criticized biblical literalism as being grounded in superstition. People who have moved to higher understanding of divine creation are often mistaken by people with lower perspectives as being faithless.

Stages of Faith

One of the most fascinating personal perspective studies is *Stages of Faith*, in which James Fowler summarized and distilled the key findings of the faith development project that his research team conducted between 1972 and 1982. The word *faith* is used to signify a person's fundamental worldview, not limited to religion.

The team interviewed 359 individuals from all corners of society, probing their values and meanings with questions such as "Does life have meaning or purpose?" and "When you are most discouraged, what gets you up in the morning to return to the struggle?" Working with the responses, Fowler traced six developmental stages of faith:

- **Stage 1: Preschool childhood.** We all begin the pilgrimage of faith as soon as we are born and in touch with the world. Surrounded by a bewildering array of sensory inputs and internal emotional flux, the infant's first God-like images may originate from the trust that the caretaker will always be there, even when out of sight. The child's slowly evolving knowledge of self also germinates an early sense of autonomy, courage, and hope. The preschooler's faith is "egocentric," fantasy-filled, and highly fluid, moved by random incidents, moods, actions, and stories of the visible faith of primally related adults.
- **Stage 2: School years.** An eight- or ten-year-old tries hard to differentiate reality from fantasy and begins to demand facts and proof of such things as Santa Claus and the tooth fairy. This leads to the curbing and ordering of the previous imaginative world. But individuals are thirsty for rich stories to provide images, symbols, and examples to help them understand the vague but powerful impulses, feelings, and aspirations that are arising within their minds and hearts. With their gathered meanings still trapped in the narratives, children are unable to draw conclusions from these sources about a general order of meaning in life. Symbols are taken as one-dimensional and literal in meaning.
- **Stage 3: Adolescence.** Triggered by the adolescent growth spurt and later by new social life beyond the family, one begins to reflect upon one's thinking, to step outside the life-stream, and to ask *why* the stories are told. Values and meanings gradually become abstract entities that can be independently articulated. But Stage 3 individuals are like Santayana's fish,[16] which cannot see the water and cannot imagine living in anything but water. The symbols and the values that they represent cannot be separated. Many individuals become stuck in this confusion throughout life, as when one cannot perceive the difference between burning the American flag (the symbol) and threatening or destroying American values or even America itself.
- **Stage 4: Early maturation.** The first phase of maturation is often triggered by an upheaval in one's personal life or

by "leaving home" and getting in touch with other cultures and ideas. Leaving the comfort zone forces one to step back and critically examine one's own value system. As a result, an inner authority is developed. This may be considered a qualitative expansion of the higher capacities of the prefrontal cortex, now capable of making "executive" decisions. Freedom from external authority enables one to see social relations not just from the interpersonal perspective but also from the impersonal imperatives of law, rules, and the cultural norm. Meanings can be separated from the symbolic media that express them. Just as a sales pitch loses its magic as soon as it is recognized as such, at this stage symbols and other social rituals can appear naïve, shallow, and worthless. There is a downside at this stage. Like an arrogant CEO who loses touch with the troops, a Stage 4 person can become overconfident in the rational self and become reductionistic and rigid.

- **Stage 5: Maturation.** At this more mature state, frustrations with the abstract self initiate a critical examination of one's position. The conscious mind comes to recognize the limit of dry reason and reaches out and seeks maximum information and insights from any source, including symbols and expressions that were rejected earlier at Stage 4. The person learns to hold multiple perspectives and embrace paradoxes. A supreme self-confidence is required at this point to accommodate the seemingly naïve or weird ideas without first imposing one's own value judgment on them. The person begins to adopt an attitude of radical openness to all input, a so-called second naiveté, which is neither a wishy-washy neutrality nor mere curious fascination with the exotic features of alien cultures. It is a "willed naiveté." The danger of this radical "see-it-all" stance is a potential loss of inspiration that can lead to complacency or cynical withdrawal from real-world problems and challenges. Individuals attaining this stage also can become isolated from others, who may not understand or be sympathetic to this point of view. As the Chinese saying goes, "The mountain peak is unbearably cold" (高处不胜寒).

- **Stage 6: Transcendence.** People at Stage 6 often sense an oceanic feeling that identifies oneself with a larger reality, or even with the ultimate existence beyond time and space. The self is almost lost in the process, becoming a disciplined, activist incarnation of the transcendental. The deep instincts of self-preservation and parochial identifications give way to larger universal visions. This abandonment of the self is very difficult to achieve since it requires a "holy indifference" to the temporal success or failure of the cause to which one has devoted one's best efforts. Detached from the specific social/cultural/personal contexts, the simplicity, elegance, and unifying power of post-conventional perspectives ultimately originate from the Tao. Stage 6 individuals are exceedingly rare. They are not perfect, especially during their formative years. Once mature, though, they appear almost like "a new species," in Richard Maurice Bucke's words. They frequently become martyrs for the visions they incarnate, and are often more honored and revered after death than during their lives.

Before we move on, let's briefly reflect on the influence of broader economic and social conditions on stages of faith. It is worth noting that material well-being and cultural stimulus seem to be necessary but not sufficient conditions for the uplifting of perspectives. Political scientist Ronald Inglehart's survey of world values broadly confirms the idea that once basic material and security needs are satisfied and taken for granted, the opportunity for self-expression and higher quality of life will be increasingly appreciated. Too many cases have shown that under grinding poverty there can be no virtue.[17] Indeed, the preliterate societies are dominated by people with Stage 2 characteristics. Living in isolated small communities, there is little chance to experience the clashes of different stories and perspectives, as well as the abstract universal laws and principles that are inventions of civilizations.

Many urban middle-class adults in developed countries are able to reach Stage 3 and even Stage 4. Committed believers who are in Stage 3 provide the mainstay and stability of organized religion or ideological movements, but they are not very reflective of the values prevailing in the society. They think there are social problems but take comfort in the understanding that the problems have always existed and will never disappear. They rely on external authorities for meaning and need external vindication about their stance.

Stage 3 is a comfortable plateau for many adults. Moving beyond this stage is far from automatic: it is usually prompted by a crisis. Moving beyond is also challenging and often leads to a deep crisis of personal identity. During brief periods of "great awakening" in modern societies, those in Stage 4 can become the most visible social force. Individuals at Stage 4 and above necessarily provide the stabilizing anchor during these turbulent periods as they carry out the dictum "I may not agree with what you say, but I'll defend to the death your right to say it."

The reason few people are able to reach Stage 5 and beyond is a complex and contentious issue. One could argue that interpersonal considerations are quite adequate for most daily decisions even in the most sophisticated societies, although some individuals will prove to be exceptions. For example, those who must face complex and novel social decisions on a regular basis may find that higher perspectives are critical in their work and personal lives.[18] Others may have such a deep understanding gained through spiritual crises or contemplation that such perspectives come naturally. However, even some who have attained higher perspectives may not be able to maintain them throughout life. As physical and mental power wanes, one may well "shrink" to lower, more comforting perspectives and allow them to guide their lives. They may retain an awareness of higher perspectives but no longer express their former concerns about the larger world beyond themselves or their own family and community.

Fundamentally, I believe the limitation has more to do with human motivations than anything else, something we will address in the next section.

The Flowering of Higher Perspectives

Although individuals rarely reach the highest possible perspective, the urge for spiritual fulfillment and the attraction to higher perspectives is an innate characteristic of human consciousness. Research on adolescents indicates that what they prefer is the highest stage they are being told about, even when they do not fully understand it.

The whole process of comprehension and understanding of multiple perspectives is probably best described as much like reading a good book: the further you get into it, the more it begins to come together and make sense. It takes a mature and reflective mind to understand that an individual does not start living until he can rise above the narrow confines of his individualistic concerns to the broader concerns of all humanity and reality. People at higher stages are happier and more at ease with their lives because they enjoy a richer sense of meaning.

8.4. Patterns of Perspectives in Society

Stop and go. Just like the biological evolutionary process, cultural evolution is always uneven, and not always moving toward higher and more conscious levels in a ladder-climbing fashion.

History does not show a relentless march forward. Like higher levels of creativity in arts and sciences, higher levels of perspective in a society materialize only when people have time to reflect and have the need for complex decision making. During the Dark Ages, the regression to barbarism triggered by the crumbling of the Roman Empire was devastating: urban life died away; commerce, currency and municipal organizations vanished; most people again lived in isolated rural obscurity. People living under these circumstances had neither the leisure nor the social stimuli to nurture and develop their higher perspectives. Their lives were much more limited to the here and now.

Europe was not alone: all civilizations have experienced periodic breakdowns and rebirth. The historian John Anthony Froude illustrates the cyclic nature of history when he says, "Virtue and truth produced strength, strength dominion, dominion riches, riches luxury, and luxury weakness and collapse."[19] Even in mature and stable places such as the United States, there seems to be a cyclical pattern with periodic "awakenings" and subsequent lost generations.[20] Given all of this, it is not surprising that historians have used the personal stages of life to describe the rise and decline of religions, nations, and civilizations.[21]

And, just as in the case of an individual, society's advancement to higher stages does not mean the earlier stages no longer exist. To the contrary, the judgments and procedures taken depend largely on the specific context of the event. In a society such as the Uunited States where most people have Stage 3 or higher perspectives, a nightclub fight may start as a typical Stage 1 dispute, where those involved were driven by raw emotions at the heat of the moment. It is only with the intervention of others, and finally the legal authorities, that the situation shifts. Those directly involved would also get a moment to reflect, engaging higher consciousness, and make rational decisions characteristic of higher stages.

It is perhaps not a coincidence that the United States Constitution, America's supreme law, relies primarily on a negative veto power just like the higher consciousness in the mind, which has to "decide" among competing motives and emotions the "right" thing to do. It is the Constitution that allows higher perspectives, for example, in giving the president the right to pardon any lawbreakers, thus vetoing the decisions of lesser institutions. In the famous exchange between the Navy lawyer Kaffee and Colonel Jessup in the 1992 movie *A Few Good Men,* Kaffee wanted "the truth"—what really happened that led to the death of a marine, and Jessup replied, "You can't handle the truth!"—the bigger reality of the military's role in a nation.[22] The court will handle the normal truth and may put the colonel in jail; but the larger truth, that of whether the nation needs someone like the colonel to defend itself, is up to the judgment of the president. In addition, the U.S. Constitution gives veto powers of one sort or another to each branch of government, not just the president, just as the conscious mind with a higher perspective would not always follow the "rational" line of thought.

A healthy society is one with a rich and open "ecosystem" of various perspectives. In chapter 4, we discussed how Western wisdom is best characterized as outward-oriented and Eastern wisdom as inward. But both cultures have stories that warn about tunnel vision, the pitfalls of single-mindedness. For example, here are two contrasting tales, both of which have to do with wells, one warning about the danger of looking upward, the other about the pitfall of staying low. In Plato's *Theaetetus*, there was "the story about the Thracian maidservant who exercised her wit at the expense of Thales, when he was looking up to study the stars and tumbled down a well. She scoffed at him for being so eager to know what was happening in the sky that he could not see what lay at his feet."[23] In a well-known Chinese tale (井底之蛙), a frog was living a happy life at the bottom of a well. One day, a turtle passing by the well on its way to the Eastern Sea looked down and had a chat with him. The turtle told him what a large and splendid world it was on the ground outside, but the frog flatly denied the turtle's suggestion that the sky was actually wider than the small round piece that he could see from the bottom of the well and saw no point in leaving his little paradise!

Religion as a rich source of perspectives. The cumulative and enriching nature of growing perspectives is best shown in the emergence of "higher" religions that have retained all components of "lower" religions. Understanding the nature of religious perspectives is very important in our attempt to make the Cosmic View as appealing and actionable as possible, which we will discuss in the next chapter.

Without exception, all major religions contain multiple, hierarchical levels of perspectives. The difficulty faced by atheist movements in their efforts to refute what they regard to be unscientific beliefs is largely due to their exclusive focus on "cold" reason; such a reliance on ultrarationalism can hardly satisfy the personal need for moral lessons and hunger for belonging, authority, and leadership.[24] In contrast, as we discussed above in the stages of knowing God and stages of faith, people at very different stages can find the same religious doctrine useful and effective, but in very different ways: that's because any given doctrine can be interpreted at literal, allegorical, metaphorical, philosophical, or mystical levels. John Dewey went so far as to define as religious anything that introduces perspectives:

> All religions have dwelt upon the power of religion to introduce perspective into the piecemeal and shifting episodes of experience. We need to reverse the ordinary statement and say that whatever introduces genuine perspective is religious, not that religion is something that introduces it.[25]

The primitive religions of the world have gradually been replaced by higher religions largely because the latter added new layers of perspectives without giving up the old layers. For example, it can be argued that the New Testament was added to the Old

Testament to form the Bible in a fashion that roughly paralleled the human "triune brain." As the old and new layers are meshed together to form a larger whole, the older ones remain indispensable; they contribute an essential set of varying perspectives to the multilevel core of the creed. The history of Christian thought experienced five major paradigm shifts (Greek Alexandrian, Latin Augustinian, Medieval Thomistic, Reformation, and Modern-Critical), but the "biblical message"—that is, our personal bond with God through Christ—and not Scripture itself, is the enduring norm.[26]

More advanced religions usually evolve at times of transitional tension but not amid a crushing crisis—for example, a community that is emerging from a warlike nomadic existence into a more settled herding or agricultural phase would be favorable to such religions. It is at these junctures that the higher religions continued to promote what can be called instinctive morality[27]—near-monogamy, and prohibition of incest, murder, theft, and cannibalism. They also imposed a reflective angle that exposes the incompleteness in all of us. Our sense of being far less than we could be—our boredom and loss of meaning after life's necessities are obtained—comes to the fore, since survival issues are no longer a major challenge. The prophets and sages articulated what the majority only dimly glimpses: the fact that in the human being there is something beyond sheer animal competence and adaptive superiority over other forms—namely, transcendence with the eternal and reflective eye.

In all successful religions, God offers both wisdom and love. To have faith in God is to have the world both explained and made secure for us. And Jesus proclaimed, "I am the Way, and the Truth, and the Life."

Multiple perspectives must be coherent and tied together to be effective, but even here there are exceptions: Genesis provides essentially two stories created by an obscure Semitic people and littered with contradictions,[28] and yet it has attracted people to come back to it again and again because it affords endlessly renewed meanings for our deepest concerns: the origin of the world and humankind, the existence of evil, and our mortality. In contrast, the Gnostics (the Gospels of Thomas, Philip, Mary Magdalene, and Judas) emphasized the abstract spiritual aspects of revelation—Christ as a revealer or liberator, rather than a savior or judge. their open-mindedness and intellectual subtlety attracted the elite, but in the end they lost the battle to become part of the mainstream Christian orthodoxy, which succeeded by emphasizing family values, developing rituals, and having their leaders as public figures rather than intellectuals who think on their own. The Gospel of Thomas, which depicted Jesus as just another wandering mystic who called for voluntary poverty and self-salvation through internal wisdom, has much less mass appeal than the canonical Gospels, which depicted Jesus as a unique miracle-performing savior who could forgive sins and assuage our fear of death by promising eternal life.

The Trinitarian orthodoxy of Anathasius ultimately won because Gnosticism was too intense and esoteric and Arianism was too bland and rationalistic. The power of the Christian God lies in the Athanasian doctrine of the Trinity—God is the Father, the Son, and the Holy Spirit in one—that effortlessly or awkwardly (depending on your personal view of the Christian faith) connects the transcendental to the most mundane perspective.[29] The "child" in us prefers a father—or even better, a grandfather—who is benevolent, forgiving, and permissive, even though the reflective mind sees the shortsightedness in this preference.

People converting to a religion always claim to be attracted by the ideology. But for most people, the main motivation is not the abstract theology, but rather the emotional attachment to the preacher and fellow worshippers. The higher perspectives embedded in the ideology are attractive spiritually, but the lower perspectives (Stages 1–3) and practical benefits are what most people can really comprehend and relate to. Today, more advanced interpretations of religious doctrine are readily available to everyone, but they remain a minority view even in free and prosperous countries. Today there is a large population of fundamentalist Christians who interpret the Bible literally in spite of overwhelming scientific evidence about evolutionary history and the roots of human morality. This has much more to do with the limits of human nature than with Christian theology per se. Even within the church, higher perspectives are readily available for those who care to look—they have been expressed by great theologians today and in distant history, such as St. Augustine, St. Thomas Aquinas, and St. Albertus Magnus, who never limited themselves to a dogmatic and literal interpretation of the Bible. The perspectives of the deepest Christian thinkers are as profound and sophisticated as those of any atheists, agnostics, or believers within other religions.

8.5. Heavy Lifting with Higher Perspectives

"Realism" resists higher perspectives. The difficulty of raising the level of the general public's conventional perspective on things is not limited to the religious and spiritual realm. In general, there are two types of outside perspectives one finds emotionally difficult: that of lifting oneself outside the instinctive system, an act of will that transcends the personal perspective; and the perspective that takes one outside of one's immediate social environment, which transcends the "me-versus-you" or "us-versus-them" group perspective.

The latent urge for transcendence represented by these two higher perspectives is seldom automatic and always requires patient nurturing. Like the initially alien nature of the scientific mindset, an outside perspective feels counterintuitive although ultimately satisfying. The limits of higher perspectives are part of the general limits of being a human, with its limited lifespan, its fixed life cycle, and its super-mammal motivational system.

Making people commit to higher perspectives is even harder than getting people to understand them. Transcendental perspectives aside, even our innate moral senses are fragile and hard to commit to consistently, as any preacher knows well. For instance, our altruistic impulses are constantly countered by an inborn suspicion and hostility to outsiders. Fundamentally, higher perspectives are as marginal as young members of the "society" of mind, with many bigger and stronger members already at the table calling the shots. The higher perspectives may receive high regard when people are discussing philosophical worldviews with nothing at stake. But the test that reveals one's true character is in the high-stakes real world when sacrifices of lower-level self-interests have to be made. All of a sudden, most people will find higher-stage perspectives unconvincing and unworthy. As we often hear, "Come on, let's be realistic!"

First and foremost, the human as a finite creature living a finite life does not enjoy any existential guarantee. A person has to focus on his or her existence in the here and now. One who is preoccupied with basic needs of food, shelter, and security will find higher levels of perspective elusive. A hungry person is not free. Weakness corrupts the mind, just like power. An insecure person lacking self-confidence is intolerant.[30] The reflective mind cannot function properly under the stress of constant fear and anxiety.

Lower perspectives are more comfortable than higher ones. Even in the absence of chronic survival pressure, most people are stuck in lower perspectives and remain preoccupied with their daily lives. For many people, the situation has more to do with desire than ability. Higher perspectives are strictly a part of our rational mind or beyond it at even higher levels of cognition, and the rational mind is too often serving the interests of other motivations rather than driving them.

Indeed, most people can understand perspectives that locate the individual in the grand scheme of things—as long as it is not about themselves. For example, we all know that the fate of any one cell in our body should not concern us since countless cells die every day. But it can be utterly depressing to realize that, in the grand scheme of things, the individual human life can be as dispensable as a particular body cell. In other words, there is no hope for *me*. A high vantage point reveals the truth about cosmic evolution, but the realization of Cosmic Being will be trumped by our mortality in the near-term, so why bother with it? It will have nothing to do with *me*.

Even for those mature individuals with a reflective temperament, the big picture is something to be suppressed, because it can render everyday human endeavors trivial and empty. For high achievers, the goals of a championship, the CEO job, or even a harmonious and prosperous society are what keep them going; yet, if those individuals are also highly reflective, they may also perceive these goals as mere temporary, distracting diversions that keep our lives interesting.

What we do not want to think about are Shakespeare's immortal words in the play *As You Like It*:

> All the world's a stage,
> And all the men and women merely players.
> They have their exits and their entrances.

Looking at the big picture directly can be painful, as typified in the lives of two great individuals who lived 2,400 years apart—Gautama Buddha and Leo Tolstoy. Both were born to great wealth and lived in privileged circumstances but became greatly disturbed at the suffering in the world, which precipitated a deep existential crisis in both of them (Buddha as a young man, Tolstoy in middle age); this in turn gave rise to a new, comprehensive awareness that existed alongside the recognition of suffering. Both subsequently lived lives utterly devoted to the truth as they saw it—the Buddha in his uncompromising realism and boundless compassion, and Tolstoy in his uncompromising adherence to his view of the true Gospel as preached by Christ, as he described in his classic work *A Confession.*[31]

Existential despair—the sudden awareness of the truth of one's existence—is an important ingredient in almost every description of spiritual awakening from ancient to modern times. In a more modest form, moments of sudden awareness of one's situation (or some aspect of it) occur in the lives of many ordinary people—including Tom Rath, the main character in Sloan Wilson's classic novel *The Man in the Gray Flannel Suit*:

> I was my own disappointment. . . . I really don't know what I was looking for when I got back from the war, but it seemed as though all I could see was a lot of bright young men in gray flannel suits rushing around New York in a frantic parade to nowhere. They seemed to me to be pursuing neither ideals nor happiness—they were pursuing a routine. For a long while I thought I was on the sidelines watching the parade, and it was quite a shock to glance down and see that I too was wearing a gray flannel suit.[32]

When the stark realization of our mortality is added to the equation, the experience can be much deeper and more terrifying than the "loss of meaning" that many of us undergo at some time in our lives. The recognition of the "must-die" outcome to our lives is something that no other animal can foresee. Although we all acknowledge that we are mortal (or at least our body is), for some of us this realization affects every part of our being in a most profound way. James Fowler vividly described his own experience:

> Four A.M. in the darkness of a cold winter morning, suddenly I am fully and frighteningly awake. I see it clearly: I am going to die. *I* am going to die. This body, this mind, this lived and living myth, this husband, father, teacher, son, friend, will cease to be. The tide of life that peoples me with such force will cease and I—this *I* taken so much for granted by *me*—will no longer walk this earth. A strange feeling of remoteness creeps over me. My wife, beside me in bed, seems completely out of reach. My daughters, asleep in other parts of the house, seem in this moment like vague memories of people I had once known. My work, my professional associates, my ambitions, my dreams and absorbing projects feel like fiction. "Real life" suddenly feels like a transient dream. In the strange aloneness of this moment, defined by the certainty of death, I awake to the true facts of life. In that moment of unprecedented aloneness experienced in my thirty-third year, I found myself staring into the abyss of mystery that surrounds our lives.[33]

The French scientist, mathematician, and philosopher Blaise Pascal,(1623–1662) expressed the prevailing sentiment of existential anxiety in his book *Pensees,* which deeply moved me as I took a lone walk and looked into the dark star-dotted sky in a summer night:

> When I consider the brief span of my life, absorbed into the eternity before and after, the small space I occupy and which I see swallowed up in the infinite immensity of spaces of which I know nothing and which know nothing of me, I take fright and am amazed to see myself here rather than there: there is no reason for me to be here rather than there, now rather than then. Who put me here? By whose command and act were this time and place allotted to me? . . . The eternal silence of these infinite spaces terrified me.

This epiphany, the out-of-the-blue terror of nihilism, is what motivates religion and all kinds of faith as we seek larger meaning than the biological self. As Voltaire put it, "If God did not exist, it would be necessary to invent him."[34] And the greater and more all-encompassing the faith, the more comfort and confidence it provides to our hungry and empty souls. That is why the cosmic perspective was born as soon as higher consciousness could articulate it!

Nevertheless, survival instincts and narrow personal perspectives have continued to dominate the mind. From the personal perspective, knowing too much could endanger our emotional comfort. It is no wonder that anti-evolution forces can be so persistent.

Historical perspective is hard to grasp. What makes history? It is truly unique and unprecedented events that paved the road for further unique and unprecedented events over an enormous length of time. Most of us cannot hope to make history, or find meaning in it, because we cannot easily identify the history-making people among us, let alone join and support them.

In other words, it is almost impossible for us to formulate or act on a real-time and real-life basis from a historical perspective. In a striking parallel to the messy workings of our body and our brain, the details of history are always messy—far less coherent and purposeful than what was written down by historians.[35] What may look in retrospect like farsighted, wise actions may be experienced in real time as utter confusion and wrenching dislocation. What leads a civilization on a path of stagnation may have been hailed by its contemporaries as the dawn of everlasting peace and prosperity.

The notion that a scientific revolution occurred in seventeenth-century England did not take hold until the 1930s. It is doubtful that any of the "revolution's" contributors were conscious of the profound historical impact they were having—otherwise, Newton would not have spent many of his best years on theology and alchemy.

Even in retrospect, it's very difficult to make sense of the past. That is why the great Greek historian Thucydides has been hailed as the greatest of all time, and condemned as the worst. He was able to distill the complex historical process into a coherent theme, but at the same time he seems to have selected the material to fit that theme in his discourse.

Abraham Lincoln's life is commonly summed up as his fight against slavery. But detailed historical records reveal a man who lived much of his life without a clear sense of what to do with slavery, his moral sentiment notwithstanding, and he certainly struggled mightily with his particular circumstances. What he lived through is far messier than even the most in-depth account of his life.

What is true at a personal level is the same at the cultural and institutional level. For example, at no point during the formative years of Athens's democracy in the fifth century BCE did any political leader wake up one day and say, "Democracy is an ideal political system that we shall create." Rather, it was a series of accidents—a very messy and uneven process.

When Martin Luther nailed his 95 theses upon the door of the Castle Church in Wittenberg in 1517, he could hardly have imagined that he would succeed in spawning a new Protestant branch of the Christian Church. He would also have been shocked if he could have seen the direction Protestantism has taken in the past 500 years, especially the role it played in creating the modern capitalist society.

But the difficulty of seeing history clearly with our limited mind and experience is exactly the reason why the higher perspectives are so important relative to everything we care deeply about. We need a holistic view—and the curiosity to go with it—in order to make sense of it all, to tie everything together. In Einstein's bold words, "I want to know how God created this world. I am not interested in this or that phenomenon, in the spectrum of this or that element. I want to know His thoughts; the rest are details."

It is nearly impossible to discern a "prime mover" in history, partly because economic, ideological, political, and social movements are seldom synchronous; but if we can see, as Robert Wright puts it, "all of history since the primordial ooze as a single creative thrust,"[36] we may catch a shadow of the Creator's thoughts and glance at our role in it.

Struggles with the long view. We are more moved by a dying child than a dying nation. We are more outraged by the bribe that a local official took than an outdated institution that deprives millions of a chance to make a decent living. We can hardly bear the sight of a live fetus being aborted, but we never object to the use of a condom as it eliminates the hopes of hundreds of millions of sperm swimming for life.

The power of our emotion is like a camera with a fixed focal length that generates sharp pictures at one distance (the personal perspective) and blurs objects either too far or too near. The aggregates that are significantly higher than the individual—species or society—are too abstract. They attract our emotional attachment only indirectly, through individual representatives (for example, a dying child). The units smaller than the individual (such as a body cell or a gene) do not attract our emotional attachment.

From an evolutionary standpoint, does the individual matter if the goal is to advance the big picture? Yes and no. The whole is made of individuals. Although the whole can survive without any particular individual, the importance of any particular individual to the fate of the whole is never certain.

We can be emotionally committed to current issues but have great difficulty putting things in a longer time horizon. Take our passion for environmental protection and "sustainable development," for example. It is immensely sensible that we take seriously the unintended consequences of our actions and set up policies that protect "public goods" and protect future generations from damage caused by our shortsighted self-interest. But, viewed from an even longer-term perspective and a larger context, we need to be proactive in another direction as well—and that involves increasing our own adaptability. Unfortunately, this urgent issue is far outside the radar of today's public debate.

The first thing we need to recognize is that, however sustainable or ecologically sound our policies and practices become, environmental crises and climate change will continue to be features of this planet, just as they have always been. That means minimizing or mitigating human impact on the environment alone is an inadequate though necessary response to long-term survival.

Earth's environment has always been in a state of flux, with periods of extremely disruptive changes over its four billion years of history—many times greater than those we are experiencing today, or are likely to experience. That much we know for certain. What we know far less about is the extent of disruption that human activities will cause. The intelligent response is to do what we can to blunt the potential impact of global warming caused by the burning of fossil fuels. At the same time, we must be more attentive to the fact that our planet will continue to create havoc on its own, with or without our "help." Knowing that Earth will not always be a hospitable environment should tell us that we need to take the long view and learn new and creative ways to adapt to our ever-changing climate and environment.

The success or sustainability of civilization depends on whether there is enough foresight *and* technology to mitigate the detrimental effects of the changing environment—whether human-caused or natural—but also to be prepared for and even take advantage of whatever present and future changes this planet will undergo, in the short and long term. Wanton disregard of our natural environment almost always spells trouble, but we must go beyond merely "not disturbing" the environment and learn creative ways of adapting to this ever-changing planet. To be free and prosper, we need to respect Mother Nature, but not worship it.

The history of our planet has been a long, bumpy ride. Abrupt changes in the environment, however devastating at the time, may not be a bad thing over the long term. Without the many "crises" in Earth's history, we would not be here today. For perspective, let us go back a few billion years to the start of life's journey. The initial flourishing of life on Earth gradually exhausted simple organic compounds. That "food crisis" eventually led not to the extinction of life but to the totally new organisms that fed directly on sunlight through photosynthesis—cyanobacteria. Oxygen, a by-product of cyanobacteria's photosynthesis, became pollution and poison to life back then because it reacted with the basic building blocks of life (carbon, hydrogen, sulfur, and nitrogen). That was actually Earth's greatest pollution crisis ever. We see it in mountains of rusty iron in Pilbara, Western Australia. Of course, the oxidized bacterial waste at Hamersley Range is now an inexhaustibly rich iron mine for us. The crisis of gradual oxygen buildup in the atmosphere, moreover, eventually led to the rise of oxygen-breathing organisms and provoked an explosion of biodiversity. We are the descendants of those initial oxygen-breathing cells.

Now let us imagine that bacteria had human-like minds 3.5 billion years ago, at the time when the available simple organic food was being depleted, and some bacteria proposed a new lifestyle of feeding on sunlight through photosynthesis. The proposal would never have been adopted. From the individual's own "long-term" perspective, this polluting way of life would have been viewed as unnatural and irresponsible. The offenders would be considered selfish, dangerous criminals to be locked up and prosecuted. That decision would have been the end of progressive evolution and the end of hope for animals, including us.

Going back to the human condition, doomsday prophecies of past decades, such as the 1972 Club of Rome report, have turned out to be widely off the mark, because they typically extrapolated short-term developments to predict long-term consequences. A longer-term analysis shows an inverted U-shaped relationship between income (GDP) and environmental indicators such as air quality, with the turning point at about $8,000 per capita.[37] There is now overwhelming evidence for the so-called Kuznets Curve: rising income increases the demand for a cleaner environment, and better technologies enable cost-effective measures to meet this demand.[38] Still, for many people, it is almost impossible to fathom that long-term progress is usually built on active responses to short-term setbacks.

Let us look at another case closer to home and our hearts. We all want our children to be better off than us. If a possible genetic change is much more likely to do harm than good to our offspring, then we would rather not introduce it. This makes perfect sense, or seems to—but weigh that against Paul Seabright's comment concerning the development of human intelligence:

> It is notoriously hard to make scientifically robust statements about something as complex and multifaceted as human intelligence, and even more so to compare the intelligence of different groups. But here are two statements we can be reasonably confident are true. First, over the course of human evolution in the 6 or 7 million years since our last common ancestor with chimpanzees and bonobos, children have been, on average, very slightly less intelligent than their parents. Secondly, over that same period, grandchildren have been, again on average, very slightly *more* intelligent than their grandparents.[39]

How can both statements be true? The straightforward explanation is that for all genetic mutations during the reproduction process, only a tiny fraction may improve intelligence that aids the bearer's survival and reproduction, and the grandchildren are the children of those who survived and reproduced. None of our direct ancestors died childless. The overwhelming majority of novel mutations are "copying errors" that are detrimental to the bearer. But without the involuntary "sacrifices" of those individuals bearing the detrimental mutations, we would not be here today to understand it!

Without exception, the same principle applies to all adaptive traits in genetic and memetic mutation on Earth. Although malfunctions caused by genetic mutations threaten the individual whose well-being is the primary purpose of the replication mechanism, genetic *errors* are *functional* with respect to another purpose (one that we might call a "higher" purpose): without such "errors," and the inevitable resultant harm to individuals of the species, there could be no evolution.

For those of us who are parents, coaches, or teachers, it is often disappointing to see children or students who flatly refuse to follow our teachings, only to repeat the mistakes we made when we were children or students. Yes, most new ideas from the young and inexperienced are bad compared with conventional wisdom or proven methods. But from the social perspective, it is exactly the continuous trying of the new that made cultural evolution possible.

We can all understand this perspective. Yet it is still ridiculous to ask parents to accept such a perspective regarding their children. We seem to be both stuck in our emotional blocks and ready to transcend them.

The emotional block of "up-close and personal." As long as the controversy is about life (and especially human life), it is very hard for most people to elevate their thoughts to a higher perspective, even with conscious reflection. Seabright's point that "mistakes make progress possible" makes sense to most of us if the subjects are industrial products, books, movies, songs, websites, shops, buildings, or roads—but not if *we* are the subjects!

The innate dominance of personal perspective in our emotional decisions seems to suggest that most people will never be able to commit to the higher perspectives, although they can understand and even be emotionally aroused by them. Please consider these questions:

- If you knew you were to lose most of your family members to a powerful earthquake, like the one that killed 230,000 in 2004, and you had the technology to put an end to all earthquakes, would you do so? Certainly, yes. And for most people, the answer is yes even when they are provided with scientific evidence that earthquakes are part of the planetary renewal process vital for life on Earth over the long run and may even be responsible for the birth of life.[40]

- If you were a hunter-gatherer, would you support the invention of agriculture, which introduces *work* (the Greek word for work, *ponos,* has the same root as the Latin *poena,* sorrow) for the first time in history (and it was mostly back-breaking labor), provides us with a lower-quality diet, gives rise to wars over the surplus food (it triggered the birth of the professional army), produces deadly infectious diseases (including malaria, probably the single greatest human killer), and institutionalizes slavery and serfdom on a large scale?
- If you knew that cities would become a hotbed for epidemic diseases (measles and bubonic plague), which would periodically kill over one-third of the population, would you allow cities, even as the centers of great civilizations, to emerge? Would you allow the Black Death to occur, which killed 30 million people, although it ended up stimulating both political and economic developments in Europe?

It is true that we have foresight, and our moral instincts are often strong enough to commit us to self-sacrificing deeds for the benefit of the tribe, the country, or even the human race. Sacrificing oneself for the good of the entirety of humanity is indeed a tall order, and a very noble one.

But let's say our primate ancestors were capable of such self-sacrifice for the sake of their species, and that they had unlimited power to act. Their decisions then might have blocked the emergence of humankind. Let's look at why this is so.

If we look back, the Golden Age for the primates is not the present time with prosperous humans as "the third chimpanzee," but rather it was the early Miocene period, more than 10 million years ago, when the tropical primates were dominated by a great flourishing of apes ranging from the size of a small dog to giants nine feet tall. Then sudden climate change wiped out much of the tropical forest and most of the ape species. From the brink of extinction, however, our ancestors adapted and gradually thrived in a time of recurrent climate changes and resource uncertainties. In retrospect, the disappearance of the tropical forest was the trigger that forced our ancestors to adapt to the dry grassland, but the wrenching transformation certainly was not in the interest of the thriving forest-dwelling apes. If the apes had been able to alter the climate to preserve their natural environment, we would not be here.

Fast-forward to recent centuries in human civilization. The capitalist system, the driving force of modern civilization, would never have been approved by those who lived through its early stages of stagnant real wages, horrendous labor brutality, and severe environmental degradation. Joseph Schumpeter describes the logic of "creative destruction" inherent in capitalism and all other dynamic evolutionary systems:

> Since we are dealing with a process whose every element takes considerable time in revealing its true features and ultimate effects, there is no point in appraising the performance of that process *ex visu* at a given point of time. . . . A system—any system, economic or other—that at every given point of time fully utilizes its possibilities to the best advantage may yet in the long run be inferior to a system that does so at *no* given point of time, because the latter's failure to do so may be a condition for the level or speed of long-run performance.[41]

This logic of creative destruction is easy to understand (we may choose to sacrifice now for the future), but it may be difficult to swallow when one's own interest is on the line, partly because a "given point of time" can be longer than the entire lifetime of a person. Witness the protectionist sentiments against the "churn" in the global economy; increased openness and competition will eventually create tremendous wealth and benefit all, but some individuals and companies in previously protected industries will find their previously comfortable lives exposed or destroyed.

We can't "pass on" higher perspectives to our descendants. From the ups and downs we experience in our lives, some of us develop higher perspectives and make a habit of evaluating things in wider contexts. However, like any acquired trait, higher perspectives are not passed on automatically. For this and other reasons, the hope for some sort of historical causality in which each successive generation inexorably progresses up the perspective ladder is unrealistic.

It is a common dream of revolutionaries that future generations will understand their ideology and continue their unfinished journey. John Adams wrote to his wife during the Revolutionary War:

> I must study politics and war that my sons may have liberty to study mathematics and philosophy. My sons ought to study mathematics and philosophy, geography, natural history, naval architecture, navigation, commerce, and agriculture, in order to give their children a right to study painting, poetry, music, architecture, statuary, tapestry, and porcelain.[42]

Unfortunately, lifetime experience, which profoundly shapes a person's perspectives, cannot be transferred as easily and automatically as genes. Listening to or reading about a wartime experience in no way matches the impact of experiencing it in person. As the Polish journalist Ryszard Kapuscinsky put it, life is truly known only to those who suffer, lose, endure adversity, and stumble from defeat to defeat.

Higher perspectives for the masses? Can the species and cosmic points of view, which first emerged at the dawn of the higher civilization and human consciousness, ever be adopted by the general public? Our conclusion is, in most cases, no. Most individuals are what Ernest Becker described as the "automatic cultural man," lulled by the daily routines of their society and content with whatever the society has to offer them.[43]

This is not to say that normal persons lead an aimless life. To the contrary, a lot of people try hard to find their "true calling," as vividly demonstrated in Bo Bronson's collection of worldwide interviews with ordinary people, *What Should I Do With My Life?* Most people do have a strong faith to live on—if we define "faith" in the broad sense of *meaning-provider*, no matter how trivial or primitive a person's lifetime pursuit may sound.

There is a story about a reporter who went to a remote village and interviewed a young shepherd about the meaning of life. "Why do you shepherd a herd of sheep?" asked the reporter. "Because I want to accumulate wealth," the shepherd answered. Then the reporter asked, "Why do you want to accumulate wealth?" The reply: "So that I will be able to get married." "Why do you want to get married?" "So that I can have a son." "What do you need a son for?" The shepherd stopped and thought for a while, and answered, "So he can continue to shepherd the sheep."

Higher perspectives are both natural and unnatural to us. We cannot be expected to worry about long-term events and futures that are simply out of sight. Yet almost everyone has at some point in their life wondered (at least for a few moments) about the meaning of life in the cosmic sense. The potential of attaining higher perspectives always exists in some people, waiting to be awakened.

The light on the hill. It is a mistake to believe that since higher perspectives have tended to materialize throughout history without the awareness of the individual, we don't need to be concerned about making them known and understood. In fact, we do need to promote higher perspectives, especially the Cosmic Vision, since we have seen how hard it is for us to understand them with our reflective mind and implement them with strong willpower. On the other hand, the lesser perspectives (short-term, personal and community interests) *can* be taken for granted, as they are driven by our deep-rooted instincts and long-entrenched cultural norms.

The cosmic perspective is about much more than helping reflective minds overcome their existential despair. Without the higher perspective, one can make a convincing argument that progress is not in the best interest of our species or our natural environment, and that every perspective is equally valid and important. In fact, a lot of people are making these arguments. The downside of a pluralistic environment is the persistence of vague science, confused vision, and dubious theology. Postmodernism in particular denies the existence of higher values and objective truth.

In light of the highest perspective, the Cosmic View, we should not worship humankind (and the natural environment that it depends on) as an end in itself. Each of us should be proud to say that I live my life not just for the sake of myself, my family, my community, my country, humanity, or even life on Earth. I live for the Cosmic Creation process because I recognize that I am born into that wonderful process and I can make a difference in it.

This transcendent cosmic sense offers the clearest understanding of what are the most important and most exciting things to do in our lives. In other words, as we get on the top of the hill with an "eternal eye" possessing the Cosmic View, everything is put into better and better perspective. How you should handle a specific issue often becomes clear and easy once it is judged in light of the big vision. I think the famous Confucius saying in *Analects*, "Men without long-term vision always suffer from short-term worries" (人无远虑, 必有近忧), can also be written as "Men without holistic views always suffer from self-centered worries" (人无全虑, 必有自忧). The best way to overcome the fear of death and the fear of change is to make one's interests gradually wider and more impersonal, until bit by bit the spiritual walls of the self recede, and one's life becomes increasingly merged into the universal whole.

$$\lim_{x - 0} 1/x = \infty$$

(Where x is the self, one can never be totally selfless, but as we reduce it, we approach infinity, which is the Cosmic View.)

Transcending our narcissistic self-preoccupation is no pinched and dismal Malthusian vision of humanity's place in the universe, but rather an expansive and optimistic outlook of human being. Humanity not only needs to have problems *solved*, but also needs to *impose* bigger and better problems upon itself as a challenge to its creative powers—for the benefit of the human spirit, our descendants, and the whole Creation.

Adding a cosmic perspective that is inevitably at odds with some of our existing values can become a cause for despair for some reflective minds; but it can also be a springboard for freedom, the means for humanity to transcend its biological limitations.

Let's consider a hypothetical situation. If the last generations of *Homo erectus* knew they would produce the first of a new species whose superior mental capabilities would soon cause their own ultimate demise, they would almost certainly have taken steps to prevent that from happening if they had such power. That conscious decision would have been good for them—they were marvelous, intelligent creatures in their own right—but it would not have been in line with the higher purposes of cosmic evolution, which is to produce more and more sophisticated minds at this stage. On the other hand, if *Homo erectus* understood that no biological species can be permanent since even the earthly environment itself does not have a permanent existence, it is likely that the opposite decision would have been taken, albeit an emotionally tough one; in this case, *Homo erectus* would have acted to facilitate its replacement by *Homo sapiens.*

I am not implying *Homo sapiens* will necessarily meet the same fate as *Homo erectus*. There is no telling how far our spirit can go beyond the flesh in which it currently resides. Human ingenuity often enables us to have our cake and eat it too. But sometimes, tough choices are unavoidable.

The power of high perspectives. For the universe, more complexity pays off; for an individual, a perspective that captures bigger complex patterns pays off. If everything else is equal, a person (or a group) who holds a higher perspective is likely to enjoy distinctive competitive advantages.

After an exhaustive review of human accomplishments in arts and sciences throughout history, Charles Murray made a strong conclusion about why a higher perspective pays off: "Human beings have been most magnificently productive and reached their highest cultural peaks in the times and places where humans have thought most deeply about their place in the universe and been most convinced they have one. What does that tell us? It is not a question to be answered with a quip."[44]

The cosmic evolutionary process for the individual is like running a maze: most paths lead to dead-ends, and at ground level, it almost seems purposeless. As often as we see promise and progress, we sense limits and decline. As Arthur Herman points out in *The Idea of Decline in Western History,* "Virtually every culture past or present has believed that men and women are not up to the standards of their parents and forebears."[45] Without higher purposes, we are "lost" in the world of petty, mundane self-interest. Yet we can look back at history and lift our sights to see the larger pattern. When I was young, I had a cynical view about popular movies—the good guys always win at the end, not because it is true, but because such an ending sells tickets. But as I learn more and more about history, I realize that in fact the good guys have won more and had more impact overall, although the bad guys always grab our attention. If the good guys were wiped out most of the time, we would not be a species with such profound moral concerns and rich perspectives.

Higher perspectives provide the stronger motivation, and the transcendental cosmic perspective is the highest possible for humanity. This is especially true when the transcendental ideal is an integrated body of beliefs that can simultaneously satisfy our lower instinctual moral and bodily desires.

I hasten to add that supposed reliance on a higher perspective is often a charade and causes more harm than good from lower perspectives. Knowing the attractiveness of a noble cause, such as saving a failing nation, political leaders have time and again urged the public for personal sacrifices in order to promote their own power and interest. As we will discuss in detail in the next chapter, on conscious evolution, one of the best strategies to achieve higher goals is to leverage personal motivations rather than running against them. It is inevitable that a mother who just wants a happy family will mourn her son's decision to dedicate himself to a larger cause. But this does not necessarily mean the mother is unaware of the larger cause, and there is no reason to believe that she *only* wants to have a happy family.

Overall, we should be optimistic in recognizing that throughout history, all the misconceptions and abuses of noble causes have not prevented higher perspectives from being realized in theory and practice. For example, the best military strategy that Sun Tzu's *Art of War* promotes is not about fighting techniques, but about the superb political and moral excellence of avoiding military endeavors altogether.

In general, the mass casualties at the individual level are part of the process that we have to live with from higher perspectives, although too often this fact has been used as an excuse for pursuing self-interest. In his famous critique of totalitarianism, *The Open Society and Its Enemies,* Karl Popper defended individualism and exposed the tyranny of Plato's idealist morality that the interest of individuals should be sacrificed to serve that of the collective state. Many people, especially certain self-styled "liberals," believe Popper was against higher perspectives. To the contrary, protection and promotion of individual freedom and initiatives can serve exactly the purpose of the state and even higher perspectives of cosmic evolution, as we will discuss in the next chapter.

Without a cosmic perspective, the question of why God won't grant His believers happiness has troubled worshippers from the beginning. The story of Job in the Old Testament is perhaps the best illustration of this bewilderment. The theologian C. S. Lewis observed, "If God were good, He would wish to make His creatures perfectly happy, and if God were almighty He would be able to do what He wished. But the creatures are not happy. Therefore God lacks either goodness, or power, or both." But rather than abandoning the concept of God, or accepting the doctrine that God has His own inscrutable reasons that are beyond our rational assessment, the solution to the dilemma is to see the nature of human happiness from a higher perspective. God's benevolence is not reflected in making us happy and free of suffering—that's a human-centric, almost childish standard. God is good by virtue of providing us with an environment that is intelligible, and by setting into motion the stream of cosmic evolution that gives us the freedom, the sense of direction, and the confidence we need to participate in our own future growth and self-transcendence.

8.6. Happiness Cannot Be the Ultimate Goal

One radical case of higher perspective is what the reflective and eternal eye sees: *Personal happiness is what we want, but it cannot be our sole aim or ultimate goal.*

Most of the happiest people do not aim at personal ecstasy, just as most of the wealthiest people do not aim at getting rich, most successful entrepreneurs do not aim at their achievements, most highly profitable companies do not aim at maximizing profit, and most highly creative artists do not aim at becoming famous. The surest way to happiness is to lose oneself in a greater cause.

Of course one could argue that sacrificing oneself in this manner is psychologically impossible, since the direct aim of our actions is always to satisfy our desires. Yet, isn't the self-sacrificing pursuit of a "greater cause" itself, ironically, one of our selfish desires? We can derive personal satisfaction even from utterly selfless acts such as jumping into an icy river to save a stranger. The issue that this line of argument raises is that "happiness" is such a catch-all word that every instinct can be said to be an instinct for attaining happiness. And don't we also have an instinct for serving or protecting others even at the cost of our own lives or well-being?

In this sense, patriotic actions (to use one example) are not truly unselfish—they simply involve suppressing or sacrificing certain primitive instincts in order to experience nobler pleasures. In this sense, there can be no irrational behavior.

However, there is a way beyond this argument. Adopting a higher perspective opens the second eye and provides us with higher motivations. It watches the dynamics of our mind and gives us a sense of detachment from our own desires. This reflective eye can recognize that bodily hedonics and immediate gratification may come at the heavy price of mediocrity, trivial pursuits, shallow attachments, debased tastes, spurious contentment, and souls without love and callings. It can realize that what is called happiness in its narrow sense comes from satisfaction of animal drives, which by its very nature can only be a transitory experience.

Conscious reflection on happiness appears to be one of the first awakening insights of the conscious mind. It was already in full blossom during the Axial Age. The Upanishads said:

> The good and pleasant approach man: the wise goes round about them and distinguishes them. Yea, the wise prefers the good to the pleasant, but the fool chooses the pleasant through greed and avarice. . . . Fools dwell in darkness, wise in their own conceit . . . I know that what is called a treasure is transient, for that eternal is not obtained by things which are not eternal. . . . The knowing Self is not born, it dies not; it sprang from nothing, nothing sprang from it. . . . A man who is free from desires and free from grief, sees the majesty of the Self by the grace of the Creator.[46]

This spiritual reflection of the human condition is still critical today. Our modern, materialistic understanding about the determinants of happiness is often mistaken, for without a higher perspective, we tend to overestimate our ability to achieve the objectives we value so much, especially happiness.[47] This is the essence of Herodotus's story of Croesus, who thought himself to be the happiest man but ended his life in utter misery.

For the Athenian sage Solon, human happiness is a function not just of health, prosperity, and familial harmony, but primarily of noble achievements and public esteem. The happiest individuals are those who are lucky enough to realize their self-worth, sometimes even at the cost of personal well-being. Confucius urged people to place learning, study, and ceremony before pleasure, profit, and power. Aristotle, in his *Nicomachean Ethics*, claimed that the pure pursuit of pleasure for its own sake is bad, especially when taken in excess, although it has an important role to play in human life. Happiness comes from reason and virtue (the Greek word *arête* also can be translated as "excellence") rather than personal amusement. This perspective demands that the individual seek excellence through the golden mean with the help of knowledge and control.

Viewing our sensual and emotional experiences coolly from a distance is perhaps the ultimate conscious mastering of our basic instincts. What we want is very straightforward, as expressed in Ecclesiastes (3:12): "I know that there is nothing better for them than to be happy and enjoy themselves as long as they live." On the other hand, there is the godlike reflective view of life: happiness can also come from suffering, torment, and grief as we strive to defy our beastlike instincts. A few years after World War II, according to a German bishop, a doctor examined a Jewish woman who wore a bracelet made of baby teeth—they were from her children who were taken to the gas chambers. The doctor was shocked: "How can you live with such a bracelet?" Quietly, she replied, "I am now in charge of an orphanage in Israel."[48] A constant reminder of a painful past is not pleasant—you may even call it self-inflicted suffering—but for that woman, it was also a powerful instrument for motivation in her mission in life.

So how shall we see happiness from a higher perspective?

- **Happiness is hard to achieve at the personal level.** Studies of identical twins and of the relationship between personality and happiness suggest that each of us has a "happiness set point" that we always seem to return to despite the normal experience of an emotional roller coaster.[49] Our brains are not trying to "achieve" maximum happiness. They are trying to regulate us and motivate us to do certain things that are useful for our survival. Indeed, the firing of dopamine neurons responds to changes and surprises only relative to our expectations. Money and material riches matter but have a surprisingly low saturation point.[50] Americans' sense of happiness has been flat since 1950 despite considerable improvements in every conceivable measure of well-being: material life, health, education, leisure time, etc.[51] In cross-country studies, national well-being and GDP per capita demonstrate a clear positive correlation, but not beyond the level of $15,000 GDP per capita.[52]

- **My happiness is often your unhappiness.** Our happiness seems far more correlated with interpersonal relationships than with material factors. We feel happy when we are in a special relationship, which necessarily means there are other relationships that are not treated in the same way. As long as an individual's happiness is a state of mind having to do with interpersonal or community relationships, our devotion to our children and kin, friends, and society cannot exist without discrimination, and this becomes the source of inherent conflict between siblings, families, and societies. A significant part of our happiness derives from status, honor, prestige, and dominance—and these are achieved at the expense of others who are in competition with us. We enjoy rooting for the hometown/national sports teams[53] and the "game" of keeping up with the Joneses. But these are zero-sum games: the happiness of the winners totally depends on the sadness of the losers.
- **Happiness is desirable but not *that* desirable.** Research and common sense tell us that being happy is not only desirable for the individual but also beneficial to society: happy folks tend to be good citizens. They are less likely to commit crimes and more likely to help others (donations, charity work, aiding total strangers, etc.). However, happiness can also mean contentment, which in turn often means avoiding challenges and settling for shrinking ambitions. Old people tend to be happy because they have either reached their goals or have given up on them.[54] The ages-old Chinese saying, "Those who are content are often happy" (知足常乐) is insightful but has unfortunately become an excuse for those who are trying to avoid tough challenges in life.[55] In general, innovation tends to stagnate unless prodded. "One may contemplate history from the point of view of happiness," observed the German phi-

losopher Georg Wilhelm Friedrich Hegel, "but history is not the soil in which happiness grows." The periods of happiness tend to be blank pages of history.

- **Negative happiness (freedom from suffering) is unattainable.** Fundamentally, negative feelings may have more survival values than positive ones. Unpleasant emotions and feelings such as pain, anger, anxiety, sadness, shame, disgust, and fear serve vital functions, such as drawing our attention to critical areas of our life, health, or environment that we might otherwise neglect, alerting us to avoid certain conditions, and motivating altruistic social behavior.[56] But a painless, peaceful life without the negative emotions is the standard popular definition of Heaven. It is unlikely that we can trounce the quandary of "no pain, no gain."[57] Experiencing negative emotions such as fear and humiliation[58] can be a great motivator. A child standing in shallow waters can never learn how to swim. Sharpening the mind critically depends on the existence of disagreement with oneself, which is unpleasant at best (disagreement with others is worse, since it may lead to strife, even violence). Even existential angst and despair is a necessary condition for our sense of freedom and self-awareness. The necessity of negative emotions in the functioning of the mind is a human tragedy, but that is true only if we insist that happiness is the goal of life and nothing else.

In summary, we must have an "out-of-body" perspective about our craving for sensual pleasures, material gains, or even well-being in general. Fundamentally, happiness is a self-centered local phenomenon, a dangerous instrument if not framed within a higher aim. We chase happiness even though we know it is nature's trick to get us going—just as we enjoy sex even though we know that it's nature's way of persuading us to engage in the reproductive act.

Life's goal is not happiness but freedom and fulfillment. Eastern wisdom coolly reminds us that animal instincts are always with us and to deny their existence is futile and counterproductive.

This irksome ability to see one's own mental states and to put one's own life in the context of the wider environment can render one's life barren and futile even when all basic needs are met. While animals can only respond to external challenges or stimuli, only humans, with our higher consciousness, are capable of stoical detachment with our eternal reflective eye. Satisfying the brain's many urges cannot be our final aim, but we do it anyhow, and we reluctantly realize it is the only way we can get things done.[59]

Our view of happiness should be similar to our view of humanity itself: our aim is not for happiness per se; yet, happiness remains the most potent motivator for us. The suggestion that happiness cannot be our aim does not imply that we have to be ascetic. To the contrary, it follows the same linked-but-detached wisdom that is reflected in the attitude of some Buddhist monks, especially the Tantric Buddhists. They refuse to accept the traditional ascetic creed of self-denial. Instead of a conventional vegetarian diet, their practices follow the slogan, "Let meat and wine pass through my digestive system, the Buddha stays within my body." With a total contempt for the hypocritical ruling elite, the iconoclastic Japanese Zen master and poet Ikkyu Sojun (一休宗純, 1394–1481) believed that one should freely shuttle between the sensual and spiritual worlds, occasionally taking a rest in between to reflect on our predicament. He fell in love with a blind singer at age 78, and the romance lasted ten years. Taiwan's famous scholar, historian, and social critic Li Ao had a collection of 100,000 academic books, but on the walls of his reading room were pictures of naked women from *Playboy* magazine—he insisted that both collections were his favorites. A person with an intoxicating spirit can get high with or without wine.

8.7. The Greatest Divide and a Leap of Faith

Our discussion of perspectives can help answer questions about "the meaning of life," but this is not what this book is really about. We are trying to address the question, "What are we really in this world for?" when we discuss the posthuman future, which also relates to such politically sensitive issues such as stem cells, human cloning, and nanotechnology.

Today, the greatest divide within humanity is not between different races, religions, or ideologies, nor is it between religious and

Figure 7: The Photoshopping of NASA Images

In order to draw popular interest, images taken by space telescopes are massaged and colored to look like nineteenth-century romantic landscape paintings. See: http://www.nasa.gov/multimedia/imagegallery

secular, literate and illiterate, scientific and non-scientific beliefs. These debates tend to be disagreements within the same level of perspective. Instead, the greatest divide within humanity is the one between the open cosmic perspective versus a wide variety of forms of human-centric tunnel vision.

To be specific: The deeply religious Christian right and the intelligent design theorists (such as William Dembski and Hugh Ross, who advocate science and don't read the Bible literally) hold the same human-centric perspective as certain nonreligious radical environmentalists (such as Earth First!), certain secular conservative thinkers (such as Francis Fukuyama and Adam Wolfson), humanists (such as Mary and Lloyd Morain), and neo-Luddites (such as Jeremy Rifkin and Pentti Linkola). Thus Jeremy Rifkin has teamed up with William Kristol, editor of the conservative *Weekly Standard*—a man who opposes him on almost every other public policy issue—to coauthor editorials calling for a ban on human cloning. Former U.S. senator Larry Pressler declared in 2000 at the United Nations World Forum: "We're embarking on a 100-year war about this issue [of biotechnology for the human genome and human being]."[60]

On the opposite side, deeply religious thinkers (such as the French Jesuit and paleontologist Pierre Teilhard de Chardin) share the same open cosmic perspective with atheists and communists (such as Leon Trotsky and J. D. Bernal), Fabian socialists (such as Beatrice and Sidney Webb and George Bernard Shaw), technologists (such as Hans Moravec and Ray Kurzweil), feminists (such as Margaret Sanger and Donna Haraway), humanists (such as Joseph Fletcher and James Hughes), and visionary futurists such as Barbara Marx Hubbard.

The human-centric camp's goal is to protect humanity from "selfish and arrogant" advocates of technological evolution beyond humanity. But their seemingly moral high ground is an illusion. As I have indicated previously, and will discuss in the closing chapters of this book, the conviction that humankind should be *the end of all ends* actually rests on the Platonic myth of an unchanging Golden Age and the Eastern closed-ended perspective rooted in an isolated self-sufficient agrarian society. In contrast, the cosmic perspective that places humanity in a larger evolutionary process is built on the Western transcendent faith in Cosmic Creation and the Eastern realistic view of humanity.

The core conviction of the Cosmic View is that nothing is permanent or absolute except the process itself. The very heart of the biblical story is not the promise of individual salvation after death, but God's intention of growing His children and gradually letting them take responsibility for and control of the world they are born into.[61] Far narrower perspectives, such as secular utilitarianism and scientific materialism, are not an adequate basis for evaluating humanity's value in the grand scheme of things, since the scope of their understanding is limited to merely human self-interests. But I argue that humanity is not, and can never be, the ultimate end of Cosmic Creation, regardless of whether there is a purposeful Creator or not.

Thus, the Aristotelian notion of human flourishing is valuable, but only in the context of what humanity can do to further cosmic evolution. Actually, the best approach to ensure the prosperity of humanity is to utilize its unique capability to ask metaphysical questions and to maximize its potential through the exercise of its forward-looking free will.

As I have advocated from the beginning of this book, in order to ensure the vitality and success of the open cosmic perspective, it is essential to blend two vital cultural influences: the Eastern realistic view of humanity with the Western transcendental vision. In other words, we should not only utilize multiple perspectives about the human condition and human potential, but we should also turn to the highest perspectives available.

I offer three pillars of support for this argument:

First, we must face the reality that human-centric drives will remain dominant and almost impossible to escape. The future-driven scientist Hans Moravec said frankly, "I resent the fact that I have these very insistent drives which take an enormous amount of effort to satisfy and are never completely appeased."[62] He is focused on developing artificial intelligence that could someday exceed the natural intelligence of humanity, but he enjoys things like food and sex just like anyone else. The scientist and extropian Sasha Chislenko could not wait to get rid of his body to become a cyborg, yet he committed suicide—not out of philosophical despair but out of a painful romantic relationship![63] Even geeks like these two are often motivated by the desire for recognition in the peer group or for winning over mates.

Second, raising personal perspectives will always require heavy lifting. Why should a person be concerned about cosmic trends other than out of leisurely curiosity? The most effective approach to increasing people's cosmic devotion is to link it to our human drives. A non-personal God that is cosmic in nature has to be blended with certain human-friendly characteristics in order to be attractive, in the same way that colors are added to the pictures taken through space telescopes (see figure 7) to enhance perception and draw popular interest.[64] People who are passionate about understanding the world for pure intellectual satisfaction are rare—they risk being viewed as misfits in this world.

Third, human morality as promoted by popular religions and enforced by laws and social institutions should not be undermined. For the most part, we are *here* to play the role of human being, not God. We are members of the species we study and try to improve upon, and we live here and now, not in evolutionary time, where the individual is all but invisible. If we start identifying ourselves with the Eternal Designer and regarding the other human beings around us as dust on the lens, we shall qualify not for the throne of the universe but for the asylum.[65] Those individuals who solely identify themselves with the universal life of the species, and regard the people around them as beneath their notice, are often those who are too lonely and timid to "jump in" to the messy stream of real life.

Nevertheless, a cosmic perspective must be articulated as "the light on the hill" to close the tremendous gap that has opened up between our understanding of the universe and our parochial focus of attention on human life. We cannot really play God as some critics claim we should do (and after all, how can we, with such limited power?), but we can play with our God-endowed gifts and initiate the creation of gods, or what I call Cosmic Beings (CoBe), a new species on the frontier of cosmic evolution that is unimaginably powerful and creative. We will not only be "interfering with nature," as we have always been doing here on Earth, but also be "disturbing the universe," in Freeman Dyson's words. Our "mind children" will be in better positions to know the true nature of the cosmos, but for us, it is perhaps best to maintain an agnostic position, as Darwin did in his mature age, because it is very hard to trust "a mind as low as that possessed by the lowest animal . . . (to draw) such grand conclusions."

What are the perspectives we need to realize human potential in creating CoBe?

First, we must recognize the absolute importance of a daring and passionate leap of faith—a leap from rationality to nothing other than being pragmatic and playful in the face of what Kierkegaard called "objective uncertainty." Today, the hope of immortality and adolescent fantasies such as time travel and resurrecting dead people are motivating, but our leap of faith is more than a personal and emotional self-indulgence. It is actually less a bet on the existence of the Creator per se, and more a learned judgment on the best motivator to gain better understanding and access to "the things of God" by a finite being. As the Protestant theologian Reinhold Niebuhr put it, "Nothing that is worth doing can be achieved in a lifetime; therefore we must be saved by hope. . . . Nothing which is true or beautiful or good makes complete sense in any immediate context of history; therefore we must be saved by faith."[66]

Second, the leap of faith is pragmatism but not vulgar pragmatism, as in, for example, telling kids there is a monster in the electrical wire in order to prevent them from touching it. Useful and practical beliefs are often not true, but true beliefs are always useful and practical. Pascal's Wager has often been mentioned as a motivation for some people to believe in God, but actually the reverse of Pascal's Wager can also be motivating: just in case there is no God, we should try our best to achieve immortality and control our own fate. The immensely popular American notion that *God helps those who help themselves,* which is often mistaken as original biblical teaching, smartly covers both wagers. This is truly what the American philosopher Charles Sanders Peirce called the Gospel of Greed[67]—a pragmatic greed to become the master of the universe as God and our self-interest intend us to be.

Third, we must keep in mind that no human belief is infallible, regardless of how well it is supported by evidence. We must be playful and not dogmatic, lest we fall into existential despair. Nothing is guaranteed, since all the cosmic patterns we have discerned so far could turn out to be mere illusions. It is probable that we are mere trilobites, ammonites, or dinosaurs—species that stayed on the top of the world for a time that seemed eternal from their perspectives but was really a fleeting moment in the cosmic scale. Bertrand Russell looked at it even more darkly, suggesting that we could be the chicken on a farm that had been fed and protected every day until the day we got our necks wrung.

Maybe the complexification trend in cosmic evolution is a local rather than global phenomenon, an illusion from our extremely confined vantage point. Maybe the fertility of the universe is quite limited, and regardless of what we do, the universe is doomed to end in a Big Crunch or a Big Chill. Just as the chickens on a farm are totally ignorant of why they are being treated nicely and what is their ultimate fate, we could be living in a setting designed by some higher being, as depicted in the movie *The Truman Show*. For us, it would be a devastating psychological blow if we were to get to know "the truth," like a painter with a terminal illness who suddenly enjoys great success, only to find out that her friends bid up the prices of her paintings to cheer her up.

As we have already discussed, cosmic history and the emergence of life and mind on Earth, if viewed from a broad span of time, does exhibit an evolutionary trend. We can never be certain that it is not an illusion, but before counter-evidence appears, this is perhaps our best hypothesis of the universe based on the faith that "God does not play dice"—or if He does, they are loaded dice.

Rather than throwing up our hands, we must push as hard and wisely as we can into the unknown, and let results speak for themselves about ultimate reality and human fate. What we can know, however limited, is the most reliable guide for our actions. We have to believe we are not misled by our encounter with the world in a fundamental way. Yes, great courage is needed in this frank and blunt picture of darkness. As Goethe put it, "without courage, all is lost": yet this sort of courage reflects the true peaceful mind of science and faith—icy, powerful, and ultimately satisfying.

Growth in the midst of darkness is not unprecedented: our biological and intellectual ancestors got us here with the same total ignorance about their future. The difference from the historical path is that we will be doing it consciously—we are aware of our ignorance—hence the need for optimism. We can anticipate surprises, but we still have to rely on human resilience, in our ability to live through the most appalling personal and public tragedies and still go on. This is an aspect that many technological optimists have underestimated.

The argument for striving for the highest godlike perspective is not just human hubris. It is what anthropologist Lionel Tiger called *big optimism,* the faith in a universe that always favors progress despite temporary setbacks. To be fully alive is to believe nothing is impossible. To be a genuine creator is to create something that is greater than ourselves. To be young in spirit is to never stop growing and to have the resilience to outlast all the skeptics.

Finally, the core of Western wisdom, the brave and selfless spiritual leap of faith, must be connected effectively with the core of human motivations in realistic ways. The important aspect of reality is that we can have multiple perspectives on nature and the human condition that can lead to the dream about becoming more intelligent, more understanding, more than just blood and flesh. Opening our eternal reflective eye, we realize that the creation of the universe must have a scope that is much greater than our daily concerns of human well-being and happiness.

Getting a dim perception of the Tao, the mind of God, we can make a conscious effort to overcome human chauvinism. This may sound arrogant and egotistical, but true hubris lies in the assertion that there is nothing for us to care for other than ourselves. And that blindness to the highest perspective carries the greatest danger for us all.

Part Four

CONSCIOUS EVOLUTION: ITS POWER AND IMPLICATIONS

CHAPTER NINE

CONSCIOUS EVOLUTION

God promises eternal life; we can deliver it.

—Philip K. Dick, *The Three Stigmata of Palmer Eldritch*

KNOWING OUR OWN limits as well as our potential, we should be propelled by the cosmic spirit manifested as natural law, and the clear direction for continuing the evolutionary trend is the growth of intelligence and its expansion beyond the Earth. From the use of fire to the domestication of plants and animals to the cultural evolution that has laid the groundwork for conscious evolution, humankind has come a long way. Now we are reaching the ultimate form of manipulation—transforming our own physical, social, and cultural building blocks at will. It took humankind millennia to recognize the pattern of cosmic evolution, but it may take a much shorter time to complete the next steps, which we will examine in this and future chapters.

The first crucial step of conscious evolution is unleashing the human genetic lineage, which has been the limiting factor on cultural evolution. The time is now, and the last great invention that *humans* will ever need to make is in sight: the autonomous intelligent cosmic being (CoBe) that is spontaneously adaptive and has a will to continuously evolve and push forward the evolutionary frontier in the universe.

The emergence of CoBe will not be the result of a simple straight-line progression. As we have discussed, natural evolution is eventful, chaotic, rough, unclean, unbalanced—it is a storm of "creative destruction." Likewise, conscious evolution will be full of unintended, unpredictable consequences as well, including calamities as well as magnificent, seemingly "miraculous" advances. Conscious evolution must follow the same strategies as natural evolution, only with more intensity and with the aid of Cosmic Vision. In this part of the book we consider key aspects of this evolution, including characteristics of conscious evolution, how to deal with various risks during the transition to the posthuman era, and why we need to move from human morality to a more universal one, a "transcendental morality."

9.1. What Is Conscious Evolution?

From natural to conscious evolution. One of the most puzzling things in the theory of natural selection is that there is no obvious (i.e., identifiable or conscious) *selector*. Evolution through natural selection appears to be an extremely blind, wasteful, sluggish process. Complex life and conscious mind emerged through self-organization and through incremental accumulation of complexity and order over time. Without a *selector* who decides what to grow and what to eliminate, perhaps natural selection is the only realistic process to make life and mind appear in the universe, never mind how inefficient and time-consuming the process is.

With the birth of consciousness, a new phase of cosmic evolution is possible that can be significantly different from the past. The invention of the scientific method added focused purpose and rigor to the earlier tradition of trial-and-error-based technological development; in so doing, it greatly accelerated technological development and knowledge accumulation. In a similar way, the conscious mind—by utilizing the emerging methods of conscious evolution that we will examine in the coming pages—now adds new impetus to the ongoing cosmic evolutionary process by also injecting newfound *purpose* and *focus*, and likely accelerating its pace even further. In other words, evolution now has a *selector*: self-aware humans who choose to align with the impulse of evolution. By "conscious" I do not mean just normal human consciousness, in which we are aware of our own thoughts and being, but an awareness of cosmic history, natural laws, and humanity's place in the ever-changing universe.

Just as the emergence of modern human beings was a gradual process, there can be no clear-cut, sudden jump from *natu-*

ral evolution to *conscious evolution*. Let us define the transitional period as *cultural evolution*. It has been the evolutionary process since the emergence of modern human beings with a reflective consciousness. Throughout human history, cultural evolution has involved making tools (which include external adjuncts—writing, printing, computers, etc.—to our individual and collective memories), creating new ideas, developing social organizations, and the like. Thousands of years of cultural evolution have created great civilizations, all within the confines of human nature. These innovations are clearly guided by human consciousness but still should be considered a transitory, or primitive, stage of conscious evolution. The "real deal" of conscious evolution begins when *we* consciously select what we want to be—and it continues as these consciously selected life forms in turn make ever more complex and interesting decisions about the future.

Conscious evolution is purposeful. Charles Darwin avoided using the term "evolution" to describe natural selection since at the time the word implied directional progress toward an ideal. New species emerged as adaptations to environmental niches, not necessarily manifesting a direction toward more complex or mindful organisms. In other words, Darwin and his successors saw adaptation as an *a posteriori* phenomenon in natural selection—insisting that there is no conscious search for improvement in an organism. Nevertheless, complexity and more cooperation and connection sometimes offer an advantage in the struggle for survival, and "survival of the fittest" has certainly meant that evolution has promoted progress—if not toward some predetermined ultimate ideal.

Natural evolution is a passive process shaped by natural laws and the environment, although the environment itself can be subject to manipulation or alteration by the biological organisms themselves. Generally, nature leaves well enough alone. As long as a species can survive in a certain environment, no great selective pressure is applied to force change. This is one reason our body and mind are full of evolutionary vestiges of our remote ancestors. For example, human blood has more or less the same salt composition as that of the seawater from billions of years ago. It must have originated back when our ancestors were fishes: the forces of natural selection must have made their blood interchangeable with seawater to prevent losing water by osmosis across their gills. Land animals do not have this requirement, but changing blood composition may entail adjustments in other aspects of vertebrate physiology. Since salt is available on land, why bother to change?

By contrast, *conscious* evolution is an active process led by conscious minds and regulated by structures that conscious minds purposely create to guide the process. Pursuing evolution consciously means there is something we *want* to achieve, some state we would love to aim for even if we cannot get there ourselves. Such a goal transcends our personal existence or even our existence as a species. Conscious evolution is purposeful in the sense that information derived from the past and hidden underneath physical appearances is *explicitly* incorporated into conscious decisions and actions, whereas for unconscious evolution, the information is only *implicit*—it is embedded in the genes and natural environment.

Being purposeful, however, is not the same as knowing exactly what the desired end point is. We must be constantly on guard against the powerful feeling of preordained destiny. Destiny is written concurrently with the event, not prior to it.[1] Our fate is not written in the stars.

Purposeful conscious evolution has to be a determined but humble adventure. Without knowledge of the origins of the universe and whether there is other life or intelligence out there, we cannot know whether we are special and whether we have a special mission in this universe. Given this deficit in our understanding, taking the position that we are significant (and may be at a unique crossroads as actors in a great universal drama) and acting accordingly is first and foremost a pragmatic measure. It is a leap of faith, albeit a leap that could be critical for our future and, possibly, the future of the universe.

While we can always strive for greater understanding, we can never be certain about the past and the future, about exactly what we really want or what we are really capable of. Still, we can use our consciousness to make fundamental improvements, culminating with the creation of CoBe after what may be many false starts. Ultimately, this noble and courageous endeavor is based on the fact that the human mind is capable of developing actionable models of the universe, even including the evolution of the entire cosmos since the Big Bang.

Conscious evolution has to be a humble adventure also in the sense that the Cosmic Purpose has to be achieved through comparatively petty human purposes. Can we transcend our pettiness and thereby fulfill the cosmic mission of evolution? Can we consciously work on bridging the gap between our ability to imagine a time span of billions of years and our individual lifespan of less than a century? For better or worse, cosmic consciousness has to rely in part on our animal motives and instincts to realize its dreams. Still, these motives and instincts are nothing to be laughed at: they represent a few billion years' worth of natural intelligence about what works and what does not, even though any given example of their expression may seem trivial.

But the question remains: Can consciousness rely on animal instincts and personal ambitions without being ruined by their narrow and nasty tendencies? That is where advanced social structures must come in, including laws and regulations. The conscious mind is slow and not in direct control of most lower-level brain activities, but it is capable of leveraging resources both internally (the instincts) and externally (our social structures).

Conscious evolution must overcome conservative barriers. Nothing in nature "wants" to evolve beyond its local environment, except for humans with their reflective higher consciousness. But during this process of evolving beyond the constraints of human nature and its attachment to local Earthly conditions, humankind must tread a fine line between maintaining its "human essence" and expanding its self-identity. In particular, if we are to go beyond cultural evolution associated with previous human history, conscious evolution in the future must overcome our conservative instincts, yet it should also at the same time respect them.

At the moment, it is considered acceptable to promote new technologies and cultural changes as long as they don't touch human nature. Bioconservatives tend to be today's value setters, and they are interested mainly in making incremental improvements to the status quo of our human lives. They tend to be satisfied with the increases in longevity and standard of living that we have seen in the last century, or in the last few millennia, and they value the hard-won evolutionary gains made over many eons of time to get us to where we are today.[2]

But there the story ends; they do not value future evolution the way they value past evolution. Indeed, an incomplete and stunted reflection on the past can lead to unfortunate conclusions in terms of evolution. Rather than elevating the human spirit and preparing us for further progress, the conservatives want to suppress additional changes and keep humanity in its present condition. To realize what I describe as the "funnel vision" in chapter 6, they tend to offer elaborate moral justifications for the human status quo. Leon Kass has even suggested that the finitude of human life is a blessing for every individual, whether we know it or not.[3] Following the same moral reasoning, the liberal bioethicist Daniel Callahan urged the medical profession to resist the "research imperative" and impose self-restraint on longevity research since there is no inherent need, only unbearable social and psychological costs, to greatly extended life expectancy.[4]

The bioconservatives have some good points, but they miss a greater point. Evolution is not just novelty-seeking. Conservation of past gains is very much part of the process of any kind of evolution, best expressed as preservation, replication, and standardization of organisms and organizations that have stood the test of time and environmental changes. Evolutionary conservatism is good, but it should not be viewed as a goal in itself, only as a necessary ingredient that makes our highest goals possible.

As finite entities, no intelligent beings can be totally free of self-preservation motives—in fact, the power and will for self-preservation have grown stronger as intelligent life has become more resourceful. But instead of defending the specific forms and structures shaped by historical local adaptations, higher consciousness should recognize the origin and limitations of its own structures and seek long-term survival and prosperity of the essence rather than appearances. For human consciousness, prioritizing essence over appearance in this sense will require the recognition of the body as a limited biological machine and the Earth as a transitory and limited environment.

Conscious evolution is powerful. Natural genetic evolution keeps an eye on being "fit" but never has any foresight for creating further evolution or optimal design. Directionless, it has relied on "lucky" accidents or random events to lead to further evolution (and at times further digression). The result is a genome or a brain that is far superior to "random" codes but is still messy and, in terms of performance, "just good enough." Francis Crick once devised a possible genetic coding system so economical and free of error that, when experimentation showed it was not nature's way, it was nevertheless called "the most elegant biological theory ever to be proposed and proved wrong."[5] In a similar way, conscious evolution adds an engineering element—design with a purpose—on top of the "natural" trial-and-error experimentations. Crick's coding system, for example, might inspire designs for artificial life that are far superior to the natural ones.

The power of conscious evolution comes from its direction, intelligence, and flexibility. Natural evolution is blind, and cultural evolution as a transitional stage of conscious evolution is restrained by genetics and cultural inertia. Conscious evolution, on the other hand, can be more powerful and purposeful without the genetic leash. We will consciously evolve by using our knowledge and artifacts in a recursive fashion. The improved intelligence and artifacts that we create will be used to develop even better intelligence and artifacts.

In a sense, natural evolution is also recursive, but it is excruciatingly slow, because according to Steve Grand, the "first law of biology" is that "nature is lazy."[6] Biological organisms adapt; they do not anticipate. When there is no pressure to change, they won't. For example, it has been estimated that natural evolution will not result in any significant change to the human species with such a large, single interbreeding community as we now have (with more than seven billion humans on Earth, many times greater than the 300 million maximum threshold for self-evolution).[7]

Conscious evolution is instead proactive. Its power is exemplified by its ability to make jumps in the organism's genetic structure and in the environment that influences genetic expression. The conscious mind can create brand new designs from a blank slate. In contrast, the predominant mechanism for a biological organism to remodel its form is by tinkering with existing genetic functions. Just a few examples: The quantum leap from the prokaryotic cell to the eukaryotic cell was not new genes, but the epigenetic ability to control the form (through editing and splicing exons, DNA segments within a gene), timing, and location of genetic expression. When air-breathing was needed as a crucial adaptation, the lung evolved from the bladder. Enlarging the brain size requires no new genes or structural changes: just more copies of certain exons in the ASPM gene on chromosome 1 appear to be sufficient.[8]

Similarly, cultural evolution rarely moves by a leap into unrelated novelty; instead, it moves by selective emphasis on certain aspects of the culture's own past. The success of a religious movement, for example, largely depends on its cultural continuity with the society at large, its moderate level of tension with society (being strict, but not too strict), and a political climate that is at least somewhat tolerant of religious unorthodoxy. Social movements grow much faster when they spread through preexisting social networks.[9]

Conscious evolution is a function of conscious experience, memory, and information technology, all of which confer the ability to pass "memes" on to any generation anytime. Genetic transmission, by contrast, allows no conveyance of knowledge across generations other than the information embedded in DNA inheritance. Natural evolution regularly produces species that soon become extinct—failed genetic structures that are removed from the gene pool forever; it eliminates the ability to expand on some potentially very important evolutionary advances.

These limitations were broken with the invention of language and writing. The later development of information technologies virtually guarantees meme preservation because the storage and transmission of memes do not have to rely on a single medium the way genes do. In turn, conscious evolution will be unimaginably more powerful than cultural evolution because it removes the critical constraint—the limitations of human nature.

As we have discussed, natural evolution is a cumulative process, with almost all of the changes involving tinkering with the basic building blocks. But conscious evolution, at least in theory, could leap back billions of years to the origins of life (life apparently originated only once, billions of years ago) in order to leap billions of years ahead (on natural evolution's timescale). It will one day alter the current biological paradigm that dictates that an organism must reproduce itself, compete with similar organisms, and face certain death in the end.

But the power of conscious evolution is certainly not unbounded. As we will examine more fully later in this chapter, conscious evolution cannot do away with casualties; it will necessarily lead to outcomes considered evil from the human perspective. It may eliminate much of the need for physical trial and error, most likely through the coevolution of simulation machines and by using designs that are generated out of the simulated process. Nevertheless, even artificial minds will never be able to provide a complete simulation of reality, thereby eliminating the need for real-world testing altogether. Adaptation by trial and error is not a phenomenon that is unique to life but is rather a fundamental characteristic of many dynamic systems in nature. Conscious evolution may not be any different.

Neither can conscious evolution be free of all rigidity. Flexibility is always a relative concept. Ironically, the rigidity of basic rules is a required condition for achieving the flexibility of the mind (if defined as configurations of patterns and information)—it is impossible to play a chess game if the rules of the game change with each step the players take.[10]

Nevertheless, by treating only natural laws as untouchable and not conferring that status on our human genetic endowment, conscious evolution can be unimaginably faster and more flexible than cultural evolution. In many aspects, conscious evolution can have its cake and eat it too.

9.2. Extreme Nature-Worship

Generally, "nature-worship" refers to the idea that the natural world is sacred, holy, or set apart as a kind of perfect eternal state. Based on this view, all things "natural" are best left as undisturbed as possible by human activities. In this section, we examine two forms of it, and I suggest that we should pursue nature-worship to an extreme that goes beyond what is currently considered acceptable.

As discussed in chapter 5, we live in an environment that was shaped by billions of years of natural evolution and thousands of years of cultural evolution—the plants, animals, tools, and social organizations have stood the test of time and are well adapted to Earth's conditions. Thus they deserve to be respected and protected. At the same time, we understand that the only permanent feature in this universe is change, and we must move forward to try new ideas and develop better solutions, and thus we cannot help but influence and change natural processes.

Although we often think of "nature" as encompassing only our outer environment, protection of the "natural state" can and often does extend to how we deal with ourselves and fellow humans. For example, many have asked whether human disease, suffering, and death should just be allowed to run their course, and have answered affirmatively, on the grounds that the natural process should not be interfered with. They follow the same principle as the ancient Chinese, who once considered cutting hair to be as bad as cutting off the head because both were given by nature and one's parents, and the tradition of the Catholic Church, which at one time opposed the introduction of anesthetics because their use would deny us our duty to suffer as imposed by God.

How we should treat conditions that we are born into can be a very emotional and contentious question. While the standards of what is acceptable are usually fixed and uniform at any one point in time, they can shift substantially over time. The issue of corpse dissection, for example, was once contentious in a way that we could not imagine today, and thus sparked what Hilary Rosner calls "the first great bioethics debate" some two thousand years ago.[11] While dissecting cadavers is now a critical component of every medical education, concerns over separation of body and soul made it difficult for medical research to obtain a steady supply of corpses throughout recent Western history.

Most environmentalists are not extremists and recognize the need for change and the tradeoffs between human progress and environmental purity. But still, the public debate about the environment tends to focus exclusively on a narrow aspect of nature, a range that covers more or less the time since the beginning of human civilization. It is a range much greater than our individual lifespan, but only an unimaginably thin slice of what has been going on for about 13 billion years.

This slice of nature is highly seductive. After all, it is this natural condition that shaped human physiology. Consider the following paragraph written by the novelist, poet, and essayist Wendell Berry, who follows the great American natural spiritualism tradition of Emerson and Thoreau and shares the ideal of the agrarian society promulgated by Thomas Jefferson more than two hundred years ago:

> I have been working this morning in front of a window where I have been at work on many mornings for thirty-seven years. Though I have been busy, today as

> always I have been aware of what has been happening beyond the window. The ground is whitened by patches of melting snow. The river, swollen with the runoff, is swift and muddy. I saw four wood ducks riding the current, apparently for fun. A great blue heron was fishing . . . A flock of crows has found something newsworthy in the cornfield across the river. The woodpeckers are at work, and so are the squirrels. Sometimes from this outlook I have seen wonders: deer swimming across, wild turkeys feeding, a pair of newly fledged owls, otters at play, a coyote taking a stroll, a hummingbird feeding her young, a peregrine falcon eating a snake. ... During the thirty-seven years I have been at work here, I have been thinking a good part of the time about how to protect it.[12]

What a beautiful landscape, a fountainhead for elevated thoughts and feelings! One that is natural yet exactly a mirror image of a pure, innocent, serene, meaningful, nurturing, and fun Utopian human society, with its happily engaged children, babies, parents, workers, farmers, and policeman. One that, based on Berry's long lifetime experience, has been there for so long and will last forever if we can stop some developer from bulldozing it.

Similarly, in a deep meditation about technology and nature, the philosopher Langdon Winner wants us to consider a picture of the gray whale juxtaposed against a nuclear reactor on the rugged California coast. Both are enormous features of the landscape, but the gray whale swims gracefully in a timeless ecosystem, offering an image of things as they have always been, while the reactor projects an image of scary and strange things as they are rapidly coming to be.[13] Again, this is a sympathetic portrayal of a great creature and landscape that few would want to destroy. But the timelessness of the whale and the ecosystem in which it thrives is only true within our limited experience. In the long view, the whale and its ecosystem came into existence relatively recently, and they will change and, ultimately, disappear—regardless of what we do or don't do. If we were around in past eons, we would have been equally attached to those environments, which no longer exist.

With our very limited lifespan, it is almost impossible for someone living in one paradigm to imagine that there will be a next one. The people of the Middle Ages didn't think of themselves as being in the "middle" of anything at all. As far as they were concerned, the way they were living was the way people would be living until the end of time.[14] But today, the scientific evidence of evolution makes it possible to glimpse fundamental changes that occurred unexpectedly for those alive at the time. Like Wendell Berry, I enjoy reading and writing at home from a breakfast-room window overlooking a pond, with flocks of wild geese passing through and squirrels running in tall oak trees. One of my favorite TV shows was *Wild America* on PBS, which depicted these types of scenes. While Mr. Berry's descriptions (and my similar experiences) stir my sense of awe, wonderment, and affection (biophilia—the love of nature and especially life—is a universal human predisposition), I know that in cosmic time, and even Earth time, this natural scene is only a passing moment and has hardly existed for a millisecond.

I have also learned that, given enough time, the entire landscape will surely change beyond recognition even without any human intervention. Recounting how the Sahara abruptly turned from a paradise for animals and human civilization into a desert in very recent history, Lee Silver concludes, "Mother Nature can be a nasty bitch."[15]

Of course, nature itself has been perceived very differently at different times and in different cultures. A traditional farmer, a modern urban dweller, an urban dweller prior to the advent of modern conveniences (like air-conditioning and plumbing), and a hunter-gatherer would have very different experiences of nature and their role in it. Even in the literature of antiquity, about forty different meanings of "nature" have been articulated, ranging from the Genesis birth to Hippocratica's body to the universe in its entirety.[16] The ardent naturalist and biologist Lyall Watson provides an interesting historical perspective:

> I was born in Africa and dream still of savannahs with scattered groups of trees near rocky outcrops which provide the sort of long, hazy views that seem to be part of the genetic memory of our species. But the truth is that both we and the savannahs are late-comers to Africa. Neither would exist at all were it not for the pre-existence of another keystone species, which has been plying its trade for some 50 million years, crafting Africa's landscape and shaping its ecology. This architect on a grand scale is, of course, the African elephant. . . .

> They need over 300 pounds of vegetable matter each day . . . killing many species faster than they can regenerate.[17]

Extreme nature-worshipping. For us, the best way to create new ("artificial") life forms and intelligence is to allow them to emerge through evolution, as nature does, only in much shorter timeframes and with much greater efficiency and variety. For conscious evolution to be successful, we need to observe and understand the *actual* workings of nature, in all its dynamism, to the point of profound reverence, or "worship." This involves everything from direct observation to a deep understanding of evolution and natural processes—thus it engages both our intellect and our deepest reflective mind. We need to grasp how the current conditions came into existence and what future conditions might look like—which we might liken to reading holy scriptures with limited eyesight—and then do our best to emulate the recognized natural ways in order to maximize the chance that the survival and prospering of life and intelligence will continue as it has for the past several billion years.

This is what I call extreme nature-worshipping. Like extreme sports, this kind of worshipping is very demanding (not physically, of course, but mentally). It won't come naturally for everyone, but for those who can handle it, it is extremely gratifying. Like extreme sports, it showcases the power of humans and their potential.

From the perspective of extreme nature-worshipping, the real significance of Darwin's evolutionary theory is not that humans descended from lower species, but that *we can continue to evolve*, and that some of our descendants will be "higher" in the sense that we are higher than *Homo erectus*, *Homo habilis*, *Homo ergaster*, or any other partially human species. Denying our past would purge our hope for the future.

As a species that shares a common ancestor with all forms of life on Earth, we are not the exception to the rule of mutability of species. What is different, however, is that posthuman evolution can be guided by conscious decisions in addition to natural forces. The more we recognize the true nature of ourselves and place it in the context of cosmic evolution, the more we realize that our species cannot be the end—and that "the end" need not be the end of our species.

9.3. Conscious Evolution Is a Wide-Open Game

The importance of variation. Evolution has three key steps: variation, selection, and replication. The conscious mind is very good at the last two. Actually, it tends to focus *only* on the last two. But the availability of variation is equally important, and to consciously maintain and foster variation is just as critical as the fostering of selection and replication. Maintaining high rates of variation requires sufficient freedom for innovation and novelty to take place, balanced with protocols that conserve new gains.

From about 2.5 million years ago until the extinction of Neanderthals about 28,000 years ago, there were always two or more hominid species. Then *Homo sapiens* became the sole survivor. During the ascent, it has had a strong hand in the termination of many other life forms, from its genocidal activities against competing races to its catastrophic impact on thousands of animal species that shared our habitat. Today, global integration aided by communication and transportation technologies presents the danger of killing off even more diversity, both in human society and in nonhuman species. When variation is stamped out in this way, it reduces the tolerance for failure and holds back risky experiments in evolution, since there is less room for error.

We should never forget the lesson of Ming China, when its centralized political system was able to shut off almost all oceanic activities with a single decree and cause the sudden death of an entire oceangoing enterprise. Thus, in one fell swoop, a vast treasure trove of technical know-how—involving shipbuilding, equipping, navigating, trading, and financing the Chinese treasure fleet—largely disappeared, as did international commerce and trade. In contrast with Europe, China was no longer open to variation in its development strategies.

The right approach of conscious evolution is suggested by the title of Freeman Dyson's book *Infinite in All Directions*, which allows both innovation and conservation of life and intelligence to manifest in as many ways as possible. Humanity's destiny is not the expansion of a single nation or of a single species, but the spreading out of life in all its multifarious forms from its confinement on the surface of our small planet to the freedom of a boundless universe.

The lesson of freedom is not just political or "geographical." When people have tried to replace freedom and diversity with detailed planning, the results have always been disappointing, from command-and-control economies to top-down strategic planning in large corporations to the planned cities of Brasilia, Chandigarh, and Canberra.

The whole idea behind centralized planning is to eliminate waste and to avoid mistakes. To that end, planning has been very successful. But we must be realistic about the limitations of planning. As Alan Turing observed, not making mistakes is not a requirement for intelligence: if a machine is expected to be infallible, it cannot also be intelligent. This is probably true for intelligence of any kind. If the desired result is to generate the truly novel and pathbreaking, perhaps there is no other way than through the process of exuberant overpopulation, maximum diversity, and a wide variety of selection mechanisms, all fostered by recognizing the central importance of variation.

A messy and sloppy business. This process of variation, selection, and replication is always messy—it is anything but streamlined and economical. Like entropy, it is a process that permeates all possibilities, the key difference being that there are more possibilities in certain directions.

Even today, the development of new technology is more dependent on trial and error than on research that is rationally guided by natural laws discovered by science. Contrary to popular belief, science did not contribute to technological development until the early nineteenth century. Galileo, for example, improved the telescope by trial and error, aided by his skill as an instrument maker rather than his understanding of optics. One of the pivotal events that moved the greatness of Rome to Paris and London during the Middle Ages, and later to Berlin and Moscow and New York, was the anonymous invention of hay: the idea, which almost certainly came from the field rather than the laboratory, of cutting grass in the fall and storing it in large enough quantities to keep horses and cows alive through the winter.

The best modern example of a totally unanticipated little invention that changed the world is the shipping container. Invented in 1956 by a self-made road-hauling magnate, it enabled cargo to be moved seamlessly and efficiently between ships, trucks, and trains all over the world and, over the next half-century, dramatically transformed the geography of global markets.

In both biological and cultural history, there are numerous cases where seemingly tiny and trivial events have led to significant changes in the development path. In so many such cases, natural selection is not the creation of a perfect fit to an ecological niche or the "survival of the fittest" in its more extreme interpretations—Darwin himself explicitly rejected "ultra-Darwinism" in his book *The Descent of Man*. Rather, it is a seemingly sloppy process that allows for the survival of the (merely) fit.

In line with the principle that "you don't have to outrun the bear, just your companion," a species is selected only for being slightly better than its direct competition. It is now well known that even within the same habitat, what is "more fit" is hard to define. The same is even true within the same species.

For seals, the battle for mating rights is ferocious, since roughly 90 percent of mating is done by 10 percent of the males. But despite appearances, brutal force is not the only way to spread the genes. While the large and aggressive dominant male controls a large territory, some males that resemble and behave like females successfully sneak in and copulate when the dominant male is not looking. For many other species (e.g., red deer and red-winged blackbirds), females have been observed to show a similar willingness to mate with less well-endowed males.[18] Even more amazing are the side-blotched lizards (*Uta stansburiana*) living in California. The males' mating strategies are like a rock-paper-scissors game without a single dominating type. These lizards have three throat colors, each representing a unique mating strategy. The orange-throated males are big and strong, which enables them to grab large territories and keep many females. The blue-throated males can hold only small territories but are able to guard them carefully. The yellow-throated males are small and weak, but they can mimic the females and sneak into the orange males' territory. When the orange-throated type is numerous, the yellow-throated males will produce a lot of offspring, which in turn allows the blue type to prosper. The cycle is played out every six years.

The prevalence of homosexual and nonprocreative heterosexual behavior among a vast array of animal species is a clear indication that animals need not be "single-minded" and efficient in order to survive and thrive.[19]

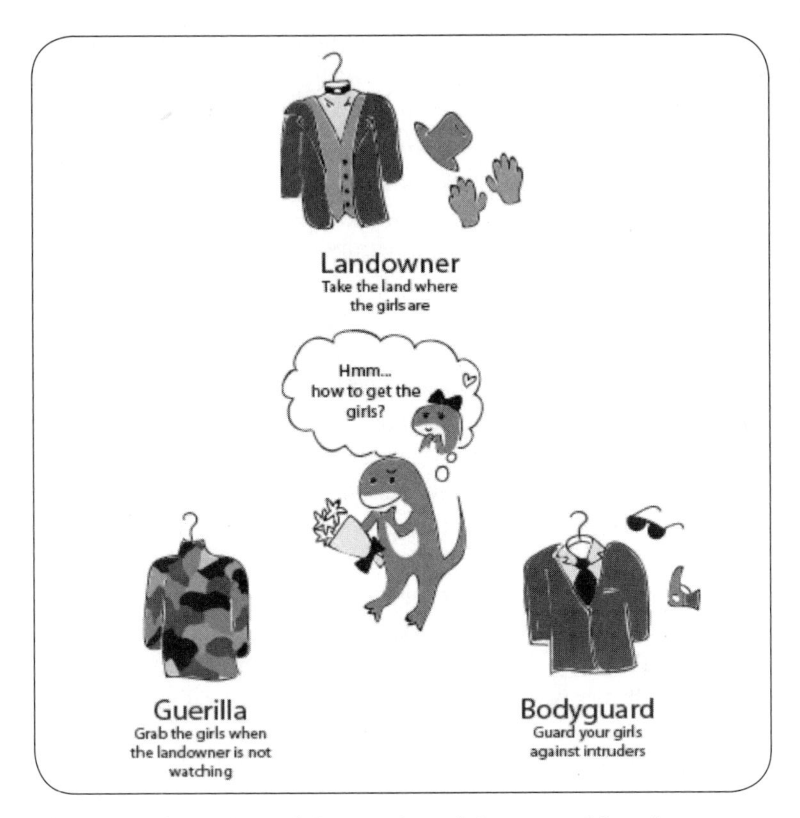

Figure 8: Three Sexual Strategies of Common Lizards

*Males of the side-blotched lizards species (*Uta stansburiana*) play a rock-paper-scissors game without a single dominating type.*

As stated earlier, this "sloppiness" has been true in cultural selection as well. A research survey conducted by the London School of Economics and McKinsey and Co. found that even in "ruthlessly competitive" markets, a surprising number of firms with inferior management practices continued to exist. The reason is not well understood, but it may have to do with the subtle niches these firms occupy and regulatory environments that prevent perfect competition.[20] Price competition as described in microeconomics textbooks is rarely the only form of competition in the marketplace. One obvious advantage of a "loose fit" is that it may allow some losers to hang on and emerge as winners when the environment changes.

In this sense, the wisdom of evolution is opportunism. The selection process itself need not follow a single pattern. Sexual selection, for example, can be said to be the "survival of the prettiest." It can favor a certain novelty on the basis of appearance only, long before the novelty shows any profitability in the struggle for survival. But occasionally the novelties result in spectacularly successful jumps in the evolution of complexity.

Ready, fire, aim. History is full of utterly unexpected twists and turns that shaped the course of events in very significant ways. Why would the future be any different? As ancient Greek wisdom had it, the most beautiful *kosmos* comes from things let loose by chance.[21] In fact, "progress" might be defined as *changes that nobody, not even the experts, anticipated.* It is the invention of new things, but more important things with previously inconceivable applications. The father of the modern computer, John von Neumann, did not anticipate the blossoming of personal computers and the computer-game industry. Not even the most vivid imagination could have conceived in the early twentieth century that quantum physics would contribute to human genome research through DNA microarray chips.

Rather than a "Ready, aim, fire" approach, in which we know where the target is and the method to get at it, the strategy we should follow is "Ready, fire, aim"—trying out different ideas rather than picking winners in advance. The only thing we know for certain is that we can try our ideas out regardless of their soundness. At the tactical level, we know that creating something truly new, even with our best thinking and simulation, will almost always result in a failure. So taking lots of shots at least increases the chance of hitting a target.

The "fire-aim" sequence is also the best way to deal with the ancient Greek question "Do you know what you are looking for?" There is a paradox here, which can be expressed by two apparently contradictory questions, both of which are self-evidently true: "If you already know what you are looking for, why look?" and "If you don't know what you are looking for, how do you know where to look?" There is usually no such thing as absolute certainty or predictability at micro levels, so the challenge is to make mistakes as fast as possible[22]—to "fail fast," as they say in the tech industry. The scientific approach to problem solving, according to John Wheeler, is to "ask anyone anything."[23]

If we do not aim, how can we get the direction right? The key insight of Darwinian evolution is that "purpose" or "direction" can be a residual fallout of this largely stochastic process. Randomness does not mean an organism or individual or organization is condemned to move without purpose. From the higher perspective, evolutionary trajectories driven by random mutations face clear constraints of natural laws and may even seem finite and "well-defined." However, this larger picture is often hidden from the individual's perspective. Planning and foresight with intelligent analysis provide the edge in competition, but they can also be overrated. In sports, great teams with strong discipline can lose to really bad teams because the bad teams err so much that they become unpredictable or because they are so desperate that they try the unthinkable. This is why the best practice in corporate strategic planning is not to preconceive well-defined strategies but to recognize their emergence and intervene when appropriate.[24]

Both nature and culture practice William Ernest Hocking's "negative pragmatism": That which does not work is not true.[25] More often than not, scientific theory follows successful practices. Theories are like roads—we build them where we have already beaten a path. Thermodynamics, for example, learned a lot more from working steam engines than the steam engine engineers ever learned from thermodynamics. It was the disagreements about the engine efficiency between Sadi Carnot and James Joule that produced the two laws of thermodynamics, long after steamboats were chugging up and down the Thames, Seine, and Hudson. Similarly, the financial markets had been running for a long time before Harry Markowitz, William Sharp, and Merton Miller built a deep and solid mathematical foundation for the analysis of investment decision making.

Only with the benefit of hindsight can we tell for sure which individuals, technologies, fads, pieces of art, literature, or start-up companies will succeed. Biographies and obituaries are rearview mirrors of significant people's lives—no journalists write about the birth of particular babies (unless they are children of celebrities), since there is no way of knowing which ones will be worth writing about. Arthur Schopenhauer once remarked, "Talent hits a target no one else can hit; Genius hits a target no one else can see."[26] The "weird" individuals are easy to recognize, but it is nearly impossible to tell the real geniuses, who are weird simply because they are ahead of their time, from those who merely stray from the norm or are just plain crazy. The record of identifying real geniuses is worse than business executives' feelings about advertising spending—about half of the money spent is wasted; the problem is that one cannot know which half.

Those who make history cannot be picked, but with a sizable population and proper environment, outstanding individuals are guaranteed to emerge, with the power to sway the path of history. The only proven way to raise masters or geniuses is to properly educate and challenge the largest possible number, thus ensuring that some potential geniuses will be in the mix. That is what was done, for example, during the Italian Renaissance, and in the great creative outburst of the sixteenth-century Momoyama period in Japan, and in the Low Countries at the time of Rembrandt and Rubens. These were periods when relatively sophisticated education was widely available to broad sections of the elites of these populations.

Surprising twists and turns. Despite its imperialistic reputation, the key strength of the Western tradition is liberty, which at its core is pluralism and individualism. The West succeeded because humanity's inherent energy was allowed, often *unintentionally*, to run wild.[27] Centralized control is the main theme of the Eastern political system, but without exception, the booming periods coincided with periods of encouraging initiatives from peasants and social elites.

What is true at the scales of the individual, the community, and civilization is certainly also true at the cosmic scale. Nobody has any idea about what will emerge next in cosmic evolution. Whether something new is on the pathfinding frontier or just another dead twig cannot be known ahead of time or from lower perspectives. Consciousness is not where the spring of creativity is.

Oftentimes, the unintended consequences or serendipitous side effects of our intentional actions lead to surprising breakthroughs. Medieval monks are credited with the invention of clocks and a variety of labor-saving devices that later contributed to the Industrial Revolution, but their original intention was merely to improve the efficiency and energy of their religious devotion. The anti-apartheid sanctions against South Africa in the second half of the twentieth century made it impossible for that country to import oil, so the necessity for liquid fuel made it the only country that developed coal-to-liquid technologies when the rest of the free world was content with conventional oil supplies.

In general, unreleased emotional power built up in our motivational system can produce seemingly wacky behavior through *redirection* and *displacement* of the original motivation, which may turn out to be useful. When animals fail to catch their enemy in a chase, sometimes they "work it off" by mock attacks on empty air or pieces of wood nearby. Frustrated humans tend to burn their energy on apparently irrelevant activities. The irrational impulsive behavior as a result of *functional change* is the origin of much of human creativity. Dissatisfaction drives humans to do what their instincts normally won't commit them to.[28] So excessive control of human behavior can hinder the release of a useful source of human creativity.

A wild and open field. To get maximum diversity in conscious evolution, the field of tinkering, discovering, and experimenting should be left wide open with as little restriction as possible—for example, it is not the job of just professional scientists and bioengineers to develop artificial life. We should also practice serendipity, which means the art of finding things one is not looking for and getting surprising findings from non-goal-directed activities.

The development of knowledge may depend on maintaining an influx of the naïve and the ignorant because accumulation of learning and experience (by the "experts") has a narrowing effect. Everyone among elite groups tends to use the same paradigms, which more or less guarantees adherence to prescribed patterns of thought and action. True breakthroughs are often accomplished by outsiders, such as Albert Einstein in science, Thomas Watson Sr. in business, and Abraham Lincoln in politics. Almost all the outstanding personalities of the Muslim renaissance of the eighth through twelfth centuries were non-Arabs (including Persians, Turks, Jews, Greeks, Berbers, and Spaniards) not yet encased by the new orthodoxy.

Maintaining a wide-open field is as much of a challenge in democratic societies as in autocratic ones. The tyranny of the majority may pose a serious risk to diversity, stifling the influence of emigrants, exiles, and refugees who carry novel ideas and perspectives. It is hard to imagine that any true innovation would ever be adopted if it first required a vote of the public. A legal system that protects free speech is an extraordinary accomplishment, but free speech includes freedom to express fear of anything new—and freedom to *not* express oneself. People often silence themselves because they defer to the crowd or powerful special interest groups, or they protect themselves against anything unfamiliar, risky, or uncertain (which almost all innovations are, at first).

Regardless of his personal motivation, Mao's slogan "Let a hundred flowers bloom and a hundred schools of thought contend" reflects the true spirit of evolution. Despite the risk of serious damage, we badly need single-minded religious zealots, political fanatics, and scientific cranks who can sustain direction even with repeated setbacks and depressing disarray and turmoil. But of course, all the mavericks and eccentrics need to be balanced with humanists, skeptics, and agnostics who can provide an environment of pluralism, tolerance, and flexibility that enables the natural spread of successful ideas through imitation without losing independent judgment. It is the messy, yin-yang environment that proves to be most fruitful and sustainable over the long term.

The power of diversity and creativity is always Janus-faced. There are many examples of failure. For example, Afghanistan for more than two thousand years has been exposed to great cultures and religions from virtually every major civilization (often through exposure to invading armies); yet it has produced neither great science nor great art of its own, but only the chronic suffering of its people on a massive scale.

In natural history as well as in economics, boom-and-bust cycles cause massive casualties, but sometimes they also characterize the most robust and creative phases of biological and cultural development. The bust portion of the cycle is an inevitable by-product of the boom, a necessary step in the natural selection process. About 65 percent of all species extinctions occurred during the Big Five mass extinction episodes. Culturally, the effective selection process usually kept only a single "flower" (or very few) alive after a hundred or so bloomed during the boom. Some of our best ideas, such as evolution, science, and democracy, suffered major setbacks and were on the verge of being totally discredited before a full revival. Thus, from a higher perspective, we should not be too concerned about losses and suffering. There is safety in numbers—that is, in permitting multiple failures without a total loss.

Accidents and chance events should be allowed to happen, since they are the essential catalyst for evolution of any kind. Counterintuitively, they also provide stability in complex systems by creating instability. A healthy heart that is flexible enough to respond to changing physical demands beats with a somewhat chaotic pattern rather than a perfect rhythm.[29] A market that allows destabilizing speculations may suffer from manias and crashes from time to time but, through a sort of "punctuated evolution," develops long-term "rationality." A successful nation-state requires a modest degree of turmoil and discontent to foster a healthy political dynamism.[30]

There is virtue in untidy rowdiness. A pluralistic ecosystem is usually one with stability and resiliency; a pluralistic human brain is highly flexible and intelligent; an interdisciplinary pluralistic research team has greater problem-solving power[31]; and pluralistic human societies tend to be ones with coherence and energy despite apparent chaos and disarray.

Diversity pays and will always pay. The diversification of life has been the most fascinating saga on Earth so far. Diversification of intelligent beings could be the next big wave. As we enter the era of conscious evolution and begin to create variations of CoBe, we may more than match the explosion of body designs that appeared in the Cambrian period.

9.4. A Greenhouse for Conscious Evolution

Now that we've seen what conscious evolution at the ground level should look like, the importance of conservation is clear, although conservation itself must be at the service of innovation.

Conservation and protection of the primitive. Darwin's theory of evolution has often been misinterpreted as representing forces of complexity and moral progress. But that's true only when you watch the surface, the frontier. In reality, to be alive is more often to be conservative and to retreat. Progress is not like climbing a ladder; it is more like accumulating a heap or expanding a circle. The upward moves get our attention, but they are actually exceedingly unusual; most historical changes are regressive in terms of morphological complexity.

For example, the animal, especially the evolution of its nervous system, is highlighted in the Cosmic View for the leading role it played. But the animal is not nature's only favored son, and intelligence is not the only competitive edge that nature offers. Animals constitute just one kingdom of life on Earth. Within the entire kingdoms of plants, fungi, and protists, there is no lineage that has shown an arrow of increasing ability to process information and become more intelligent. Even within the animal kingdom, only one among twenty-five to thirty major branches (or phyla)—the chordates—developed real intelligence.

Evolving a more complex genome, body structure, or even intelligence cannot be a sure sign of progress from the long-term global perspective. For example, some fish species have evolved four eyes (two to see above water, two below), and some reptiles have a third eye (pineal eye) on top of the skull. These innovations did not catch on for various reasons, while simpler versions continue to prosper. The "survival of the fit" is really a special case of "the survival of the stable." From the individual organism's perspective, changes are forced, and are considered as accidents and mistakes.

In general, low-level niches are more plentiful—and this is still the world of microbes. By far the most plentiful life-form on Earth consists of single-celled bacteria and archaea, not just by number but even by weight. They are also more stable and resilient. No life other than these simple creatures can carry out its life cycle in places where the prevailing temperature is greater than 160°F. While the overall trend is no doubt toward greater complexity, there is danger in chasing progress and advancement single-mindedly, only to end up on a limb. Major mass extinctions and collapses of complex civilizations showed us that complexity has its costs and fragilities. The most complex is usually the first to go when facing environmental change or new competition. In general, organisms can become too well adapted to a *particular* environment or too advanced in a single direction.

One important consideration for persistent growth in biological and cultural complexity is to build stable lower-level subsystems so that when the top level collapses, it does not collapse all the way down. It may be a combination of luck and life's inherent resiliency that none of the five major mass extinctions in history set life back to a Bacteria Age. And none of the collapses of major civilizations wiped out the Axial Age wisdom or even basic agricultural technologies entirely.[32] The U.S. stock market lost 90 percent of its value during the Great Depression, but the foundation of the political economy survived, and the market returned to its 1929 pre-crisis peak in 1954.

In comparison, the disappearance of the Mycenaean palace-based "original" Greek culture around 1200 BCE was a horrendous collapse. The surviving people fell to subsistence farming and became illiterate for four centuries. The fall of the Roman Empire led to the long Dark Ages, during which time much of Greco-Roman technology and culture was completely forgotten in Europe—for example, the use of legumes in crop rotation to restore the soil; the mining and smelting of iron and manufacturing of iron tools; and the harvesting of honey from hollow-tile hives doubling as garden fences. Western Europe in the ninth century was a source of "eunuchs, slave girls and boys, brocade, beaver skins, glue, sables, and swords" and not much else.[33] In the eyes of Muslim travelers at the time, the Christian world was like today's Chad or Somalia.

To handle the inevitable booms and busts in the creation process, conscious evolution must work on conservation and protection of diversity. Complex systems, whether they are molecules, biological organisms, technologies, or social organizations, are inherently less stable and more fragile than simpler ones.

Protecting and nurturing creativity. We need a "greenhouse" for conscious evolution. But protection is not just about securing the survival of existing lives, organizations, and any other entities—the richness and liveliness of diversity make it unnecessary to take any *specific* form of life too seriously. Instead, the most important aspect of the evolutionary greenhouse is protecting and nurturing creativity and novelty. And variation is the essential first step in doing this.

The first order of business, in terms of protecting the fountain of creativity, is *the distribution of real estate*. In *On the Origin of Species,* Charles Darwin pointed to the role that land and sea barriers played in isolating populations, which over time led to the diversification of species. Human traveling and shipping around the world have broken down this natural isolation in many locations, and invading foreign species contributed to the current wave of mass extinctions, the sixth in history. Geographical isolation played the same role in promoting cultural diversification. As *Homo sapiens* spread out of Africa 150,000 years ago, these wandering groups became isolated from each other and gradually adapted to various local environments. This phenomenon of far-flung local adaptation was critical for the cultural evolution that eventually led to the rise of the great civilizations.

Although the hard geographical barrier is not the only isolation mechanism for all speciation,[34] some differentiation of environment is required. Formation of new cultures also requires environmental differentiation. If Africa were an isolated continent with a homogeneous and stable natural environment, cultural evolution could have been very limited. Without some barriers, societies that are powerful—with better technologies, more effective social arrangements, greater population densities, more energetic traditions, etc.—will either swamp their neighbors or force their neighbors to copy them in order to keep up and maintain their autonomy. Already, rapid advances in communication and transportation in modern times have speeded up cultural transmission, as people have a natural tendency to imitate each other. Testing of various ideas certainly requires isolation. Without the Iron Curtain, the social experiment of the two Germanys and two Koreas would have been aborted early.

The globalization process is not building a greenhouse for conscious evolution in the sense that large populations within a single society tend to become more protective of the status quo and resistant to change. This appears to be true from the institutional standpoint. The rising power of social groups and vested interests that don't have the broader evolutionary perspective at heart often makes mature societies intolerant of radical reform and bold social experiments, especially in liberal democratic societies.[35] In contrast, relatively isolated peripheral groups without entrenched special interests often exhibit greater flexibility and courage for the untested.

Ernst Mayr referred to the ability of isolated founder populations to undergo profound genetic restructuring as *speciational evolution*.[36] In the same way that variation flourishes for isolated groups in the biosphere, we need to see pioneer human populations operating and innovating in relative political isolation. Cultural evolution needs to give room for heretical ideas to escape the tyranny of the majority, the way Buddhism managed to prosper in the Far East after its dismissal in India, and the way Christianity spread to the world after being rejected by mainstream Judaism. The colonization of the New World was pioneered by those who wished to create a utopian society in the wilderness free from the persecutions they experienced in Europe.

In this light, we must carefully evaluate the pros and cons of a single global liberal democratic super-state. We may need not one Supreme Court but many independent Supreme Courts to come to different verdicts on the controversial issues raised by the advancements of technology, and we may need many independent political systems to enforce different regulations with regard to unconventional ideas and ideologies.

It is unnecessary and impossible to bring everybody to the cutting edge of conscious evolution. Whether the innovation consists of the electric light bulb,[37] high-speed travel on a train, or test-tube babies, history has shown time and again that the key to resolving many controversial issues and uncertainties does not lie in the endless heated debate characteristic of a consensus-seeking system, but in the experiments carried out by fearless and faithful individuals or small groups of individuals.

One day intelligent beings will spread into galaxies and far-flung communities that can be separated by hard physical limits. But during the transition period, when all experiments are conducted on Earth or "within earshot," we will likely face an extremely fragile phase of cosmic evolution. It is becoming nearly impossible to isolate the damaging impact of violent extremists on one hand and to escape the tyranny of the majority on the other.

However, isolation for the sake of diversity has a price: after all, genetic and cultural recombinations (through symbiosis, sex, cross-pollination, or other mechanisms) usually operate much faster than random mutations or natural "drifts" that occur with populations in isolation. DNA evolves not only with random mutation at the base level, but also through crossover, with chunks of genetic code swapped—a sort of mega-mutation similar to the wholesale importation of culture from more developed civilizations.

The advantage of access when evolution is occurring in multiple places simultaneously is that one need not build everything new from scratch. In fact, the most significant inherited variation came from symbiosis after life languished at the bacterial stage for about one billion years. The merger of fungi and algae produced a great variety of lichens with surprising characteristics found in neither fungi nor algae.[38] The root growth of the first forests benefited from the alliance between fungi and plants.[39] Likewise, cross-pollination and stimulation have been equally critical for cultural evolution. In this singular case, the contact and DNA crossover between previously isolated populations was critical to evolutionary progress.

Permanent, complete isolation reduces or even destroys the chance for learning and stimulation, which can be a fatal outcome. An elaborate web of interlocking cycles in a cell, an ecosystem, and a market economy all rely on interconnected and interdependent diversity to ensure error tolerance and resiliency. Furthermore, specialization by splitting out is not the only means of diversification. From the symbiosis of the first eukaryotic cell to the human imagination, mixing up has always fueled creativity and complexity of all sorts through positive feedback loops and non-zero-sum games, by which I mean both sides benefit from interaction and cooperation.

The second order of business in terms of protecting the fountain of creativity is *granting privileges to the talented and the motivated.* Let those who are willing and capable of going the extra mile or traveling off the beaten path be themselves. For example, schools serve indispensable functions of educating and training the younger generation, but they are not suitable for all. Strict conformity and standard procedures may be appropriate for most students but stifle the originality of creative minds by suppressing variety, surprise, chance, and excitement. It is unlikely that Ansel Adams would have become one of the greatest photographers in history had his father not taken his energetic and "hyperactive" son out of school.

The third order of business for fostering creativity is what might be called the *controlled allowance of behaviors that might be inappropriate or dangerous under certain circumstances*—for example, drinking alcohol is permitted, but driving while drunk is not. The mind is subtle and powerful because it is at once "occupied" by the Greek gods of Apollo (the god of tranquil calm and conservatism) and Dionysus (the god of ecstasy, abandon, and creative destruction). The liberal democratic capitalist system is dynamic in the same sense: it indulges (and stimulates) human desires of *all* sorts (including those emotions that traditional religions, cultures, and societies frown upon, such as greed and envy), even as it is regulates them by a clear set of rules (especially restrictions laid down by law) and ruthless reality checks.

It is in this sense that conscious evolution cannot be sternly unidirectional. Evolution gives everyone elbow room to do the things near to his or her heart, a chance to show what can be done on one's own. It is a process of wild exuberance and extravagance without hard caps on efficiency or waste.[40] Sexuality, the key mechanism for biological exuberance, was itself initiated as a kind of auxiliary function of reproduction, a superfluous gadget for a bacterium.[41]

The fourth order of business is what might be referred to as *the occasional "pardon" of the less fit.* Paradoxically, one crucial piece of progressive evolution (increase of complexity) is what can be called randomized relaxation of selective pressure—occasions in which defective individuals can linger alongside fitter neighbors. The following example illustrates this:

The emergence of eukaryotes could be traced to the odd situation where mildly deleterious genetic mutations in the proto-eukaryotes were "allowed" to survive in an environment of smaller populations.[42] Eukaryotes are big, compared with prokaryotes. Their genes are segmented into exons, which are interspersed with functionless segments of "junk" DNA known as introns. The eukaryote cells have to delete the introns when they copy a gene for use in building a protein; the failure to edit out the introns could result in a defective protein. Had selection pressure been strong in proto-eukaryotes, all introns would have been eliminated. But instead, in the somewhat loose environment of a large eukaryote cell, introns and duplicate exons lingered. The cell managed to cope by developing an enhanced ability to edit. As the editing was applied to exons as well, eukaryote cells ended up with the capacity to make different proteins from the same gene. It is this flexibility that has given eukaryotes great versatility to build different kinds of cells, tissues, and organs with a limited number of genes. As a result, we can say that our existence can be traced to the accidental introns.

So, take a deep breath and relax: there is nothing wrong, or to be ashamed of, in pleasing as many cravings in your head as possible, even those that may seem regressive or less strictly useful, healthy, or adaptive. The political genius of the American Founding Fathers is reflected in the phrase *the pursuit of happiness*. Without articulating what happiness means, these words confer the freedom for people to choose different ends for their efforts. This flexibility and open-endedness allow the maximum expression of human aspirations, which also leave room for the Cosmic Vision to emerge.

Loftier goals can emerge from lower motivations under the appropriate social environment. Reading the succinct and self-confident United States Constitution, one can hardly believe that it was a creation of bickering and compromises, a product of both deep political wisdom and narrow self-interests—but it was.

People behave like children. Human beings have a specific nature that drives their actions, just as a cat must hunt and a fish must swim. Yet, human nature is multifaceted, and its expressions are dependent on interactions with the environment. We should have the confidence that we are the only species on Earth that can rebel against the tyranny of the instincts, but we should also understand that the most effective means of rebellion is to allow the selfish genes and prevalent memes to have their day. The world's greatest leaders and managers have arrived at the same observation: people don't change that much—they tend to uniformly pursue self-interest or some singular passion that drives them. Don't waste time trying to put in what was left out. Try to draw out what was left in—that is hard enough.[43]

If someone wants to spend a fortune on cloning a pet or creating a unicorn, the person should not have to listen to a lecture on how the money would be better spent saving starving children in Africa. The Eastern wisdom of *wu wei* (无为)—which values the feminine fertile state of *presence* and *being*, or the concept of getting things done by being receptive and without intention—teaches that conscious evolution will be best served if it moves with human motivations rather than against them.

Like kids, all people need pressure as well as freedom to move. When small groups of people moved to isolated tropical paradise islands, they lived so effortlessly that they did not bother to keep previously cherished technologies and culture. Fortunately, thus far survival and social pressures have always been there for the vast majority of humanity. We noted earlier that the First Law of Biology is the idea that nature is lazy; however, it is also true that we humans always feel that what we want is just slightly beyond our reach and that we have a chance to get it if we try. Yet these innate drives can be suppressed under certain circumstances. One characteristic of many societies, communities, and industries that have stagnated or declined has been the provision of comprehensive and often excessive benefits to individuals (from common workers to CEOs), resulting in disincentives, the stagnation and paralysis of creativity, and reduced satisfaction and meaning in one's own life.

One totally mistaken prediction one hundred years ago was the coming age of leisure.[44] The prominent economist John Maynard Keynes predicted that the human race's economic problems might be solved within a century, assuming 2 percent average growth of real per capita income, and that people would give up the austere virtues of the traditional work ethic and start to enjoy living for the moment.[45] Yet today, people are busier than ever. At the individual level, Keynes's prediction did not materialize, partly because he underestimated the importance of the fact that the human is a social animal that places a high value on status and other measures of relative well-being. Status is a permanently scarce resource. The insatiable urge for relative well-being keeps people on the "hedonic treadmill" long after their basic needs are met. Baker's Law of Economics states that one never wants what one already has or what one can afford. In its encouragement of "arms races," the capitalist system mimics the ecosystem.

In sum, the pressure is on, even for the wealthiest, and it is good. Maybe Nietzsche was too harsh about the satisfied bourgeois and utilitarian "last man." We should dread the possibility of what Gunther Stent called the "new Polynesia" in *The Coming of the Golden Age:* an endless life of social and intellectual stasis and universal leisure being replicated with each generation, and being centered on sensual pleasures and a satisfactory social life. Such a life does not necessarily imply idleness, but the theater of struggle and challenges (as well as recreational opportunities) is largely self-referring—perhaps personally interesting, but without larger meaning or significance. In this possible future scenario, we will carry on this engagement in a childlike fashion, like the young shepherd we mentioned in the last chapter whose lifetime objective is to have his son be another shepherd, until some event makes us realize it is too late to move beyond the human.

Leaving room. One of the crowning achievements of the Enlightenment is the principle of tolerance, especially of moral and spiritual diversity. This might even extend to the extreme pronouncement made by John the Savage in Aldous Huxley's *Brave New World*, that people should be allowed even to demand "the right to grow old and ugly and impotent; the right to have syphilis and cancer; the right to have too little to eat; the right to be lousy; the right to live in constant apprehension of what may happen tomorrow; the right to catch typhoid; the right to be tortured by unspeakable pains of every kind."

Primitive pastoral life is hard, but not impossible, if given enough room to roam. Modern technologies are not necessities for human life. The isolated state of the self-sufficient peasant life has been cherished not just by illiterate farmers but also by thoughtful intellectuals such as Thomas Jefferson, Mohandas Gandhi, and E. F. Schumacher, who all regarded it to be sustaining and satisfying. Leon Kass has a right to demand a "right to ignorance" and freedom from "the benevolent tyranny of expertise."[46] Following a rich tradition of early American "come-outers," Daniel Quinn has a right to urge teenagers to move "beyond civilization" (the title of his book; he was actually suggesting that we sink beneath it) to what he called the "Tribe of Crow"—the idea that within a small community, people can lead an aimless life in relative poverty but without the stress to "get ahead" or "make something."[47]

There is no need to expunge the argument that the primitive lifestyle in these communities represents an ideal way of human life—we are confident that few would choose it after experiencing what modernity has to offer. There is also no need to wipe out living-fossil societies such as the Hottentot tribes in Africa, the tiny Hunza Kingdom in the extreme north of Pakistan, the Semai tribal people in West Malaysia, or the tribes in the Brazilian and Colombian forests who are so isolated that they have never heard of their country.[48]

We should also be relaxed about the possibility that some social and technological advances will be reversed. Throughout history, there have always been attempts to get rid of higher consciousness and higher perspectives. As the great English poet, essayist, and lexicographer Samuel Johnson quipped, "He who makes a beast of himself gets rid of the pain of being a man." In a similar sense, the appeal of the Eastern realistic view is its self-perfection and existential satisfaction, a view that focuses on wisdom rather than on knowledge; on self-control rather than on power; on excellence rather than on success. In every civilization, including our own,[49] there has always been a romantic dream of primitive barbarian nostalgia. The back-to-the-land communalist movements have emerged again and again despite the fact that few of these experiments have been able to sustain themselves, let alone spread and grow.

Leaving room means not only tolerance for diversity but also protection of cultural isolation. We've noted that the enthusiasm for further integration of global civilization is dangerous, as it wipes out so much diversity.

What is more, we should never lose sight of the fact that protecting diversity is the means, not the end. And given limited resources at any point, the selection process must always follow the diversification process to give opportunity for the winners to prosper. In history, the elimination process gradually trims down diversity after periods of radical, explosive change. The number of basic biological body types (phyla) in the Cambrian period was far greater than it is now. The number of languages and religions was also far greater in ancient times. By the same token, many kinds of new life and intelligence we create will most likely be trimmed back at some point; of course this selection process may be emotionally taxing and far from error-free.

While all the above propositions may be true, let's still keep our perspective: we should not accept propositions we find laughable just because some great ideas were once laughed at; nor, for example, do we want to label as art every pretense at art just because in the past we failed to recognize some masterpieces. We have suggested that the ultimate selection criteria should be in the service of cosmic evolution; this requirement should be placed above the myths and value systems that were created before the emergence of the Cosmic Vision. After getting ready and firing the initial random shots, we must aim, and aim high for the next round.

9.5. Get Ready for Failures, and Lots of Them

Allow failures. In mutation (diversification) and selection (elimination through competition, the struggle for survival, and other mechanisms), conscious evolution follows the same principle as natural evolution. Evolution can never be safe and uneventful. The paths of both natural and cultural history are littered with periodic explosions and implosions. The natural way is to allow accidents to happen, and sometimes they can even be fatal to the participants. A cancerous cell can kill a person eventually; a virulent virus attacking key species can destroy an entire ecosystem. Nature can afford it because of independent multiplicity.

Thus, it is essential for understanding the evolutionary process that we think in terms of whole populations—and that we think statistically. Unfortunately, these kinds of thinking are at odds with our intuitive sense, since they require a perspective higher than the personal one. As self-centric beings, we find it cold (and even *mean*) to consider the fact that the fate of the specific (whether it is the individual atom, cell, person, firm, civilization, or, in the future, planet Earth) has never been important in the grand scheme of things.

Life and mind, and maybe all complex systems and communications networks, depend on safety in numbers. "An organic being is a microcosm," observed Darwin in 1868, "a little universe, formed of a host of self-propagating organisms, inconceivably minute, and numerous as the stars."[50] Population dynamics are paramount for natural selection, which is a process akin to playing the game Go: winning the game is all about the overall situation (*da shi*, 大势) made possible by the individual pieces, but it has nothing to do with the individual pieces themselves.

Mutations of genes are the spontaneous generators of diversity in biology, just as curiosity and play are the generators of spontaneous changes in cultural evolution. In both processes, great explosions of new entities must entail great casualties. Failure is the chief experience in biology and in culture. This evokes for me a quotation by Winston Churchill: "Success is the ability to go from one failure to another with no loss of enthusiasm."

Almost all natural "experiments" fail. Only a single gene survived unchanged in the last billion years—the one that codes for the protein Histone H4.[51] To a first approximation, *all* species are destined to become extinct—out of roughly thirty billion species, only thirty million survive today, a 99.9 percent failure rate—and this does not consider all failed mutations at the *individual* organism's level during the development of new species.

This failure rate is comparable to that of cultural evolution. Most fads eventually fade away. Most new styles eventually go out of style. Most new religious sects and cults fail to last beyond the founder's life. Most start-up companies either go bankrupt or are bought out by others.

In fact, the frequency and percentage of extinction of companies between 1912 and 1995 closely resembles those of biological species in the last 550 million years.[52] Within the first forty years of the S&P 500 index of the largest U.S. companies, more than 85 percent of the companies (426) on the original list dropped off the list. They either failed to grow or disappeared entirely (went bankrupt or were swallowed by other companies). The casualty rates of *small* companies are much higher—about 80 percent disappear within the first three years.

Few, if any, political entities have survived for a long period of time. For all of human history, no dominant state has managed to maintain its position over the long haul. The classical ancient Greek political culture that first originated democracy was a lucky survivor among the numerous ancient societies, most of which vanished without a trace. And during the rise of the Greek polis, there were more than two hundred city states (*polis*) at the height of political vitality that eventually perished. Just two survive in popular memory today: Athens and Sparta.

But these high mortality rates are not necessarily a bad thing. For example, from the perspective of the market, there is not much virtue in the longevity of a firm itself. The purpose of a firm is to do something remarkable,[53] not merely to survive. It is not worth spending a fortune just to keep a firm alive.[54] In fact, the greatest shortcoming of Japan and the emerging Asian economies has been their reluctance to let failing firms fail. Although far from an absolute rule, keeping too many outdated businesses alive can be destructive of the progress of a country's economy and the well-being of its people. As Joseph Berliner pointed out, it is not the invisible hand but rather the invisible foot of competition—the foot that kicks out inferior firms and business practices—that is the greatest source of innovation and thus of progress and economic growth.[55]

There is no reason to believe that things will be different in the future. The churn rates of human organizations, whether they are firms, associations, churches, or political states, largely depend on the competitive environment and frequency of external shocks, regardless of how hard people are making conscious efforts to prolong their organizations. Indeed, the greatest social and economic advances always come during the periods of greatest disruption and turmoil. However, we are in the process of modifying the "natural variation" part of natural selection, although, as I have emphasized repeatedly, we must always keep in mind that the precise path of evolution can never be mapped out in advance, let alone predicted. Today's science and technology are developed with the conscious mind and increasingly aided by computer simulations of reality, yet most breakthroughs follow the same pattern of Françcois Jacob's *principle of tinkering,* or poking in the dark.[56]

Let the winner rule. Conscious evolution has to allow the winner to emerge, while keeping natural barriers in place to maintain diversity—a tricky balancing act, without a doubt.

As with natural and cultural evolution, the path of conscious evolution will necessarily be a "wasteful" process with many dead ends that cannot be anticipated. Lawrence Krauss quoted the biblical words "Many are called, but few are chosen" to describe the incredible odds against the survival of each of the subatomic particles that initially emerged (1 in 1,000,000,000) in his book *Atom,* but odds have been just as incredible with all evolutionary events. We have already discussed the Cambrian explosion and the explosion of human wisdom during the Axial Age as odds-defying milestones. The following much more recent (and better documented) technological leaps round out the picture:

- Shortly after the first automobile was invented at the end of the nineteenth century, hundreds of automobile companies offered a bewildering array of products of all shapes and sizes. Cars were running in the streets powered by electric, steam, and gasoline engines. Nobody could predict which car company and what type of propulsion system would dominate the automobile industry decades later. Eventually, a handful of automakers who relied on the gasoline/diesel internal combustion engine proved to be the winners. Other types of power trains disappeared from the market (but luckily, not from our collective memory).

- In the 1920s and 1930s, hundreds of inventors in small start-up companies were building and selling airplanes to intrepid pilots and fledging airlines. The number of different varieties of airplanes that appeared during this period was estimated to be 100,000! Many of the pilots crashed and most of the airlines went bankrupt. Out of 100,000 types of airplanes, only about 100 survived to form the basis of modern aviation. Largely because of this rigorous and "natural"[57] selection process, today's airplanes are astonishingly reliable, economical, and safe.[58]
- The invention of the Internet and the World Wide Web triggered the dot-com boom at the end of the twentieth century. In the gold rush of the virtual world, a wide variety of "business models" were dreamed up and tried by tireless young people with the support of venture capital. Within a few years, a ruthless elimination process forced by the NASDAQ stock market crash quickly sorted out what works in e-commerce and what does not. Those companies that quickly ran through the remaining cash from their IPOs burned themselves out.

Again, if records had not been kept, future historians could infer divine interventions for the miraculous emergence of near-perfect transportation and information technologies. In all these cases, the market system encouraged and funded speculative ideas with even the slimmest chance of success because the potential payoffs were believed to be enormous. Each individual person or company in the race has high confidence in their final success, although from the social and market perspective they are irrational: a vast majority will fail, just like genetic mutations in biological evolution.

As Eric Hoffer observed more than a half century ago, it is the unique glory of the human species that many rejected individuals do not fall by the wayside but become the building blocks of the new, and that those who cannot fit into the present often become the shapers of the future.[59] The key for this process to succeed is that nobody is allowed to pick winners ahead of time.

Get ready for failures. The first step in the creation of new technologies with conscious guidance is called *design.* While the aim is to minimize failure, we cannot avoid it. In his classic study, *To Design Is Human,* Henry Petroski wrote,

> Though ours is an age of high technology, the essence of what engineering is and what engineers do is not common knowledge . . . I believe that the concept of failure . . . is central to understanding engineering, for engineering design has as its first and foremost objective the obviation of failure. Thus the colossal disasters that do occur are ultimately failures of design, but the lessons learned from those disasters can do more to advance engineering knowledge than all the successful machines and structures in the world.[60]

First published in 1982, Petroski's fascinating case studies concentrated on mechanical and structural failures, but his insight really applies to all human design endeavors, including future technologies. Although all evolutionary processes are wasteful and cruel to the most stupefying degree, we are unlikely to find a better method of creating the new. Success is grand, but failure can be more insightful. Falling down is part of growing up; it is an inevitable consequence of a risky and uncertain "free" world.

If we are able to recognize the inevitable wastefulness of any evolutionary process in a world dominated by free will and randomness, then we will be better able to adopt a higher perspective on socially sensitive issues such as the human misery that may chance to accompany unexpected or passing events. At the local level and within a shorter time frame, the path is usually characterized by a pattern marked by periods of relative stability punctuated by intense bursts of change and turbulence—plus the transitional suffering of groups affected by contingent events.

Seen in this light, what is often missing in studies of the future is the basic understanding that evolution almost necessarily creates crises and reversals.[61] Progress is only demonstrable when taken as a whole—as the full summation across all the individual actions and events during any period of time, a summation that is orders of magnitude greater in scope than any given individual efforts and actions themselves. Cherry-picking a few points in a messy history littered with relapses to show a smooth progression path can be delusional and misleading. Take, for example, the table of human life expectancy in a book promoting the transhuman future:[62]

Life Expectancy (years)	
Cro-Magnon Era	18
Ancient Egypt	25
1400 Europe	30
1800 Europe and USA	37
1900 USA	48
2002 USA	78

This chart seems to portray a nonstop steady progression, but we all know that major lags or reversals in life expectancy have occurred over these millennia in almost every generation, and certainly in particular localities. As Wendell Berry rightly suggested, to naively trust "progress" or our putative "genius" to solve all the problems that we cause is worse than bad science; it is bad religion.[63] *Any major evolutionary change can be viewed as a failure and setback from some local perspective or smaller time frame.* This includes, for example, biologically, the mass extinctions; culturally, the bloody and brutal Western colonization and exploitation of the world following Columbus's exploration; financially, stock market crashes and currency crises; economically and socially, the initial deterioration of economic well-being and tearing up of the social fabric in the early stages of agricultural and industrial societies.

There is nothing wrong with feeling morally outraged by today's setbacks and the resultant human misery, but ignorance of what it took to get us here can easily lead one to the unwarranted conclusion that there was little or no "collateral damage" during the evolutionary past. We will continue to make things better with our best efforts, but it is different to say we want to avoid all potential setbacks and moral outrages regardless of what is going on in the larger scheme of things.

Unit of selection. The early stages of the posthuman world could be even crazier and messier than previous eras, but the logic is the same. Now largely dictated by technology, the proper unit of selection will continue to evolve, and that unit may have to include the individual person. ("Unit of selection" refers to units that are subject to the variation and selection process.) Up until Old Testament times, mass killing was a normal part of the selection process among competing tribes, clans, and families. Much more recently, two brutal world wars in the twentieth century made it clear that mass killing is no longer acceptable. Then the invention of the atomic bomb made large-scale wars unacceptable. The unit of selection had to be lifted from person to social organization—we must change from eliminating an organization by wiping out its people, to eliminating an organization in a peaceful way (via rule of law) so that it would not hurt people. That is exactly what has happened, as we discussed above.

Even at an organizational level, the selection process is emotionally taxing—people develop emotional attachments to their firms, villages, and communities and do not want these organizations to be eliminated or replaced. The additional impediments for selection at the personal level, impediments that we shall address in the next two chapters, are the emotional and moral aspects of life: we are no longer talking about dumping just machines or abstract organizations.

In summary, natural evolution is eventful, chaotic, rough, unclean, unbalanced—it is a storm of "creative destruction," not a walk in the park. Conscious evolution will be full of unintended, unpredictable consequences as well, including calamities as well as magnificent, seemingly "miraculous" advances. Conscious evolution must follow the same strategy as natural evolution, only with more intensity and with the aid of Cosmic Vision. Going slow may be tactically correct at times but not strategically sound. To take new initiatives is to risk disaster, but to stand still is to render disaster inevitable.

CHAPTER TEN

RISKS AND FEARS

To every man is given the key to the gates of heaven.
The same key opens the gates of hell.

—Overheard by Richard Feynman while visiting a Buddhist temple in Hawaii

We are as gods and we might as well get good at it.

—*Whole Earth Catalog,* 1968

IN THE LAST chapter, we discussed the differences and similarities between natural evolution, cultural evolution (an early stage of conscious evolution with human nature as its anchor), and future conscious evolution (a time when all things are subject to variation and selection except for natural laws). Like natural and cultural evolution, conscious evolution will still be unpredictable and volatile, but its power will be much greater because new intelligence and new motivations will be used to generate even more advanced intelligence and motivations.

While many readers may believe the Cosmic Vision makes perfect sense, they may be very worried about grave risks in the posthuman world, when enhanced humans or fully autonomous robots and cyborgs will emerge amid "normal" human beings during the transition period, perhaps raising existential risks for humanity or even for all life on Earth. In this chapter we address the most common concerns, starting with the reasons for fear, why fear can be helpful, and how we should deal with various risks in an uncertain future.

10.1. What Fear Is For

The value of fear. Fear is adaptive in the natural world, where dangers are so omnipresent that chronically afraid, hypersensitive individuals have a better chance of survival than the less fearful.[1] Stress is the pump that raises the selective pressure and stimulates proliferation of genetic and cultural mutations. Monopoly and hegemony usually bring stasis. Without fear and tension, stagnation usually sets in for both biological and cultural evolution. The path of least resistance for evolution is downward, not upward. The majority of species have degenerated and become extinct, or, what is perhaps worse, gradually lost many of their functions. For example, the ancestors of oysters and barnacles had heads, snakes have lost their limbs, and ostriches and penguins have lost their power of flight. Humanity may just as easily lose its intelligence.[2]

Indeed, peoples that moved to some isolated small tropical islands in the Pacific Ocean lived so comfortably that they lost both technical skills and spiritual aspirations. Andy Grove, former head of Intel, summed up the value of fear in *Only the Paranoid Survive*:

> The most important role of managers is to create an environment in which people are passionately dedicated to winning in the marketplace. Fear plays a major role in creating and maintaining such passion. Fear of competition, fear of bankruptcy, fear of being wrong and fear of losing can all be powerful motivators.[3]

As a highly successful CEO, Dr. Grove experienced an intense fear of these factors while living in his native Hungary before moving to the U.S. His personal experience may help us to understand how that small country produced such an exceptional crop of scientific geniuses between two world wars, among them John von Neumann, Michael Polanyi, Leo Szilard, Edward Teller, and Eugene Wigner. As we discussed in chapter 8, happiness is not our end but a great motivator for achieving our goals. But fear is also a great motivator, maybe even greater than the attraction of happiness. The great pressure brought on by existential fear produced a disproportionately large number of outstanding individuals. Likewise, fear of the posthuman future will create many of the geniuses of the future.

Although William James and other scholars have argued that the ascendancy of man from beast created a living condition that requires less fear, the opposite may be true, especially during periods of drastic change. As the ethnologist Irenäus Eibl-Eibesfeldt noted, the human is perhaps one of the most fearful creatures, since added to the basic fear of predators and hostile conspecifics come intellectually-based existential fears. Existential philosophers found dread, angst, and anguish at the core of our unique human existence. One can find evidence of fear lurking in the background of many kinds of emotions that on the surface might seem to be the antithesis of fear. Courage is the ability to overcome fear. In the posthuman world of advanced artificial life forms, we certainly will need more courage when existential fear inevitably increases in the face of uncertainties and perceived threats.

The fear bias. Like all biological organisms, we tend to gravitate toward fear when it comes to handling the tension between continuity and change. Fear can be overcome by many other motives—otherwise the virtue of courage and our admiration for heroism would not exist—and almost all human cultures have implored people to overcome fear. But because of its survival value, fear remains the most electric emotion: it concentrates our attention, sharpens our perceptions, and quickens our reactions as no other emotion can. The following facts are illustrative:

- Fear always remains. As Joseph Conrad wrote, "A man may destroy everything within himself, love and hate and belief, and even doubt, but as long as he clings to life, he cannot destroy fear."[4] Fear arises from the neural networks that link the amygdala with the orbitomedial frontal cortex, the anterior cingulated cortex, the central gray matter, and the hypothalamus.[5] It is very difficult to silence.
- Fear is a defense mechanism as automatic, biological, and enduring as can be. Its evolutionary root is escape from pain. Nociception—the sense of pain—is the first function of the nervous system; the flinch from a noxious stimulus is within the ken of creatures with one nerve cell.[6]
- Traumatic memories are easier to recall than happy ones.[7] We obsessively remember traumatic or painful events that

> we would much rather forget,[8] possibly due to the fact that *once is enough* for animals facing deadly predators. Consequently, villains grip the imagination with a force no hero can hope to match.[9]

The extinction of conscious thought in intense sudden fear is often so complete as to reduce us to automatism.[10] The most traumatic phobias and flashbacks are stored in a special ancient organ—the amygdala—for hair-trigger influence on our behavior. We can conclude that humankind possesses an innate bias toward fear because of its overriding influence in evolutionary selection.

Fear of change. In particular, fear triggered by *change*, or by the threat of change, is one of our deep-seated instincts because changing the self is experienced internally as the equivalent to self-destruction, and changing of the environment always brings uncertainty and risk. We may even fear any change in our *understanding* of the world, for it may cause psychological damage, such as denting our self-esteem.[11]

It is natural for us to stick to the tried-and-true. Ernest Becker writes:

> Philistinism knows its real enemy: freedom is dangerous. . . . Kierkegaard had no illusions about man's urge to freedom. He knew how comfortable people were inside the prison of their character defenses. Like many prisoners they are comfortable in their limited and protected routines, and the idea of a parole into the wide world of chance, accident, and choice terrifies them. . . . In the prison of one's character one can pretend and feel that he *is somebody,* that the world is manageable, that there is a reason for one's life, a ready justification for one's action. To live automatically and uncritically is to be assured of at least a minimum share of the programmed cultural heroics—what we might call "prison heroism": the smugness of the insiders who "know."[12]

It is also natural for us to protect our self-identity first before doing anything else. Without enduring self-identities, our lives would lack coherence. As individuals, we automatically cling to peculiarities that define ourselves. Social evolution supports this search for personal coherence: A working society evolves mechanisms that stabilize ideals—and many of the principles each of us regards as personal possessions are really "long-term memories" that our cultures have learned across the centuries and passed down to us in the form of "wisdom."[13]

That said, the complexity of our human nature means that we are not always risk-averse. Most of us understand, at least during moments of calm reflection, that our lives would be meaningless and worthless if there were no risk and nothing ever went wrong. In our daily life, we have an intuitive sense of seeking a risk homeostasis. Take our driving behavior, for example. Drivers sensibly slow down after dark or when the road is wet, even while advanced safety technologies such as ABS prompt drivers to be more careless in their driving. Sometimes, we even *seek* risk in the impulsive search for the thrill of the dopamine rush. We are descendants of cavemen who dared to explore uncharted territories—driven by the "novelty-seeking" gene and population pressure—rather than of those who cowered in their caves.[14]

In the book of Genesis, the first emotion experienced by humanity is not desire, not shame, but fear.[15] From the story of Adam and Eve to the legend of Faust and the modern parables of *Doctor Strangelove* and *Jurassic Park*, the theme that humanity's attempts to gain knowledge and power will result in a critical loss of value and lives has haunted Western thought and acted as a powerful force against the Western idea of progress.

Fear and ignorance. In dealing with complex social and cultural changes, conscious reflection may be unable to reach a quick or easy conclusion about their meaning—and as a result we tend to prefer the status quo. As Livy (59 BCE–17 CE) claimed, "We fear things in proportion to our ignorance of them."

Take the case of the highly controversial and emotional history of the test-tube baby. It seems hard to believe today, when the procedure is so routine that it is usually covered by medical insurance, that in 1973 in vitro fertilization (IVF) was thought by some to threaten the very fabric of civilization—marriage, fidelity, the essence of family; our sense of who we are and where we're headed; and what it means to be human. If IVF were to be allowed, some said, all the stabilizing threads of our human existence would unravel.[16] Those who are against cloning today are repeating the same arguments. How quickly people can forget how their earlier positions changed over time!

It has often been said (though no one seems to know who said it first) that all new and truly important ideas must pass through three stages, which go something like this: first they are dismissed as nonsense, then they are rejected as being in opposition to religious or other dogma, and finally they are acknowledged as true and obvious. (The basic three-stage theme has been stated in many different ways, but the fundamental pattern—beginning with dismissal and threat, and ending with nonchalant acceptance—is the same.)

This three-stages-to-acceptance pattern shows up in a host of areas, from science and technology to social and cultural issues (for example, the trajectory leading to racial and sexual/gender equality) to acceptance of everyday consumer products ranging from the automobile to the microwave oven and the cell phone (some of which may have no obvious risk factors at all).

Risk and uncertainty are ubiquitous, unavoidable, and conceptually uncontrollable. For most of us, the assessment of risks is not always rational but biased by several factors, including whether we have subjective control, whether the risk is man-made (which automatically triggers the "fairness instinct"), and whether it portends a potential catastrophe (perceived or real).[17] Unfortunately, new technologies usually sound loud alarm bells based on these criteria.

Excessive caution by the public is easily made worse by the scare tactics of the opponents of new technology—but also by the seeming arrogance and elitism (and woeful ignorance of the art of persuasion) on the part of many scientists and innovators. In cases like IVF and cloning, debate tends to degenerate into a mish-mash of irrational fears and extremely polarizing rhetoric.

It undoubtedly helps if the public is provided with historical examples in which emotionally charged opposition has given way to broad acceptance. IVF is an extreme example of this pattern in recent decades. In such cases, opposition is often absolute and couched in religious or even apocalyptic language. But it is really facts and results that change our minds; results can quickly and decisively defeat our fear bias and alter public opinion. High-flown rhetoric notwithstanding, most of us tend to be pragmatists; let the results speak for themselves. When asked what American book he would place in the hands of every Soviet citizen if he could, Franklin D. Roosevelt reportedly answered, "The Sears, Roebuck catalog"—perhaps because the Sears catalog (which was *the* catalog of the mid-twentieth century) would demonstrate the results of the American economy and lifestyle far more than any discussion of belief or ideology ever could.[18]

A few examples from science and technology illustrate how seemingly intransigent opposition, even in the face of contrary evidence, can give way to acceptance with increased familiarity:[19]

- In the case of in vitro fertilization, a 1969 Harris poll found that a majority of Americans believed that producing test-tube babies was "against God's will." But test-tube babies were born without suffering any apparent ill effects. As a result, less than a decade later (1978), more than half of Americans said they would use IVF if they were married and could not have a baby any other way.[20] Even the prominent bioethicist Leon Kass, who once strongly opposed the practice, reversed his position in 1979.
- A few decades ago, cosmetic surgery was relatively rare and viewed as almost immoral in America. Today, most public figures (especially women, and especially actors) either have had it or are suspected of having had it.
- Many women in 1900 were shocked by the invention of the X-ray and swore that they would never allow such a gross intrusion of their bodily privacy. Later they became used to the idea and demanded regular X-ray checkups.[21] Today, far more intrusive medical tests (such as amniocentesis, the testing of the amniotic fluid surrounding a fetus) are routinely conducted.
- When the newfangled Great Western Railway steam-engine train left London in 1844 at forty-five miles per hour, commentators expressed horror at such unnatural speed, while physicians urged passengers to avoid trains, warning the public that anyone moving so fast would surely suffocate.[22] Now, of course, people enthusiastically embrace high-speed rail with speeds approaching 200 mph, and there is discussion of the feasibility of much faster speeds. And whatever we think of air travel, nobody complains that passenger aircraft can cruise at up to 600 mph.

10.2. Overcoming the Fear Bias against Conscious Evolution

Lessons of the past. Given the limited appetite for risk, the best approach to encouraging society to accept new features of conscious evolution is to reduce or even eliminate risks as much as possible. Of course, innovations always require educational efforts to promote acceptance. Visions and motivations, such as the Cosmic View, are so important because they prioritize our risk-management efforts. Likewise, it is critical that we have sound social institutions, as faith or trust in them will provide a basic level of security for all while maintaining maximum room for experimental "play."

In addition to public education and demonstration efforts, the public acceptance of new technologies often requires a regulatory framework more stringent than is scientifically necessary. For ideas as to how to proceed, we should look to the regulatory mechanisms of modern capitalist society itself, including property rights, social safety nets, and the concept of limited liability in bankruptcy that, for example, enable entrepreneurs to chase their business dream. Let's consider several general lessons to be learned from the modern economic system.

First, providing a social safety net encourages more risk-taking by liberating us from the most dire consequences of failure. The sense of fear is intensified when we are weak and cannot afford the negative consequences. When birds are starving and desperate for food, you might expect them to take everything they run across that looks edible. But, in fact, hungry birds shy away from unfamiliar food. A creature in genuine need cannot take a chance on being immobilized a day or two by food poisoning.

More often than not, the really novel ideas come from *play*—when nothing is at stake—rather than from work; and they come from those who have extra time and resources to kill rather than from those who are the most needy. For example, Islam's golden age came when it was the most powerful, politically and militarily, which afforded its elites opportunities to engage the cultures they conquered with an open mind. Later, when faced with waves of barbarian invasion and economic decline, Islam quickly retreated from its position as a free, tolerant, and inquiring society, resulting not only in oppression of its people but in intellectual sterility.

Second, an environment that heightens the sense of change hardens people's attitude toward it. It is no surprise that the weak and the losers tend to be excessively cautious, which in turn makes them even weaker in competition.

Much of human history over the last ten thousand years has been about technological change that brings people closer to strangers or that forces people to face entirely unfamiliar situations. Such an arrangement is loaded with uncertainty and risk—especially given the inherent moral deficiencies in human nature. The Industrial Revolution drastically quickened the pace of change (and the population's mobility) and spurred widespread psychological distress and rebellious movements such as the Luddites. After numerous attempts at structural reform, the successful economies in today's global environment are those that manage to soften people's attitude toward change with a firm "Change is not only inevitable but actually good for you" mentality.

Third, embracing change is much easier when there is a greater sense of control. The increases in social/economic integration due to policy changes and advancements in transportation and communication technologies have eroded people's sense of control over their lives, which is directly linked to their sense of economic security. Modern technologies have greatly reduced many kinds of natural risks, but at the same time they have created many artificial ones. As science turned from being a "natural philosophy" to being a social institution focused on technological application, it became clear to the public that some new technologies could bring devastating unintended consequences that were utterly unpredictable. The empowerment of the individual is critical in addressing the issue of control, which means allowing individuals to play a "wide open" game as we discussed in the last chapter.

Fourth, practical policies must recognize that our risk tolerance varies widely. Individuals in a society typically exhibit different kinds of biases toward new man-made risks: the *fatalist* believes we are powerless and face hopeless odds against the elite or profit-driven corporations; the *hierarchist* believes the solution is strong top-down regulation and control; the *individualist* happily embraces new technologies with an understanding of the risks necessarily associated with them; and finally, the *egalitarian* abhors all "unnatural" products and practices, and also fears technological dependence and concentration of political and economic power.[23]

Most of these approaches tend to be focused on man-made risks (rather than natural risks), and our alertness of risk tends to be associated with emotionally charged social judgments—especially where they relate to cheating, equality, justice, exploitation, and manipulation. In other words, our moral instincts and emotions make it hard to take a balanced view of the risks, which demands detached rational analysis and a long-term, holistic perspective.

The social/emotional dimension is one of the most important reasons why people's resistance to new technologies has increased in modern society despite a significant drop in the overall level of risk. Plenty of real dangers are always present regardless of the state of technology. Whether we pay attention to them is another matter. No doubt the water in fourteenth-century Europe was a persistent health hazard, but it became a public preoccupation only when it seemed plausible to accuse the Jews of poisoning the wells. The heightened sensitivity to moral defects of institutions such as large corporations, as compared with natural dangers in which there is no one to "blame," explains why, for example, asbestos poisoning is seen as more fearsome than fire. Asbestos was developed to save people from burning; asbestos poisoning is a form of industrial pollution whose toll of deaths by cancer justifies a particular anti-industrial criticism more strongly than does loss of life by fire (except, of course, in cases where a fire is a result of a product defect or other human or corporate negligence). Similarly, there is no obvious way in which the incidence of skin cancer caused by leisure-time sunburn can be mobilized for criticism of industry, and so we hear less about it than about lung cancer caused by smoking.

The general apprehension about genetically modified organisms (GMOs) in food—which is of interest here because we are moving toward a future of genetically modified *beings*—is another example of public distrust of corporations that lead and profit from technological innovations. Just as people instinctively dislike greedy businessmen, corporations that try to make a profit through growing food in ways suspected of being unhealthful, inhumane, or exploitative are not likable. On the other hand, profit motives are not incompatible with important or even necessary innovations, although the processes by which such innovations come to market must of course be regulated. Here Adam Smith's insight of the invisible hand is critical—motives can be selfish and still serve the public good. Contrary to some beliefs, Smith did not believe in absolute laissez-faire, and he insisted that harmful excesses should be controlled. Prudent regulations regarding public safety and environmental impacts are needed, and corporate social responsibility should be promoted. Sound judgments are needed to decide what is the proper level of regulation, and rational policy analyses are needed to address our emotional and political-sensitivity biases.

In the case of GMOs, as with other new technologies, it is important to have open discussion among all stakeholders, such as scientists, media, and the public, as well as thorough testing and prudent application. The public is correct about the need to thoroughly test GMOs before introducing them into our diets, but this applies to all genetic variations, man-made or not. For example, many plant species have an incentive to develop poison in order to avoid being eaten. It is interesting to note that when seed companies found it nearly impossible to change public perceptions, they had to adopt a more costly approach, which was to use the same gene maps as discovered in advanced GMO research to speed up the traditional breeding of new varieties based on natural genetic traits in wild species, thus avoiding the stigma of GMOs.[24]

The introduction of advanced biotechnologies and nanotechnologies inevitably brings with it a social perception of high risk. These risks can easily attract a public backlash, in sharp contrast to the relative public apathy toward natural risks of much greater magnitudes. Not infrequently, even the most brilliant minds fail to take into consideration the complex feedbacks between technological developments and the human societies that bear their impact.[25] Real risks and suffering due to lack of new technologies and natural disasters, on the other hand, are often unperceived. We tend to forget that before the introduction of modern agriculture and medicine (an uneven and risky process), life expectancy was half of what it is today, and millions perished when there were major weather disruptions or contagious-disease outbreaks.

New challenges in the posthuman world. Human history shows a trend toward faster change and greater risk, but up until now these changes have been by and for humans. However unpredictable humans are, their nature is universal and unchangeable across time and across cultures. In Adam Smith's words, "Every faculty in one man is the measure by which he judges of the like faculty in another. I judge of your sight by my sight, of your ear by my ear, of your reason by my reason, of your resentment by my resentment, of your love by my love. I neither have, nor can have, any other way of judging about them."[26] As Israel Baal Shem Tov suggested, to carry out the seemingly counterintuitive Torah command "Love your neighbor as yourself" (Leviticus 19:18), the emphasis should be placed on the words "as yourself."[27]

But now, for the first time ever, we face the prospect of dealing with unknown intelligence with a different nature—beings to whom we fear we may one day cede control of our own fate. This is a change of unprecedented magnitude and unpredictable—even unimaginable—consequences. In the face of this prospect, the fear of change as a powerful social emotion will only escalate if it is not dealt with appropriately.

Such an unprecedented challenge calls for unprecedented new mind-sets and institutions. However, institutional resistance to change can be equally as strong as public resistance, and also more complex. Throughout history, deliberate actions to hold back progress and suppress the new have been the norm. For vested-interest groups, it makes total sense to protect the status quo. Favorable social environments for change have rarely been nurtured intentionally. Throughout history, they most often have come as the ruling powers were weakening and losing control.

Indeed, even in the West, with its long intellectual tradition of the idea of progress, the wide acceptance and embrace of progress—involving a belief that the present is better than the past and the future can be made better than the present—came about only very recently, when positive, technologically induced changes transformed the lifestyles of individuals within the span of a single lifetime. The concept of "progressive degradation," on the other hand, has been enduringly popular and can be traced all the way back to Hesiod's Five Ages and even Homer, who evoked a past that was better than the present.[28] In Plato's creation story *Timaeus*, the entire evolutionary trend is the exact opposite of the Cosmic Vision. In the beginning, there were only the gods. Men, the highest of animals, were generated by the gods; women were the result of degeneration of certain men—and were cowards and villains. All the other species devolved from men by a process of corruption and degeneration. Birds came from harmless but too easygoing people who trusted their senses too much; land animals came from men who had no interest in philosophy; fishes, including shellfish, degenerated from the most foolish, stupid, and unworthy of all men.[29]

Although Judeo-Christian theology emphasizes the linear, progressive aspect of history, one of its central features—the idea of humanity's fall resulting from the events in the Garden of Eden—is clearly regressive. Actually, it's a bit more complicated, for the Bible speaks not only of the fall but also of a return to the Edenic state, a time when "the wolf shall dwell with the lamb and the leopard shall lie down with the kid" (Isaiah 11:6). But this hope of "return" is itself regressive, idealizing a past state of perfection; therefore it contrasts sharply with the Western idea of progress and with the Cosmic Creation vision contained in the Bible itself.

What is most needed to overcome the fear bias in the posthuman era lies in the modern concept of progress itself—the fundamental belief in open-ended change leading to better things, based on today's understanding of higher human potential. Ultimately, progress in this sense is essential to realization of the Cosmic Vision. Today, most people no longer see social change as leading to corruption, decay, or degeneration, but the results of our current shift to a modern, evolutionary worldview are still largely mixed. We are in a long transitional period to more public acceptance of progressive change, because this shift will also inevitably bring with it chaotic ideas, paranoid anxieties, hybrid solutions, and regressive tendencies. However well educated, we are still human with a certain "speed limit" on accepting change.

In general, the environment and technologies one has known since childhood are usually accepted unquestioningly, and incremental changes are usually also accepted once examined. Paradigm shifts, on the other hand, are often perceived as threatening. We have a very limited capacity to believe or accept things that are drastically different from what we perceive in our daily life. But there is much we can do to enhance a society's capacity for unprecedented change and tolerance of risks; here I will make several suggestions.

First, we should protect and promote the pioneers. It is unrealistic to expect most individuals to take risks and embrace change without seeing results. For most new technologies, the willing early adopters are always a tiny minority. They are willing to accept the tradeoff between excitement and the thrill of new ideas and the risks and potential failures. Hence the efforts of trailblazers play a critical role in alleviating public fear of change. Those who take experimental brain-modifying drugs or subject themselves to genetic procedures today play the same role as the daring pilots in early aviation history.

Nearly all of us appreciate those pioneers who contributed to past advances in our technologies and lifestyles. But most of us do not appreciate those who are making history right now. As a Chinese proverb goes, "The great man is a public misfortune." Most of the time, we resent those who disrupt our deeply ingrained values and worldviews. They make the rest of the crowd feel unworthy. The sure sign of a true genius is that the "confederacy of dunces" is against him.[30] Thus, pioneers need all the protection and promotion we can give them.

Second, we should highlight the risks of the status quo. One of the best approaches to fighting the conservative forces in democratic policy debates is to emphasize the risk of refusing change. We need to ask those who are resistant to change, "Where does the status quo come from?" Everything we know was utterly new at some point in history. The new always appeared to be odd and crazy as measured by the prevailing conventional wisdom. Once committed to and tested, however, the impossible or the extreme can become reasonable, normal, or even inevitable in retrospect.[31]

Yesterday's status quo often becomes today's backward enterprise or discredited idea; and many of yesterday's unquestioned, "commonsense" beliefs have become untenable today. But these changes don't come easily. The status quo within a given culture can become so hardened that it takes something like a foreign invasion, or wide exposure to development in other nations and cultures, to break through the resistance to new ideas and technologies. Every country with a long history has suffered periods of antiprogress stagnation and paid dearly as others advanced.

Third, we need to highlight the dangers of remaining merely "human." Today's status quo is to remain human. The popular imagination seizes on the dangers of our robotic creations, or of aliens from space; but it is static human nature that makes the growing power of knowledge in modern society a real danger. The more advanced our social organization, technology, and intelligence become, the more brutal and efficient our methods of killing. The ever-widening gap between our progress in creating more efficient technologies—including weaponry—and the stasis of our primitive aggressive (all-too-human) urges may lead to our collective self-destruction. One key lesson of the twentieth century is how quickly the dreams of utopians in the Victorian era (and since) were turned into the disasters of two world wars.

The romantic intellectuals always had a misplaced faith in the mutability of human nature. Unfortunately, as long as we remain human, ideological fervor will rise when the political climate is right and utilize the power of modern technology and social organization to serve our animal instincts. So, perhaps counterintuitively, the *least* risky long-term strategy is to transcend the "human" in the face of technological advancements. And that includes expanding technology to address and transcend human nature itself. That is the great challenge that transhumanism places before us.

10.3. Addressing Frequently Voiced Fears

To emphasize the danger of remaining human is not to deny the risks in the posthuman future. In the following quesstion-and-answer section, I address the most frequently voiced fears—the questions and concerns that arise when we consider the possibility of a posthuman world. First, a note of caution: it must be made clear that nothing can be ruled out in the future, and the range of possibilities tends to be wider than our wildest imaginations. But here we will look at why the risks and concerns that we can imagine at this moment, when examined in full context, are risks worth taking, and as I mentioned above, the greatest risk of all may be the risk of doing nothing.

Is there a strategy for a smooth transition to the posthuman world?

The simple answer is no. As we emphasized in the last chapter, the game plan for conscious evolution is to leave the field wide open under evolving social structures. Yet, when things are left open in this way, as in the real universe, there will be huge swings as there have always been—witness the centuries-long regression to the Dark Ages within relatively recent history.

Just look at the path of life and (in more recent time) cultural development on Earth. Despite the fact that our planet sits in an extraordinarily tranquil "habitable zone" in the galaxy and boasts many self-stabilizing mechanisms, its planetary environment has at times behaved like a switch rather than a dial during periods of abrupt climate change that have led to devastating (but also, in retrospect, highly stimulating) impacts on life. The planet has experienced five mass-extinction events, and there have been several periods in natural history and cultural history when the evolution process came to a complete halt and the entire enterprise of life and culture seemed to hang in the balance.

Thus, we can be sure that the transition to the posthuman world will also be a very bumpy ride, but there has never been such a thing as "costless and painless progress" from the individual perspective. But then there has never been "costless and painless" stasis either. The choice is a rocky path to CoBe or a seemingly comforting status quo that will eventually—and inevitably—be proven unsustainable (and even rockier).

Are we creating a deadly competitor in the cyborg or superhuman of the future?

There are many unknowns when discussing future scenarios, but as I will discuss below, deadly competition between humans and posthuman beings will not be in either our interest or theirs. We should always proceed cautiously, but we can also be confident that, in the end, it will not be the human competitor that will prevail in cosmic evolution.

Unlike cultural evolution in recent history, when human nature is fixed and we all pretty much want the same stuff, evolution beyond the human species is about divergence in motivations and values. What we highly prize may not be that attractive to a different kind of being. As I have argued, those posthumans that are most likely to survive and thrive are those who transcend the worst traits and motivations of human nature, although nobody can guarantee that all "versions" of posthuman beings will be designed in this way or that any given design will be reliably nonviolent.

Deadly competition between superhumans and human heroes is what makes us fearful. This is the focus of numerous sci-fi movies and novels, but it is actually not the kind of competition that superhumans who are unimaginably stronger than humans would enjoy or find necessary. That is because much of the fear concerning deadly competition is the result of a misunderstanding of Darwin's theory of evolution, which is much more complex and subtle than a single-minded "struggle for existence." Although deadly power is often wielded against equally matched competitors, the ultimate will to power is the will to overcome the limitations of the self. Nietzsche's "superman" is not a super-brute, but one who is strong enough to lift himself above "all-too-human" needs and desires. Further, even the Nietzschean will to power is empty and sterile without what Viktor Frankl called the *will to meaning* and the virtues of objectivity, courage, and a sense of responsibility beyond oneself.[32]

Thus the ultimate will to power is a godlike will to forgive: indeed, it was Jesus Christ's power to forgive, even more than his performance of miracles, that shocked people and gave Christianity its enduring strength.[33] How is Jesus's "turning the other cheek" possible, given our natural instinct of retribution and self-defense? It requires superior mental and physical strength. This is a characteristic of Zarathustra in Nietzsche's poetic tale; he avoids death and wins over his enemy with a high mind nobler than most people can imagine:

> One day Zarathustra had fallen asleep under a fig tree . . . an adder came and bit him in the neck, so that Zarathustra cried out in pain. When he had taken his arm from his face, he looked at the snake, and it recognized the eyes of Zarathustra, writhed awkwardly, and wanted to get away. "Oh no," said Zarathustra, "as yet you have not accepted my thanks. You waked me in time, my way is still long." "Your way is short," the adder said sadly, "my poison kills." Zarathustra smiled. "When has a dragon ever died of the poison of a snake?" he said, "but take back your poison. You are not rich enough to give it to me." Then the adder fell around his neck a second time and licked his wound.[34]

Jesus, Buddha, Zarathustra—almost every civilization has stories about the sage or hero that is some kind of superlative idol we want to be but cannot be. What are the characteristics of a super-being that are likely to succeed in the long run? To use the analytical framework developed by New York University's historian James Carse, we can conclude that beings of the future who gain immortality would play an open and infinite game with dramatic *strength,* which is the courage and resilience to prevail against all odds, rather than theatrical *power,* which is the force to bend or defeat others in a closed and finite game. Strength must be applied in an unselfish fashion because any "self" has a finite and closed existence. True power is not the "theatrical" ability to eliminate, but the ability to connect and grow with others, utilizing the will to forgive and to overcome the self.

The urge to love, to embrace others, is not a human invention and can be a successful survival strategy. While aggression is necessary in the animal world—at least in the form of self-defense—competition for survival is not always "red in tooth and claw." What we find in natural ecosystems is an array of interspecies relationships: (1) predator/prey, (2) host/parasite, (3) master/slave, (4) symbiosis—intimate physical association that benefits both, (5) interspecies cooperation—such as bees and flowering plants, and (6) mimicry, among others.

Deadly aggression always gets our attention, but it is actually a fairly limited activity in the totality of animal and human life. Cooperation's competitive edge is evident from the beginning of complex life and human civilization. Tracing back to the origins of multicellular life, symbiosis has been as important as competition for both interspecies and intraspecies relationships. An environment of limited available resources can make "eat or be eaten" the survival rule, but it also increases the pressure for cooperation. It is a common misconception that competition occurs only at the personal level; thus we may falsely assume that the happiness of one individual necessitates the injury of another.

Indeed, one of the most important insights of the modern era is that wealth and power are best built through promoting peaceful commerce (economic cooperation) and international trade rather than violent conquest and plunder of weaker nations. Paradoxically, violence between peoples has led us to create better and better institutions that ensure peace.

It is hard to imagine, but given this growing tendency of cooperation in natural and cultural evolution, the brutal "eat or be eaten" competition may well give way to symbiosis-like evolution at some stage in the posthuman world.

How can we prevent transhumans' violent behavior toward humans?

The science-fiction writer Isaac Asimov famously proposed three laws for all robots[35] to follow: (1) a robot may not attack a human being or, through inaction, allow a human being to come to harm; (2) a robot must obey the orders given to it by a human being except where such orders would conflict with the first rule; (3) a robot must protect its own existence as long as such protection does not conflict with the first two rules. These "laws," though they sound just and logical, are utterly impossible to implement if the autonomous robots are to be intelligent and able to reprogram themselves.

However, we should ask why robots—or genetically modified humans—would have to copy our social instincts. I have argued in this book that human nature and human instincts are limiting. Those transhuman designs that for some reason do emulate or inherit exactly the same human instincts are unlikely to be successful in the vastly complex posthuman world. Robots that are able to reprogram themselves would recognize this likelihood as well.

We have often heard the argument that autonomous intelligent beings guided by different values could be even more dangerous than humans. Presumably, they could kill people mercilessly since they are not "one of us" and do not share our moral instincts. If they could replicate easily, they may not even value what we value the most—self-preservation—which almost all law enforcement mechanisms are based on (by successfully invoking the fear of death). Locally, it is quite conceivable that intelligent life could become extinct as a result of interspecies power struggles. Historically, during the global spread of *Homo sapiens*, all of its close relatives were wiped out within a short period of time. The only survivors were the great apes that lived in dense jungles inaccessible to humans. It may well be that wholesale extermination was the necessary price for the exceedingly rapid pace of human evolution over the past half million years.[36]

However, *the nonviolence scenario, while not a certainty, is far from wishful thinking. In fact, the whole of cosmic and human history tells us that this is the most likely of all scenarios.* Fundamentally, different minds with drastically different motivations don't compete. According to Gause's Law, which states that maximum competition is to be found between those species with identical needs, transhumans are extremely unlikely to be close competitors with humans.

We also need to recognize that the murderous instinct serves the purpose of eliminating rivals for resources and mates *in a shared space*. However powerful, it is fundamentally a local phenomenon. For example, while surplus food made raids far more attractive at the advent of agricultural civilization, it has been noted that many Neolithic villages seem to have been undefended, which suggests that in the great age of Neolithic expansion, tensions were relieved by emigration. Once a given frontier was closed, however, war became a chief instrument of Malthusian dynamics to hold population growth in line with available resources, which led to the creation of professional armies.[37]

On the other hand, technological advancement has continued to expand the quantity of available resources, rendering Malthus's laws no longer operable. If the world had somehow remained at the level of the hunter-gatherers, it would have been impossible to envision that the Earth could comfortably support a population of seven billion. Yet, because of a myriad of technical achievements as well as the evolution of the democratic market system, there is still plenty of potential left even on this tiny planet, not to mention outer space. For example, our total *annual* energy consumption is the equivalent of only about one hour's worth of sunlight reaching Earth's upper atmosphere. We will continue to expand our ability to utilize natural resources, but transhumans will be able to do it much faster. By incorporating superior technologies, including capacities designed for inhospitable environments, transhumans almost certainly will not covet the ecological niche that humans currently occupy, just as we no longer have to fight with wild animals for food.

Contrary to popular opinion, crime and exploitation has always been a function of need *and* weakness. The rise of the West was characterized by brutal conquest and mercantilistic exploitation by the Spanish empire, which was inherently weak, lacking a dynamic domestic economy. That inherent weakness was one of the motivations for the empire's imperialist adventures, but it eventually led to the empire's downfall. Today, developed countries around the world have no need for more land grabs and forced labor to fuel further growth.

For much of the world, our relationship with less-evolved beings—wild animals—has also changed fundamentally since the Industrial Revolution. In the old days, "there [were] no compacts between lions and men."[38] Now wildlife is seen as something fragile and precious as environmental awareness takes hold. If it is possible to develop a new intelligence that makes today's greatest geniuses seem like simpletons, *we* will be among the less evolved beings; but we should not thereby feel threatened, because the new sentient beings would almost certainly eliminate any reason, advantage, or necessity for mistreating or killing humans.

Apart from the arguments we have already given against the likelihood of deadly violence from transhumans, observations based on natural and cultural history show further ways to minimize deadly competition: trade, physical separation, and differentiation. First, fighting makes sense only when it is less costly to grab than to trade. As the economist Frederic Bastiat noted, "Where goods do not cross frontiers, armies will."[39] Second, separation in space reduces, if not eliminates, tension. Expansion to oceanic or desert environments or outer space is an obvious option and much easier for new beings evolved and designed for that purpose. The third observation is the most fundamental: transhumans will evolve new values and motivations.

When we think of examples of strong and weak, we tend to think of the relationship between predators and prey—say, eagles and rabbits, or lions and zebras. Yet in reality, the eagle is just as weak and incapable as the rabbit in its occupation of a fixed small niche in the same ecosystem. The strength of the transhuman is extremely unlikely to be its ability to overpower the human. When the technological and intellectual distance between the transhumans and us reaches that between us and the ants, transhumans may not even share what we consider our fundamental universal values, which we refer to by such words as "love," "freedom," "happiness," "justice," and "faith." While these values can point to a transcendent dimension as partially revealed by our highest wisdom traditions—a dimension that is likely to survive because it can further cosmic evolution—they also have large human-specific characteristics, mainly to counterbalance and overcome our animal instincts.

Based on the above discussion, shouldn't we still fear the emerging transhuman?

We cannot help it, even if we have a positive attitude toward the transhuman. But, as pointed out previously, fear has a positive evolutionary value: our fear will actually be one of the driving forces behind a positive outcome, just as our fear keeps us from harm in our everyday lives.

We have no fear of small children, but we have a natural fear of adult strangers and an innate tendency to be antagonistic toward other groups. The motivational complexities and unknown personal background of adults make them unpredictable to fellow humans. But as complicated and unpredictable as human beings are, we can still trust that they have the same mental apparatus, motivations, and social instincts. We take it for granted that when we see a person for the first time, we can get a rough idea of the person in a split second: gender, age, skin color, health condition, mood, etc. We have special neural circuits for facial recognition and a sharp sense about subtle differences in physical appearances, which can be unconscious expressions of emotions that cannot be concealed or faked despite our best conscious efforts.

The transhuman, however, could be totally unpredictable and not easily understood by any human measure or standard. We may not be able to handle the onslaught of artificial "downloadable" personalities and designer genetic enhancements.

By comparison, the imagination of many popular works of science fiction is rather tame. The character E.T. from the Spielberg film of that name, despite its superficially weird and alien appearance, was primarily human in its attributes, with cute exaggerated baby features and quite ordinary human emotions. The fearsome Golem of middle-European Jewish folklore, Frankenstein's monster, and the Terminator are all cultural archetypes of cold-blooded mechanical intelligence based on the naïve belief that only "nature" can provide the vital force that is the soul of humanness.[40] Such wooden, one-dimensional machines are not the real threat.

We may also fear transhumans because they don't present us with a "human face." This may be exacerbated by the fact that we are already feeling the loss of face-to-face assurance in the electronic world—"on the Internet, nobody knows you're a dog." We complain about faceless corporations and institutions that we have to rely on in various ways. Corporations, in turn, have developed and nurtured brands as abstract but reliable personalities that we can trust just like some old friend or neighbor. By providing legal protection of commercial brands, our social institutions alleviate fear of having to deal with strangers. And through massive expenditures on image, advertising, and publicity—and of course by creating products or services that are reliable over time—these brands come to represent the appearance of a trustworthy personality.

To be successful in gaining humanity's trust, the transhumans may also have to evolve similar abstract personalities and reliable signs and track records, just like corporate brands.

As we will discuss in detail, the whole issue of personal rights and identities has to be addressed when fledgling transhumans become autonomous members of our society before they move on to form their own. The fear factor is a natural barrier for transhumans to be allowed to mingle with humans, and like our early ancestors, they will seek the path of least resistance for expansion. They are most likely to spread into currently unoccupied territory rather than taking over the existing environmental niches that sustain natural life.

While we tend to focus on discussing the differences between "us" and "them," the differences among "them" in all their various types will surely be many orders of magnitude greater. New social institutions and cultures will surely evolve to help humankind to differentiate and discriminate among the proliferating transhumans, exerting selective pressure on both sides. How the new institutions and mind-sets manage to reduce our anxiety is likely to be the single biggest factor during the transitional period. The psychological burden of giving up our control to transhumans could be similar to—but much more disruptive than—any elder generation's ceding control to the next generation.

If we create our successor species, does that mean that humanity will not be needed in the future?

Yes, but it is true for humanity only in its current form. Humanity will no longer be able to assume a preeminent place in the known universe, and we will no longer play the leading role in cosmic evolution.

"We will not be needed" is actually a concern primarily for the elite. As Alan Turing pointed out, "An unwillingness to admit the possibility that mankind can have any rivals in intellectual power occurs as much amongst intellectual people as amongst others: they have more to lose."[41] Computer whiz Bill Joy's article in the April 2000 issue of *Wired* magazine, "Why the Future Doesn't Need Us," hit a raw nerve among many. He argues that we should relinquish certain genetic, nano, and robotic technologies now before their development drives us into extinction. Mr. Joy is not the first to predict that humanity someday may be irrelevant, nor is he the deepest thinker on the new technology's implications to humanity. Lee Thayer wrote in 1976 that given our fixed human nature, "each increment of growth and progress now costs some further loss of our human relevance."[42] Indeed, from a higher perspective, humans are a transitional species. As with any other animal, humanity's lasting value may not be much more than as "living fossil" in the distant future. The human species is extremely precious at present, but it may not be so after it spawns countless new beings that are far more "human" than us. They will not be masters of us humans, but they will be masters of the universe in ways we can never be. Our spiritual (if not fleshly) offspring will first be our servants, then our partners, and finally gods.

Recall that as great a figure as George Washington, once having achieved his spectacular, world-redefining victory over the then-existing status quo, only wanted to retire to his estate in Mount Vernon and leave the future to his heirs—the "unborn millions." Surely, we humans can have this same enormous sense of personal satisfaction as someday we leave the future to our spiritual heirs—the unborn trillions of trillions of Cosmic Beings!

Recall, too, that the first thing the general and future U.S. president George Washington did at the Constitutional Convention of 1787 was to ensure that the very first article of the U.S. Constitution had a clause that would protect and encourage the widespread creation and dissemination of technology and innovation. Like no one else anywhere in the world, he and a few other American founders foresaw the critical role of innovation in the future well-being of those unborn millions of Americans.

In the end, long after we have relinquished our place on the frontier of cosmic evolution, our offspring—the "unborn trillions"—will have a say in *Homo sapiens'* continued existence. However, one thing is worth noting. While few miss *Homo erectus* today, if they were found to be surviving in some remote area, I am sure the developed world would make a great effort to protect them.

Will transhumans be a threat to human dignity even if they do not destroy us?

The answer depends on how we define "human dignity." Those who make the argument against transhumans base their idea of human dignity on our supposed sacred position within God's creation (or within nature). In their view, the preservation and enhancement of currently existing humanity is an end in itself. That is the premise, for example, of the statement in *The Decree* by the Council of Europe on Human Cloning, which states, "The instrumentalisation of human beings through the deliberate creation of genetically identical human beings is contrary to human dignity and thus constitutes a misuse of medicine and biology."[43]

With the Cosmic Vision, however, humanity's greatness lies in its untapped potential, which includes its potential for self-transcendence as a species. The human is a transitional creature like anything else in this universe. The fate of humanity is neither a fixed salvation nor enslavement at the hands of transhumans. To the contrary, human dignity is reflected in our willingness to recognize our value and position in the grand scheme of things beyond that which is "merely human." The sense of dignity or self-worth is ultimately a sense of calling. In this light, transhumans are not a threat to human dignity, but its greatest expression.

Human dignity is an important moral and philosophical issue that will be discussed further in a special section in the next chapter.

Rather than move forward to the next step in evolution, shouldn't we instead get rid of human civilization since it is such a danger to nature?

Some individuals look nostalgically to the prehuman past rather than forward to the posthuman future. For example, many have argued that it may not be a bad thing for humanity to become uncivilized beasts again. They say we must avoid our cultural chauvinism and realize that humankind has become so maladapted and destructive that our demise could be a good thing. Followers of the "deep ecology"[44] movement believe nature has supreme value in itself regardless of its utility to humankind. Anarcho-Luddites such as Derrick Jensen and Kirkpatrick Sale (as well as the "Unabomber," Theodore Kaczynski) believe humanity in its precivilization original state is the most natural and desirable. These people regard technological civilization as the enemy of their cherished ideas of "nature" and raw humanity, and are sure that the transhuman would spell the end of us.

The reality may well prove to be the opposite. All one needs to have is a longer-term perspective. In chapter 8 we described the seemingly counterintuitive fact that industrialization eventually helps preserve the natural environment and reduce man-made pollution. Let us consider another once-counterintuitive idea: that new labor-saving technologies improve the welfare of workers. In Dickensian England (in the early nineteenth century), what sparked the violent Luddite movement was the belief that factory-based automated textile machines were permanently replacing the weavers' family-based manual work. Those rebels who followed the legendary machine-wrecker Ned Ludd truly believed they were trying to save the weavers' lives from ruin. As it turned out, after their exact way of textile production was eliminated by superior technologies, the standard of living and working conditions turned out to be far better, and more jobs were created as a result of the higher income and demand.

The resulting prosperity did not just benefit the few. While one can never trust machine-owning capitalists to place worker welfare as their top priority, the workers' welfare had actually improved so much that few longed to return to the old ways of family-based operations. By then, the workers' movement started to emphasize employee rights in the factories and society rather than destroying the factories and labor-saving technologies, and by the late twentieth century many economists realized that the best way to enjoy economic security and prosperity is to embrace new technologies by investing in education and training, and to share in the wealth they create through investment, philanthropy, and entrepreneurship.

The same logic can be applied to address the concerns of today's worried naturalists. Perhaps one can never trust transhumans to place protection of the environment and the relatively primitive way of human life as their top priority, but it would not take long before earthly resources, including the collective human brain power, became trivial relative to their rapidly expanding capacities. For them, there may be a strong urge to maintain their heritage, and it may require only a little effort to preserve the natural environment.

Is it possible to proceed peacefully into the posthuman future by respecting everybody's opinion?

Most likely not. The small but vocal opposition to the coming posthuman era, including the blank-slate "Romantic Naturists," the technophobic nature "guardians," and the magicalistic New Age "spiritualists,"[45] can effectively damage a society's capacity to adopt new technologies and to reallocate resources in response to evolutionary pressures. While dissenting voices and actions must be tolerated and even protected in an open society, we should be alarmed and fight fiercely if "the train"—the creative destruction process that began as soon as we become truly human—is entirely stopped or slides backward. To be creative, change or destruction in some form is inevitable. It is in this sense that Daniel Dennett calls the idea of evolution wonderful but dangerous. It is also the reason that Fukuyama calls transhumanism the world's most dangerous idea.

At times of heightened tension between creative bursts and destructive worries, it is worth keeping in mind that nothing fundamentally new can be expected to materialize without a fight against the chauvinistic conservative forces and their comforting mysticism—whether from the left or the right of the political spectrum. "Think not that I have come to send peace on earth," declared Jesus with unusual candor, "I come not to send peace, but a sword." Seemingly inconsistent with Jesus's other sayings, the point is not to advocate war and violence, but to make clear the inevitability of conflict in certain situations and encourage us to face it head-on.

If history is any indication, there are almost certainly many zero-sum games to be played, when some individuals or groups gain and some lose at a certain point in time, even though the overall evolutionary pattern is non-zero-sum. I suspect that in many cases, tough political decisions will have to be made that will not please everyone or benefit everyone. Progress is rarely uniform, seldom driven by consensus, and in most cases not recognized as true progress by the public until a later time, since people first tend to focus on the dangers, damage, and losses caused by change.

As a general principle, dissenting voices are encouraged in free societies, but the world is not a debating club. The potential brutality of the selection process is most clearly shown in Western civilization's global dominance over the last several hundred years (although it certainly has existed in other times and other parts of the world). The Western way of thinking (monotheistic religion and rational science) prevailed not because of peaceful debate but by forceful exposure of weaknesses that had managed to survive (sometimes even thrive) in isolated conditions. At the same time, false beliefs that cannot stand the empirical tests die out as they inflict fatal wounds on their holders. I am not suggesting that the bloody past will necessarily be repeated in the future—most likely it will not be repeated on a broad scale, since eliminating individuals for their ideas is generally a thing of the past—but many firmly held beliefs and newly generated ideas will be proven wrong and will be dropped during the evolutionary process.

Can we trust some individuals or groups to develop the Cosmic Being for us?

No, we cannot entrust to any individual, group, or organization the responsibility for development of CoBe.

Within each person there is a dangerous destructive psyche that is so deep-rooted that a certain alertness and suspicion are necessary. We always look for exemplary individuals to be our leaders, but we are always wary of their selfish sides. Power-hungry people who claim to embody a higher vision are likely to have their character flaws exposed: "The higher a monkey climbs, the more you see of its behind."[46]

History shows again and again that those who regarded themselves as carrying out the orders of God, or the mandates of destiny, were acting under a delusion, often with disastrous consequences. What we tend to forget is that all "deities" reside in the human beast. For those rare persons who truly appear to be above the crowd (most notably the founders of religions, such as Jesus and Buddha), a common trait is that they do not seem to care about their personal well-being *at all*.[47] The empirical research on stages of morality shows that we have few, if any, living Jesuses or Buddhas who operate from the highest moral perspectives and possess pure, selfless motivations.

The traditional religious or moral authorities cannot be blindly trusted to manage conscious evolution, either. Like any other institutional leaders in the social sphere, they will be facing their own conceptual challenges and must adapt to new realities in the post-human era. When human nature gets "upgraded" or expanded to include higher perspectives and greater wisdom, traditional teachings of human morality must also evolve and adapt to new reality.

The bottom line is that any socially acceptable mechanism for creation has to be transparent and open to criticism and adjustment. In law and social institutions, we find unchanging, transparent, and somewhat abstract rules superior to individuals and organizations. On the other hand, even decisions and actions guided by noble motivations or higher perspectives are no guarantee of positive results. Some of the recent attempts to radically change the course of history under the guidance of "scientific understanding of the laws of history" left a terrible track record. From the Enlightenment to Marxist revolutions, it is heartbreaking to see untold numbers of revolutionaries attempting the betterment of humankind but ending up giving their lives for a mirage, a chimera.

We might then ask if it is better to trust nobody and leave everything to the "natural flow of things," hoping they will work out eventually. There are several reasons for not adopting this fatalist attitude. First, we are part of nature and we represent the frontier of evolution on Earth, if not in the entire universe. Consciousness emerged through natural and cultural evolution, and it is a result of the "natural flow of things"; to separate nature from consciousness is to create a false dichotomy. Second, there is no indication that unconscious "nature" is better at avoiding mistakes than conscious beings. Quite the contrary, as we discussed in the last chapter. Third, even though the dominant concern for most people does not extend much beyond their daily life, most of us do worry and care about the long-term future and human destiny. That is something we can rely on in motivating and shaping the conscious evolutionary process.

Is it possible to restrict ourselves to developing only the "good" technologies?

Unfortunately, no. The yin-yang reality is that no power is single-edged. We need look no further than our own body. Every mechanism that helps us survive has its potential downside. Stress is an effective mechanism for dealing with emergencies, yet chronic stress leads to cardiovascular and other diseases.[48] Our body's ability to generate a strong inflammatory response to fight disease organisms in childhood increases the likelihood of heart disease, diabetes, stroke, and cancer in late adulthood.

Those who see only the dangers of modern technologies and innovations should look again at the dread, damage, and suffering once brought on by some of their most cherished values and tools. Nobody in the West today is against freedom of speech or fails to recognize its role in the ascent of our culture over the last five hundred years. Yet, as Freeman Dyson suggests in response to the dread of machines-run-amok scenarios expressed by Bill Joy and many sci-fi writers, "There is an analogy between the seventeenth-century fear of moral contagion by soul-corrupting books and the twenty-first-century fear of physical contagion by pathogenic microbes. In both cases, the fear was neither groundless nor unreasonable."[49] In the mid-seventeenth century, both church and state in England had legitimate concerns about the unrestricted freedom of unlicensed printing, since it would irreversibly spread dangerous ideas in a time of long and bloody civil and religious wars in Europe. The poet John Milton, however, urged the Parliament of England to refrain from prior censorship of books because we can never assess their full impact beforehand:

> Suppose we could expel sin by this means; look how much we thus expel of sin, so much we expel of virtue: for the matter of them both is the same; remove that, and ye remove them both alike. This justifies the high providence of God, who, though he commands us temperance, justice, continence, yet pours out before us even to a profuseness all desirable things, and gives us minds that can wander beyond all limit and satiety. Why should we then affect a rigor contrary to the manner of God and of nature, by abridging or scanting those means, which books freely permitted are, both to the trial of virtue, and the exercise of truth. It would

> be better done to learn that the law must be frivolous which goes to restrain things, uncertainly and yet equally working to good, and to evil.

In the end, Milton expressed the Western faith in the God-given intellectual vitality that Dyson believes we should inherit from patriotic and proud seventeenth-century England.

Nuclear technology is another important case of two-edged technology. The bombing of Hiroshima and Nagasaki during World War II pales beside the almost inconceivable destructive power of the arsenal of nuclear warheads amassed during the Cold War—enough to destroy human civilization and much of life on Earth countless times over. Even peaceful use of nuclear power generation has the potential for enormous environmental damage if not managed appropriately, as the experiences of Chernobyl (1986) and Fukushima (2011) have shown. Yet we have found many other applications of nuclear technology without risks, and nuclear power has for the most part proven to be a cost-effective way of supplying reliable energy without adding greenhouse gas to the atmosphere.

Many new technologies for the posthuman era will likely to be controversial as well, most notably nanotechnology and genetics. Just like free speech and nuclear technologies, the right approach is not outright ban but controlled application to take full advantage while limiting potential downside risks.

By creating transhuman robots, are we playing God?

Exercising God-given capabilities is a way to honor God; to refuse to exercise them is a form of dishonor. Even if we wanted to, we could not play God, and the act of advancing our technology does not mean we are usurping divine prerogatives, even if we are creating posthuman beings. In earlier times, the scope of what "only God can do" was vast; now it is shrinking rapidly, and it will continue to shrink. Therefore, the notion that we can "play God" is based on a false assumption of what our natural, or God-given, capabilities are. We should also consider the possibility that we truly are godlike in our ability to be consciously creative, even co-creative with divine evolutionary purposes, and that there is no strict boundary between the human and the divine.

On the other hand, if we stick to traditional definitions of "playing God," then we might find that we cannot help but do so by our very nature. If, for example, "playing God" means "tampering" with nature, then we already have a long and proven track record of doing it—for example, with healing and medicine. While some cultures still hold that illness is a form of divine punishment or that curing illness should be the exclusive province of God, we have long tried various ways to cure diseases, from the premodern use of medicinal herbs to the modern inventions of drugs, surgery, and vaccines.

Breakthrough technologies that violate existing norms or conventions have always been looked upon as "playing God." The introduction of the smallpox vaccination in the eighteenth century was opposed by various sects for over 30 years on the grounds that we were interfering with God's intentions. Similarly, Benjamin Franklin's invention of the lightning rod led some of his contemporaries to accuse him of playing God at a time when lightning bolts were considered to be a form of judgment sent down by a disapproving God.[50] It was believed that praying during thunderstorms or ringing specially "baptized" church bells would keep the bolts of lightning away. In the early years of motorization, Henry Ford's Model T also raised a few religious objections: it was said that if God had intended humans to go forty miles an hour, He would have provided us with wheels instead of legs. Today, the taboo subject is cloning, which indeed carries great risks at its current stage of development, but eventually will be accepted as normal and unproblematic in the same way that in vitro fertilization is today.

If "playing God" means making conscious choices (versus being driven by instincts), then we all play God countless times in every day of our lives. Even the evolution of the human species was executed by our conscious selection of sexual mates and our decision to have babies. In a sense, parents who selected certain types of people for their children to marry were genetic engineers. The only difference is that in the past, people have tried to engineer based on phenotype (expression of genes in a certain environment), whereas we are developing the capacity to alter and manipulate the genes themselves.

If "playing God" means consciously eliminating, modifying, and creating new species, then we have been playing God for many thousands of years. Neolithic breeders started this endeavor with the domestication of plants, animals, and microbes some ten thousand years ago—cross-breeding for unnatural characteristics that violated species integrity but satisfied human needs and desires. In the meantime, species that competed or preyed on our food stocks were eliminated.[51] *Homo sapiens* was able to outlast other hominids partly because of our direct ancestors' ability to transform an entire series of wild plants into wheat, peas, lentils, and so on. And today, most people do not feel it is unnatural to enjoy their favorite seedless watermelon, sweet corn, and countless other highly "unnatural" farm products. Creating totally new organisms in the new field called synthetic biology is not radical from a historical perspective.

If "playing God" means controlling one's environment and destiny, we are also not doing anything new in that regard. Even the primitive aboriginal Australians managed to transform the ecology of their continent through systematic burning. Immortality was once thought to be a defining characteristic of God or gods, but we have been trying to find the magic pill or practice that produces longevity since the dawn of civilization.

If "playing God" means acting from godlike perspectives instead of with human motivations, then we can and already do play God, and with His blessings. As the *Tao Te Ching* puts it, if we can put our selfish instincts aside, like a sage ruling from the purest motives and relying wholly on quiet and inner peace, then we can do things for the whole of the world; and if we love the world, then we are ready to serve it. And if we are thus behaving in a godlike fashion, so be it.

Those who object to "playing God" for fear of offending our Creator should be reminded that there is nothing to fear in our creativity if in fact we inhabit a universe that is created by an all-powerful God. It is absurd to suggest that we have become so powerful in manipulating nature that it is time to give back to God some rights and protections from our own intrusions that, after all, are possible only because of our divine endowments. If God is omnipotent and omniscient, then God was aware of and could handle all the future possibilities of intelligent conscious beings—His own creatures—when God created the evolving universe. So how could God be offended by anything we do? To presume that human technological interventions in the natural environment violate God's rule is to worship Mother Nature,[52] not the Creator God.

If God could not anticipate the growth of human civilization when God first created the world, then God is not omniscient and is not truly God. Actually, human beings can only carry out the possible consequences of God's creation, the ever-evolving universe and its natural laws.

The natural environment and biological species are not sacred. There is no single, ideal ecological "balance" to be maintained or restored. In fact, living things, from their very beginnings, have been "playing God" by manipulating their environment—not just their local environment, but the global environment. They do it unconsciously but "naturally." In a sense, we are incapable of anything "unnatural." In another sense, we are already practicing the art of building various "artificial universes" when we create rules and laws of chess games or political systems. Whether someday our progeny will acquire or create the ability to alter natural laws or to create baby universes is a fascinating open question. But again, whatever happens should be considered as a consequence of God's creation. The question of why God created (or is creating) this universe is perhaps forever beyond our comprehension, since we are forever trapped inside of it and unable to gain an outside perspective.[53]

The Cosmic Game is in a sense similar to a basketball game: there are clear rules that are strictly enforced, but the referees allow any outcome as long as the rules are not violated. They are never expected to participate in the game by suddenly jumping up to block a player's shot. It's a game that naturally evolves as new conditions and new players come along. So our attitude should be what the *Whole Earth Catalog* declared in 1968: "We are as gods and we might as well get good at it."

As we create our successor species, can we bear the psychological burden of being gods?

I recognize that many will worry that playing God in this way and altering human nature will drive us insane. But followers of Western religions have long heard demands of radical self-transformation from their prophets. In the past, our comprehension of such demands was limited. However, what is now technically possible continues to change exponentially, rendering these prophecies of human transformation ever more capable of fulfillment. Yet, despite this legacy, Christian fundamentalists have been among the strongest voices against evolution. But if they would literally follow the biblical teachings as they claim, these Christians would be ardent supporters of the posthuman world and would be able to equate the biblical Spirit with the Cosmic View, as in Romans 8: 6–8: "To set the mind on the flesh is death, but to set the mind on the Spirit is life and peace. For the mind that is set on the flesh is hostile to God . . . and those who are in the flesh cannot please God."

More generally, the concern about playing God points to the unbearable psychological burden we face, given the terrible mismatch between our freely creative inner selves and our finite and predetermined body and instincts. But the mind has an amazing ability to adapt to its own inventions. To our ancestors, flying across oceans, moving about in "magic carriages," and having a conversation with someone thousands of miles away would have seemed like acts of demigods. There is little doubt, moreover, that transhumans will go about their business neither more nor less in awe of their status than we are of our own. And why shouldn't we and they feel like the demigods that we really are?

Of course, this recognition will at first be limited to a relative few. We must recognize that our psychological capacity for absorbing change is limited, especially in the short term, but over time, we typically manage to go through three aforementioned psychological stages after the initial shock: total instinctual rejection, followed by refusal to accept through rational thought, and finally denying it is anything new. By then we come to believe that we have intuitively known and understood it all along.

Imagine that a Westerner visits an isolated primitive tribe with news that some people in a distant land have landed on the moon. What will be their reaction to the news? Nothing less than total astonishment. However, if the tribe worships the moon, then there could be outrage or a deep sense of the moon losing its "dignity." Such psychological pain would be intense but may not last for long. The tribe would soon adapt by creating comforting justifications. People in impoverished countries did not suffer from long-term depression even after they became aware of the existence of the far wealthier Western civilization.

Was the eugenics movement in the early twentieth century a lesson we should forget?

No, history should not be forgotten, but we must be careful about what kinds of lessons to draw from emotionally charged historical episodes.

First of all, the early-twentieth-century eugenics experience is simply one of the latest human manipulations of our genes. We've already noted that intentional or unintentional human eugenics has been practiced throughout history: genocide was practiced in tribal conflicts, and weeding out infirm infants was common in numerous cultures (it was especially well-known in Sparta). Without in any sense endorsing these practices, we can acknowledge that they are natural characteristics of our species.

Although the idea of achieving human perfection can be traced at least back to Plato,[54] who in *The Republic* considered correct mating arrangements vital for society, the first fully conscious eugenic efforts were initiated by Sir Francis Galton. Eugenics became a powerful social movement in the late nineteenth century and reached its peak in the early twentieth century.

In retrospect, the fundamental assumption of eugenics, that certain segments of the human population are congenitally inferior to the rest, is mistaken but not unique to the eugenics movement. Many Victorians, for example, believed that the more advanced status of the white race justified slavery and racism. Similarly, Ernst Haeckel's cosmic vision and his pantheistic beliefs of organic progress were rooted in the intolerant Aryan ideal—an ideal that appealed to pseudo-educated minds who had sought an authoritative yet simple account of modern science and a comprehensible view of the world.[55]

In the early twentieth century, this crude understanding of progress and the source of human diversity led to some laws that called for involuntary sterilization of "inferior" races or groups in thirty states of the U.S. as well as in Canada, the Scandinavian countries, and Germany. In 1907, U.S. President Theodore Roosevelt signed an immigration act that, on the grounds of eugenics, excluded "idiots, imbeciles, feebleminded persons, epileptics, insane persons" from being admitted to America. Negative eugenics, with its aim to get rid of the "biologically unfit," eventually provided "scientific" support for the Nazis' murder of millions of

Jews, Gypsies, Slavs, and homosexuals. The Holocaust was perhaps the single most traumatic event that ultimately led to the complete discrediting of eugenics after World War II, although at the 1963 Ciba Conference held in London, several Nobel Prize winners proposed various versions of eugenics to counteract the problem of genetic deterioration in modern civilizations.[56]

What Nazism and Marxism shared was a misguided conviction that they had found the iron logic of history and, by following that logic, they could reshape humanity. For Nazis, the preordained logic of human history is a racial struggle, with progress based on the superior Aryan race eliminating all the other, presumably inferior ones.[57] For Marxists, human history is a history of class struggles, and human destiny is the inevitable victory of the proletariat.

As we discussed earlier, the idea that "progress" is made only through the strong violently eliminating the weak is a tunnel-vision conception and has little to do with how evolution actually works. In fact, the conviction that there is a simple logic of evolution of any kind is erroneous.[58] The initial successes of both the Nazi and Communist movements were not due to their understanding of historical trends but rather were due (at least in part) to their firm grasp of intergroup human psychology: if you can divide people into groups—even artificial groups decided by a coin toss as done in psychological studies—it is easy to fan hostility among them.[59]

Another useful lesson from these episodes is that the rise of a new system is always a turbulent process with its impact uncertain except in hindsight. When market capitalism hit its low point during the Great Depression, it was hard to argue with its enemies who believed that the system had fatal deficiencies. Millions of people were thrown out of work with vast production losses, human misery, and dire political consequences—the Great Depression led directly to World War II. The real terror of the Great Crash was the failure to explain it. It is only with the hindsight of several decades of multinational research that economists have come to realize that the Great Depression was *mainly* caused by macroeconomic policy mistakes, although the exact dynamics are still debated.

Getting back to eugenics, there is a world of difference between the eugenics practiced earlier in the twentieth century and what is practiced now. Eugenics then was about setting up a command-and-control system to make people breed for the state;

today, genetic technologies are driven by private individuals' own choices. The "Big Science" approach that totally disregards the basic evolutionary principles was a product of National Socialism and Communism, and hopefully it will stay in the dustbin just like the political systems that incubated it. The principal lesson we should draw from the earlier eugenics movement is not that its seemingly lofty goals inflicted great human misery (although they certainly did), but rather the evolutionary principle that freedom works better than coercion.

Another lesson is that we should aim for maximum possible diversity, rather than focusing on trimming the "unfit." There is no room for arrogance in the Cosmic Vision. We shall never know who the unfit really are, although we tend to know more over time. The failure of the early eugenics movement is a warning about the dangers of new technologies, as well as a painful reminder that most new ideas fail. But that should not prevent us from trying, failing, and trying again until we prevail in the end.

Given potential dangers, shall we withhold new technologies until they have been proven to be absolutely safe and harmless?

This is the argument known as the *precautionary principle* (PP), at least in some of its definitions and practices. The PP essentially prohibits any new technology or activity unless it can be scientifically proven that there will be no resulting harm to health or environment.[60] The PP is psychologically comforting, but it is more wishful thinking than a practical approach.

In principle, we should handle the risks in new technologies the same way as we handle other kinds of risks. The PP has been used by those with a hidden agenda to halt creative innovations, since no amount of testing can guarantee the safety of something new. It follows from this that an absolute ban on risk does more harm than good. We should always weigh the pros and cons of waiting for further evidence. When the negative consequences are irreversible and catastrophic, a bias toward caution and inaction is warranted. But there is a critical difference between calculated caution and unthinking inaction. Moreover, history is made up of countless acts that had potentially irreversible world-ending consequences.

Imagine a debate circa 1492 on the eve of Columbus's historical trip based on the PP, which can be summed up as "control and regulate first, investigate later." Some might have asked, what if Columbus might run into a superior power in a strange land that could then come and destroy us? Since nobody could prove that such a power did not exist, the trip would never have been allowed.

Then there was the famous "Maybe we'll ignite the atmosphere" argument back when the Los Alamos scientists were building the atomic bombs during World War II. Few believed the chance was that big, but nobody could totally rule it out, either.

The PP, when implemented in full as a hard legal or regulatory constraint, would leave society paralyzed. It is a very peculiar form of power, what Jean Baudrillard called *deterrence*: "what causes something not to take place."[61] Since nobody has perfect foresight, the PP is essentially a doctrine of "never do anything for the first time." Concerning newfound genetic knowledge, the British writer Bryan Appleyard warned us about the moral of Leontes's suggestion in Shakespeare's *The Winter's Tale*: when you drink your tasty wine from a cup, how can you know if there is a spider at the bottom of the cup?[62] We cannot. The seemingly rational approach of the precautionary principle suffers from tunnel vision, which prevents one from realizing that the biggest risk lies in doing nothing at all.

Risk does not disappear simply because one is averse to it. It is a necessary evil of existence. Selective, temporary delay can be a smart policy, but attempts to eliminate risk will ultimately bring greater danger over the long run. This is why higher perspectives such as that provided in the Cosmic Vision, aided by faith and courage, are needed when one is facing great uncertainty but also great potential.

What we need is what Max More calls the "Proactionary Principle"[63] in active management of risks. The best way to deal with risk is to enhance our tolerance of failure with diversification and preparation. The precautionary principle is irrational because it ignores dynamic feedback from shocks and unintended consequences. As Joel Garreau pointed out, "If someone in the 1970s had correctly forecast the ubiquitous presence in the 21st century of computer viruses viciously and constantly attacking the brains of our most sensitive systems, any sensible person would have concluded we were doomed. It might have seemed laughably Pollyannaish to believe that an immune system could co-evolve to match this problem."[64]

Should we wait until we become technically adept and morally wise?

The short answer is that we can become technically adept not by passively waiting but by active trial and error. Our wisdom evolves as technology improves, and our instinctive moral sense cannot get an upgrade until we alter human nature.

The sinister side of human nature is in part a reflection of human rigidity in the face of rapid technological change; and again, to the extent that we turn away from improving technology, we fail to grow in moral wisdom. To quote Pogo, "We have met the enemy and he is us." The more we delay the improvement of human nature through advanced genetic and robotic technologies, the worse the consequences of the mismatch will be: during the twentieth century, 86 million people were brutally killed in armed conflicts and a nuclear arms race brought global civilization to the brink.[65] Improvements are long overdue!

If evolutionary history is any indication, our posthuman endeavors will go through some traumatic bottlenecks that could drastically trim diversity at some point, just as the initial release of oxygen decimated most life forms on Earth. There is no denying that we humans are now on a rope over an abyss, tied between beast and superman. But not knowing exactly how we can cross over doesn't mean it can't be done. Numerous species were eliminated in the brutal process that led to our emergence. But what has worked throughout evolutionary history is to create even more new species, not protecting existing ones at all costs.

The experience of a "J-curve," where things have to get worse before they get better, is nothing new for us, whether it is the growth of economic activities causing a period of worsening pollution or the transition from dictatorship to democracy causing a period of political chaos. Still, it cannot be emphasized enough that these local, transitory pains have given rise to cosmic joy and comprehension of evolutionary laws that govern the universe; and that has provided us with an ever-changing, open-ended future that has invariably brought with it a cleaner, safer, and more prosperous world.

In the end, we must try our best but be ready for the worst, take our chances and see what comes at the end of the tunnel, and enjoy our God-given freedom with a humble acceptance of fate.

Concerning the risks, do we know too much for our own good?

Let us use this question to summarize the discussion in this chapter. Certainly the technical and moral challenges on the posthuman frontier are daunting, and the existential risk is potentially considerable. Yet there is a big difference between "we should be cautious" and "we should just wait." In pursuit of the totally new, humans are always torn between caution and curiosity, between accepting our place on Earth and striving for transcendence.

Today's questions in regard to the danger of human knowledge and foresight have to be put into historical perspective. They have been asked and answered numerous times by some of the sharpest minds in history, as literary critic Roger Shattuck's book *Forbidden Knowledge* indicates, but we must return to these questions again and again. Beginning with the earliest creation myths, such as the biblical story of the Fall and the Greek mythical characters of Icarus, Prometheus, and Pandora, people have been enjoying the fruits of knowledge and power while remaining deeply wary of their dangers. In John Milton's famous words, humanity must be "lowly wise" and dream not of other worlds.[66] Reflective thinkers like Montaigne were keenly aware of the danger of "man's attempt to rise above himself and humanity" due to his suspect faculties—we are animals after all. Yet despite his deeply held skepticism about the power of reason, Montaigne never stopped using his rational thinking in his reading and writing. The frankness with which he dealt with these contradictions speaks directly to us today.[67]

Echoing Montaigne, Voltaire's modest proposal in his work *Candide*, "Let us cultivate our own garden," also advises us to live within our reach or range. But Francis Bacon astutely observed the virtues of a dynamic approach. He differentiated three types of philosophers: presumptuous dogmatists who think they know the truth, despairing skeptics who believe nothing can be known, and persistent inquirers who keep asking questions in order to extend imperfect knowledge.[68] Bacon's choice of the last type has been more or less the road that the Western pioneers have traveled, and a road that we shall strive to take.

The dangers of knowing too much are less than the dangers of knowing too little. Trying to conceal and regulate things because you think they might be difficult to deal with can be a patronizing underestimation of the public. In a pluralistic world, caution usually predominates, as it should. Only the energetic, overconfident hypomanics are willing to throw caution to the wind—most of them end up as losers, but all of them should be our heroes.

In the end, I believe that the best mind-set is the Proactionary Principle,[69] which follows Mao Zedong's principle of being "strategically fearless, tactically paying full attention" (战略上藐视, 战术上重视). Technological risks are not fundamentally different from other types of natural or man-made risks that humanity has always dealt with—and only cool-headed, rational assessment can help us discriminate and effectively deal with them.

CHAPTER ELEVEN

THE MORAL ARGUMENT

We still don't know how to put morality ahead of politics, science, and economics. We are still incapable of understanding that the only genuine backbone of our actions—if they are to be moral—is responsibility. Responsibility to something higher than my family, my country, my firm, my success.

—Václav Havel, address to the United States Congress, February 21, 1990

Ethics (which can be defined as a theory or system of moral values) involves at least three fundamental, interrelated issues:

1. For what end should we live?
2. What fundamental principle, if any, should guide our actions?
3. Who should profit from our actions?

Since antiquity, moral philosophers have debated numerous ethical issues and established many sets of moral standards and guidelines. Practically, ethics is about our attitudes toward other people, and the most fundamental moral principle calls for us to treat people as we would like them to treat us, the so-called Golden Rule. However, we may have to address ethical issues from a totally different perspective when we take actions that result (directly or over time) in changes to human nature itself—such as the creation of cyborgs and autonomous robots.

As we discussed in the last two chapters, conscious evolution is most likely to be messy and unpredictable. As far as we can tell, there is no way to insure against a posthuman transitional period that may entail realignment of social interests, mass casualties, or even "near-death" experiences for the entire civilization. Hence, deep concerns about bioethics and other moral challenges in a posthuman world would have many dimensions, including fair and humane treatment of people, respect for God or nature, political rights for nonhuman intelligent beings, and whether it is OK to treat humanity not as an end but as a means. Concerns about the posthuman future center around these issues.

The central argument of this chapter is that without a transcendental morality, we could be forever trapped in a human-centric morality without clear answers to the ethical questions related to the posthuman future. The first step to developing a transcendental morality is to see the humble origins of, and profound limits to, human morality. The second step is to ground our posthuman morality in the Cosmic Vision—the vision of humanity's pivotal but transitional place in an evolving universe. Just as human morality has served a vital function of strengthening and reinforcing our moral instincts, a transcendental morality will serve as a guide for us to treat humanity and forthcoming advanced intelligent lives as they should be treated, under the vision of furthering cosmic evolution.

11.1. "Is This Moral?"

Although cloning humans seems to be one of the defining moral issues at the moment, almost every major new technology—from the plow to the nuclear reactor—has raised grave (and often seemingly unprecedented) moral concerns.[1] Medical innovations that we consider routine today, such as anesthesia, blood transplants, vaccinations, birth control pills, and organ transplants, have all faced fierce resistance on alleged moral grounds. Often they were considered inhumane or offensive to God or to our "higher" nature.

One of the most widely accepted moral principles today is "First, do no harm." This is clearly a sound principle for handling human relationships. But within the context of cosmic evolution, can we treat it as an absolute moral principle? Might it be possible that, in that broader context, there is actually some *harm* in "doing no harm to our fellow humans"? Again the question comes down to whether our ultimate aim is human happiness and well-being or fulfilling our ultimate potential and our responsibilities toward cosmic evolution.

It has often been argued that only the conscious human can commit evil acts because only a human knows what he or she is doing. But, as I've previously argued, this unique reflective consciousness is a mere add-on to our animal motives and instincts, as are our moral instincts of empathy and justice, all of which exist and operate independently of our conscious thoughts. It is naïve to believe that the conscious mind has full control of the "society of mind" that forms our central nervous system. The reality (though counterintuitive) is often the reverse—our inborn instincts and motivations utilize our rational and reflective mind to justify actions that satisfy their urge. Our "free will" is not as free as we subjectively feel, and it is often unknowingly hijacked to serve an instinctual agenda. The source of our moral dilemmas—defined as our innate drives to do things that our consciousness finds unacceptable or harmful from the perspective of others—can be traced back all the way to the beginning of animal history. Murder, or intentional killing, started as soon as animals appeared on Earth some 600 million years ago—the capacity for movement *is* the capacity for war,[2] for movement opens up competition for territories and resources and forces individuals to confront unaccustomed creatures and habitats. Most animals have no problems with killing members of their own kind as a means to survive and to reproduce. Humans, however, have evolved stronger-than-usual moral instincts in the course of living together as a highly social animal. The development of reflective consciousness further enables us to "put ourselves in others' shoes" and strengthen the empathetic sentiments. Our greatly strengthened moral sentiments make moral dilemmas more pronounced, but we are certainly not the first to "experience" them—our belief to the contrary is human hubris.

Darwin said that there is grandeur in the way of life that is produced by natural selection, but he also admitted that his moral sentiments were violated by the natural-selection process that he discovered. He wrote to his friend Joseph Hooker in 1856: "What a book a Devil's Chaplain might write on the clumsy, wasteful, blundering low and horridly cruel works of nature!" Cruelty seems to be woven into the fabric of biological life, and we are very much a part of it. Natural selection is a process of exuberant creation *and* ruthless elimination. Similarly, the best practice in science and culture is a kind of negative pragmatism: find out what is not working and get rid of it.

Killing is perhaps the most efficient selection mechanism in the evolution of all varieties and in every sphere. Since all life is in the position of potentially being killed (individually and as a species), we might ask why nature bothers with birth and rebirth at all. But waste seems to be unavoidable and even virtuous. Under the second law of thermodynamics, order and structure cannot emerge without dissipating heat (disorder and waste) into the environment. Life does not emerge and persist in defiance of the second law. On the contrary, life, as an ordered and dissipative system, maximizes the rate of entropy production in the universe (or, from the thermodynamic perspective, *gradient reduction* in a system characterized by 5800 Kelvin incoming solar radiation and 2.7 Kelvin temperature of outer space). This is what Rod Swenson has proposed as the law of *maximum entropy production*: the more life, the faster the energy in the universe is degraded.[3]

In this light, the "sinful" act of killing or taking things away from others is far more "original" and universal than the emergence of conscious humans. This observation relates not just to predators and exploiters: nothing that has a distinct structure from its environment can escape the predicament of being "selfish," of gaining something at the expense of others, and of polluting the environment. The second law is a built-in feature of the universe. The constant struggle between selfish and cooperative forces is such a fundamental yin-yang dynamic for evolution that it may encompass not only all life on Earth but all life in the universe.[4]

With insight into how natural and cultural evolution work, we can understand how messy and ungainly these processes are if viewed from the narrower perspective of our "First, do no harm" standards. When we marvel at intricate biological organisms or elegant scientific theories, we must realize that a huge amount of chaff has been discarded during the biological and cultural creation process. In the absence of destruction, the process of creation—improvement on and advancement of the species—is simply impossible: "No pain, no gain," as the saying goes.

Psychologically, however, we wish no harm to be done, not just to other human beings but to other lives and the natural environment—and that is why supernatural beliefs of painless creation and frictionless progress are so emotionally attractive. These forms of creation would involve neither trial and error nor struggles for survival that involve killing, cheating, exploitation, stealing, and so on.

In this chapter I intend to show why attempts to extend human moral standards beyond human relations are unwise, impractical, and inappropriate, and that we instead need to create a new moral standard that transcends the narrow requirements of human morality.

Indifferent nature. If we define God as the Creator of the universe, then God's concern about any given individual cannot be the end, even though each individual is a part (however small) of the whole process. As we discussed extensively in chapter 8, in order for evolution to work, higher perspectives must be "given" higher priority, and this often contradicts our limited human priorities. If we were at the beginning of the natural evolution process, and we somehow came to the realization that over 99 percent of the species to be created during the process would be wiped out, often in violent ways, then we might well have concluded that the Creator is evil.

Again, the only straight answer is that the creation of any particular single species, including our own, is not an end in itself. And we cannot use human morality to judge *that* Creator, whether we believe the Creator is a conscious being or not, because the creation of life on Earth is a much wider endeavor than human interactions.

Moving closer to recent history, at the beginning of primitive humans' ascent from a group of primate species, we might ask who made the decision that only *Homo sapiens* would survive. Does God decide from minute to minute the fate of each individual person?

No, only a nursery-school god would do that. It is only by putting ourselves into some kind of resonance with the principle that governs all, at all times, that we can begin to perceive the deep significance of a *cosmic* God. Nature seems to be *neither kind nor cruel but indifferent* to the fate of the individual.[5] This reality is more evident in the animal kingdom than in the human world, largely because with animals we can use the "other eye" to observe from the outside. For example, in the animal world, both "evil" and "moral" behaviors play a role in the game of life. There are numerous examples of animal kindness and altruistic behavior, but also numerous examples of "evil" acts—of which the following are a bare sampling:

- A wasp raises its young by finding a cockroach, immobilizing it with a sting, then digging a hole in the ground, dragging the cockroach into the hole, and laying a single egg on the roach's body. Her child then eats the cockroach alive from inside out over the course of some days and emerges out of the hole to repeat the cycle. We can marvel at nature's ability to develop such a complicated set of instincts, but think about how mercilessly the cockroach is treated! In Darwin's words, "I cannot persuade myself that a beneficent and omnipotent God would have designedly created [it]."
- The queens of a particular species of parasitic ant have only one remarkable new adaptation—a serrated appendage that they use to saw off the head of the host queen.
- A cuckoo sneaks into the nest of a bird of another species and lays her egg. She then takes away another egg, so that the nest owner will not notice the difference when she returns. The cuckoo egg is hatched early, and it pushes out all the other eggs in the nest laid by its foster parents so that it can get all the parental attention and food.
- For the ovoviviparous sand shark, the young eat each other up in the usually protective confines of their mother's oviducts until only one well-fed shark is left to emerge supreme.
- At up to twelve feet long, the Komodo dragon in Indonesia is the world's largest lizard. Its lethal weapons are shark-like teeth and virulent bacteria in its saliva. It attacks a deer or pig and leaves it with messy, lacerated wounds. Then it waits around while the animal suffers an excruciating slow death from blood poisoning over the next few days.

Should we eliminate the above "immoral" species from nature based on our moral standards? On first thought this very idea is absurd, since our moral standards don't permit killing a species en masse just because it does cruel things—and seemingly cannot help itself! In reality, when our own survival is at stake, we never hesitate to do whatever we can to control and eliminate other species, such as malaria-carrying mosquitoes.

On top of all the miseries inflicted by predators and parasites in interspecies competition, the members of a species often show no pity to their own kind. Infanticide, fratricide, siblicide, and rape are routine practices in many species. Not all mother animals are kind nurturers and fearless protectors of their young—in fact, nature abounds with mothers that defy the standard maternal script in a raft of macabre ways. There are mothers that zestily eat their young, mothers that drink their young's blood, and mothers that raise one set of their babies on the flesh of their siblings.[6] In addition, the coldhearted culling of the young is a centerpiece of the reproductive game plan for many species, including the adorable giant panda.

Most Westerners are either indulgent in self-supremacy (animals are immoral because only man is made "in God's image") or shocked by nature's cruelty, partly because they are immersed in the tradition of a personal and benevolent God. Although Buddhist monks are famous for their extreme mercy for all forms of life, overall the Eastern view is far more realistic and sober. Lao Tzu said, "Heaven and Earth have no morality; they treat everything like worthless *Chu Gou* [a dog-like article made of grass, used in ritual ceremonies and thrown away afterward]. The sage has no morality; he treats everybody like worthless *Chu Gou*" (*Tao Te Ching*, chapter 5). Chuang Tzu said, "In ancient times, the highest wisdom is the view that the world is a deep void. This is the apex, nothing can be added to it. The next level of wisdom is the view that there are things in the world, but there is no clear distinction among them. The next level view is that things can be clearly distinguished, but there is no good and evil. The next level, which could hardly be called wisdom, is the view of good and evil. With that bias brought into the mind, the Tao is damaged" (*Qi Wu Lun*, 齐物论).

Obviously Lao Tzu and Chuang Tzu were able to grasp the Cosmic View intuitively. In many mundane areas of life, we have no trouble thinking this way. For example, "neither kind nor cruel but indifferent" is the attitude we take to computers or TVs when we unplug them for good. We do not miss them as long as we can get better ones—in fact, this is the most likely reason that we dump fully functional TVs, PCs, mobile phones, or cameras. Our emotional attachment is sometimes extended to personal items such as a sports car, but in general we tend to be rational and indifferent concerning inanimate objects.

We are far more emotionally involved in a political revolution, which destroys human rulers, than in a technological revolution, which destroys reigning machines. On the one hand, we may hate terrible rulers and want to "see justice served" on them; yet we are emotionally involved in other ways as well—to some extent we see ourselves in them, because we share the same human instincts, and because we know that "there but for the grace of God go I." And we know that destruction and revolution affect many people and involve very complex and varied human motivations, all of which are human. Therefore it is hard, emotionally, to commit to "creative destruction" at the societal level. It is psychologically unbearable to discard human beings like old and outdated machines just because better alternatives with higher performance and cheaper costs become available. Those who do so are said to be "cold-blooded" or even evil.

Yet, in sports, coaches are expected to adopt the attitude of indifference concerning which player to select for the team or which player to send onto the field. This is also supposed to be the mind-set of movie directors, business managers, and military leaders. The reflective conscious mind is fully capable of sacrificing short-term gains (or gains for a part of the whole) for the sake of long-term gains (or gains for the whole). But in human relations, the attitude of indifference is meant to be restricted to the professional realm. Coaches, directors, and managers must be trained to put their emotions aside when they deal with people professionally. Most voters, on the other hand, seldom select political leaders purely on the basis of professional qualifications.

The obvious reason we treat people and machines differently is that we instinctively feel people are qualitatively different. But if we dogmatically cling to this attitude, there will be numerous moral challenges in the posthuman future, not unlike the challenges those who cling to a literal interpretation of the scripture face in the age of science. For example, the boundary between human and machine will become less distinct, and we will face entirely new questions, such as: Should we extend our ethical and moral standards to machines that have feelings and conscious thoughts? If the machines express pain or distress, will we try our best to hang on to them longer than we would have on the basis of purely rational considerations?

We have already seen that expressions of human morality change significantly with social and technological change—for example, in regard to slavery, child labor, women's rights, and so on. When lines between humans and machines blur and the "obvious" reason that we treat humans differently becomes not so obvious, we will see more clearly that human morality is a special, not universal, principle.

Morality is precious. Placing morality into a higher perspective by no means implies that we should not care about personal suffering. Cruelty, inequality, and physical harm must continue to count as costs in any social and political decision, especially those made in the name of a grand vision. It is important to recognize that indifference is not the same as callousness. To the contrary, compassionate love is a vital part of "indifference." To get the best out of their players and team members, coaches and business leaders often show genuine concern and care toward their players and workers. The love that many feel about their "personal" God is not totally groundless—God is indifferent in the sense we have explained above, but we can also say that God cares about and "loves" all of His creation, even though He "allows" it all to be destroyed.

In this context, we can see why Nietzsche's famous critique of Christian morality is either mistaken or—perhaps more likely—misunderstood. He was reacting strongly to the Christian doctrine of love when he suggested that nature is "boundlessly extravagant, boundlessly indifferent, without purpose or consideration, without pity or justice, at once fruitful and barren and uncertain." He cried out that Christian love is the exact opposite of true living, which is "valuing, preferring, being unjust, being limited, endeavoring to be different." He argued that it is wishful thinking to try to dictate human morals and ideals to nature.[7] But it has often been argued that his verdict is not a complete description of nature. We must say that Nietzsche was wrong if in fact he thought that nature is simply cruel; at least that's what many of his readers believe was his position.

As pointed out previously, nature is both kind and cruel; for, while the instinct to kill is deep-rooted, the instinct to cooperate and be good to others is equally so. The origin of life can be traced to the merging of two complementary structures: proteins (the hardware) for metabolism, and nucleic acids (the software) for replication. Symbiosis between different biological organisms kicked off the diversification of species. Other forms of cooperation for mutual benefit have been some of the strongest forces behind natural and cultural evolution toward higher levels of complexity and diversity.

It could not be a cosmic accident that multicellularity emerged independently many times, that most if not all social animals such as ants, bees, and humans are immensely successful, and that complex human societies and higher "selfless" perspectives emerged again and again in different cultures. In other words, the richness of life and culture may depend as much on cooperation as on competition. The very definition of being *alive* is the ability of vast numbers of seemingly independent entities to support each other, resulting in a lively *whole*.

Although Malthus's doctrine of callous population dynamics was one of the inspirations for Darwin, so was Adam Smith's doctrine of emergent cooperative behavior in social interactions. The evolutionary process is far more complex than the strong exploiting or dominating the weak. In fact, Darwin himself held a strong belief that the evolution of humanity and strong morality go hand in hand:

> Two classes of moralists: one says our rule of life is what *will* produce the greatest happiness. The other says we have a moral sense. But my view unites both and shows them to be almost identical and what *has* produced the greatest good or rather what was necessary for good at all *is* the instinctive moral sense.[8]

The moral instincts in humans are invaluable because they counterbalance some of the more primitive instincts, such as aggression. This implies that acting in one's own self-interest does not necessarily mean behaving selfishly. And these are not recent revisionist interpretations of the original Darwinian theory. Thomas Huxley, one of Darwin's greatest supporters, explained in 1893 that in the phrase "the survival of the fittest," the word "fittest" has a connotation of "best"; and about "best" there hangs a moral flavor.

We are the best and most intensely social animal if we consider the power of our cruelty *and* our kindness. "Good" and "evil" can both be winning strategies for the individual in the struggle for survival. Furthermore, our moral impulses can be very destructive when they are hijacked by our animal instincts; for example, wars and other destructive mass movements can be fueled by self-righteousness and sadistic retribution disguised as social justice. Nevertheless, we need our moral sense more than ever in our dealings with social issues. And luckily, there is no indication that our moral sense and commitments have diminished with modernity. The Noble Savage is a myth: primitive people were unable to commit grave immoral damage to the invaders of their spaces and to their environment mostly because they lacked the means, and this inability had nothing to do with any lack of motivation or superior morality.[9] Some hunter-gatherer societies were scrupulously egalitarian, but only because there was violence or the threat of it to anyone who sought to "get ahead" of the others.

Overcoming moral bias. If we understand that human morality is not perfect but arose out of the dynamic interactions of competition and cooperation in a particular social environment, then it is possible to recognize potential pitfalls in following our moral sentiments dogmatically. Morality is not an abstract good; it cannot be isolated from its context. Morality evolves and is always contingent. To recognize and overcome our moral biases is not easy, especially in the West, with its deep-rooted idea of an absolute dichotomy between the human and all other species. It requires conscious effort and the courage to go against societal taboos.

Because we are social animals, part of our moral sense derives from our natural hypersensitivity to suffering inflicted by or on others directly and intentionally. Any act of enriching oneself at the expense of others is universally condemned. But a higher perspective on the entire process may show a different story.

We should be wary of saying that what makes an act good is that it is unselfish, since the determination of what is "selfish" is a human judgment from a certain perspective. Good intentions do not automatically equal good results—for example, protecting the weak by insulating them from competition may, more often than not, end up making them weaker and thus eroding their self-confidence. The "justice bias" is a result of treating equality as the end rather than as a means to achieve better results for the society as a whole. Politicians who instinctively side with a weak group without thinking through what made them weak may create solutions (via programs or assistance) that exacerbate rather than solve the root problem.

As another example, we should also be aware of the distance bias in our "automatic" moral judgment. Many of us feel compelled to spend five dollars to help a sick person on the street, but much less so when the same money could save starving children on a distant continent.

The question of why we honor moral sentiments must be answered in the same way as we answer the questions of why we enjoy sex and higher status—all complicated but not inexplicable issues. Our moral sentiments are not ready-made ideal forms but historical products handed down by our Paleolithic ancestors as a part of our behavioral repertoire, and to some extent they are shared by other social mammals.

11.2. A Realistic View of Human Morality

We have seen that our moral sentiments have tremendous value in balancing our animal instincts. They transcend race, sex, religion, culture, and any other characteristic that differentiates humans from each other. They may embody elements of principles that will prove to be universal, such as compassion. For humankind, ethical standards such as the Golden Rule and the Hippocratic Oath are timeless.

However, as a whole, *human morality is human*. It is not divine; it does not transcend our species. No abstract moral principles exist outside the particular nature of an individual species. Our moral laws are dependent on human nature and are far more "local" than natural laws. This perspective is difficult to appreciate within a society that is strictly human. But it is easy to conceive of some other intelligent species (whether earthly or extraterrestrial) evolving moral principles that are repugnant to us, such as cannibalism, incest, or parricide, to name just a few.[10]

There can be no absolute moral judgment even within human society, although the homogeneity of human genes implies that broad moral principles should be universal. The subjective judgment of how a person is affecting others' well-being is made by a "society of mind" in the brain—each part of this "society" competing for conscious attention—rather than some single, one-pointed entity that we subjectively feel and imagine our brain or mind to be. Recent brain imaging studies show that in quick judgments, our moral sense is controlled by emotion and intuition, and moral reasoning is often a post-hoc justification for what unconscious processes already have told us is right or wrong.[11]

Morality can only be understood in terms of such human-specific inclinations as emotions, desires, urges, drives, and instincts—there is no universal Kantian "categorical imperative" for all rational beings, if "rational" is defined to be inclusive of nonhumans, aliens, and posthumans. Human morality is conditioned by human nature—when the latter changes, so must the former. Nietzsche's most famous words, "God is dead," are a proclamation that there are no divine or eternal underpinnings to human morality. It is not a typical atheist's assertion of God's nonexistence, but rather of the humanistic God's vanishing from this world that is now equipped with science, a world ripe for the "superman." Nietzsche knew that developing such a superior being was largely wishful thinking at his time, and that he was speaking to people not yet born.[12]

There is nothing inherently good or evil when one person cuts open the body of another with a knife: intention, or rather the external interpretation of it, is critical since the cutter can be a surgeon or a murderer. Moral demands, which hold the reflective consciousness responsible for behavioral consequences, are *only* the best available control mechanisms, since we do not yet know how to remove evil motives from the brain.

The prerequisite of strong conscious control is the primary reason why human morality—with its idea of equal rights—does not apply to animals or small children, nor would it be considered applicable to prehistoric humanoids, nor should it necessarily apply to any posthuman species. Pets and other animals, young children, and individuals of very limited mental capacity can be deadly (just as we can be), but we do not judge them by our normal moral standards. We know that the more effective way to control their behavior is to address them at a level that they can respond to. Every existing animal or nonhuman species has its *own* nature, its own hierarchy of instincts—in a sense, its own set of virtues and vices.

Just as slavery was considered a presupposition for a normal society and viewed by many as perfectly moral in classic Hellenic civilizations and even up to nineteenth-century Europe and America, and the caste system in India was considered essential for social harmony and the individual's accumulation of karma for four millennia, the high status of freedom and equality as inviolable human rights should be considered a product of the modern Western civilizations and nothing "eternal"—even though it is built on universal human instincts such as empathy. It is both a strength and a weakness for the Western religious tradition to anchor its ethics in supernatural beliefs—a strength in that it fosters the rule of law, a weakness in that it converts historically derived moral principles into absolute imperatives.

Viewed from the lens of an individual lifetime, most of our moral instincts appear absolute. But from the historical perspective, new types of ethical standards are constantly emerging. This process is unlikely to stop; in fact, it is only likely to accelerate. Modern ethical beliefs are an indispensable part of human culture, but they are still a particular historical product.

Moving beyond human morality. Perhaps the first stage of expanding human morality, with or without transhuman insight, involves deepening universal moral principles such as the Golden Rule.

In the famous passage in the Talmud (Shabbat 31a), Rabbi Hillel was challenged to recite the whole of the Torah while standing on one leg. He replied, "What is hateful to you, do not do to your neighbor; that is the whole Torah, all else is explanation. Go and learn this." The Golden Rule is often stated positively as "One should treat others as one would like others to treat oneself" or something similar.

Independently, the Golden Rule has been discovered many times in history and around the world by sages, prophets, theologians, atheists, and moral philosophers and theorists. It is widely regarded as the most important moral standard. Numerous ethical codes can be seen as concrete expressions of this general principle.

Nevertheless, the Golden Rule might be taken to imply that since we humans are the same, we should be treated the same. How might we expand such a statement to apply to those who are radically different from ourselves? I believe the common phrase "*Put yourself in the other's shoes*" is a better fit for the posthuman world, where all kinds of intelligent beings jump into our ever-more-complex moral equations. The perspective-based concept provided in the Cosmic View stretches the reciprocal-based one in several dimensions:

- It not only covers the Golden Rule and all its other variations—such as the Silver, Brazen, and Iron rules, all of which incorporate the perspective of the "other"[13]—but also offers the flexibility of jumping into the perspectives of all kinds of collective entities we can think of, from family, tribe, society, and species all the way to the Creator of the universe. It asks one to break out of not only the personal perspective, but also the limited collective perspective. For example, Miles's Law (created by Rufus E. Miles Jr., assistant Health, Euducation and Welfare secretary under three U.S. presidents) states, "Where you stand depends on where you sit." Miles's Law attempted to broaden the limited collective perspectives of organizations by encouraging outside perspectives.
- It breaks the constraints of space and time. For example, it can be extended to people who are long dead or not yet born, who cannot react to what you do. John Rawls's "veil of ignorance" thought experiment follows the same principle; so does the surprisingly enduring slogan "What would Jesus do?" coined in Charles Monroe Sheldon's 1896 novel, *In His Steps*.
- It is applicable to the human mind itself by asking the conscious mind to understand multiple motives in the "society of mind" within our brain, including unconscious motives. Knowing what is behind these motives helps us decipher puzzling questions such as why we are capable of committing both good and evil acts.
- It relaxes the requirement of symmetry in reciprocal relations—a requirement that can be implemented only in relations between human beings, but does not work in relations between species or between biological and artificial intelligence (or humans and transhumans).

- It resolves the tension between positive freedom (liberty) and negative freedom (safety and security). Personal, civic, and "sovereignal" freedom for a person, a community, and a nation may seem conflicting, but they can become equivalent once we move beyond our personal viewpoint. From the holistic perspective, positive freedom cannot be enlarged without also enhancing negative freedom, and vice versa.

This stretched Golden Rule (shall we call it Platinum Rule?) could be the bridge between human and transhuman, the first step toward establishing interspecies morality for conscious intelligent beings. It asks us to take an "impartial spectator" position, to use Adam Smith's phrase.

To move beyond human morality, we must put the doctrine "we should always treat humans as an end" into proper historical context. Although that doctrine is gaining in global popularity and acceptance, this seemingly lofty ideal—championed by Kant as a "categorical imperative" and followed by numerous moral philosophers—is in fact peculiarly Western. It is, at the end of the analysis, a historical product of the Enlightenment and another form of human hubris. The instinct of self-preservation and self-gratification is powerful and extremely valuable, but from a higher perspective nothing is good in and of itself. This is the wisdom expressed in the book of Ecclesiastes, which in its cry "All is vanity" (insubstantial and impermanent) emphasizes the fact that human life and human goals, as ends in themselves, are futile and meaningless.

Just as we cannot pursue knowledge for its own sake, preservation and even perfection of the self can never be an end in itself. The absolute end is the entire Cosmic Creation. Just as human morality is not absolute, the same is true for human rights and personal autonomy. Human rights as we know them are the political rights of humans vis-à-vis other humans. Humans can be treated as the principal end when we are the highest known sentient being, which was taken for granted when Kant wrote his treatise, but we must always be prepared to relax this moral imperative when humans are no longer the "highest."

Robert Nozick identified three distinct kinds of moral status in his 1974 book *Anarchy, State, and Utopia*:

- **Status 1:** The being may not be sacrificed, harmed, and so on, for any other organism's sake.
- **Status 2:** The being may be sacrificed, harmed, and so on, only for the sake of beings higher on the scale, but not for the sake of beings at the same level.
- **Status 3:** The being may be sacrificed, harmed, and so on, for the sake of other beings at the same or higher levels on the scale.

Prevailing Western human morality clearly places animals in Status 3 and humans in Status 1. But as Nozick suggested, even as we consider human affairs, humans may occupy Status 2. He asked:

> Do ordinary views include the possibility of more than one significant moral divide (like that between persons and animals), and *might one come on the other side of human being* [i.e., one placed higher than human being]? Some theological views hold that God is permitted to sacrifice people for his own purposes. We also might imagine people encountering beings from another planet who traverse in their childhood whatever "stages" of moral development our developmental psychologists can identify. These beings claim that they all continue on through fourteen further sequential stages. . . . However, they cannot explain to us (primitive as we are) the content and modes of reasoning of these later stages. . . . Do our moral views permit our sacrifice for the sake of these beings' higher capacities, including their moral ones?[14]

We have not been obliged to think about these questions so far—presently Status 1 and 2 yield the same moral conclusions as far as humans are concerned, since in the known universe we are the "highest" in terms of capabilities and perspectives—but this will no longer be a hypothetical question as we enter the transhuman world. We *will* create these now-imaginary Nozickian beings that may be superior to humanity, morally and otherwise.

The mainstream Christian point of view, which perceives a deep-seated dichotomy between humanity (and angels) and other living creatures, has caused Western thinkers to have great difficulty in accepting humankind's deep historical and biological connections with all life on Earth. It has been an emotionally and intellectually grueling task to accept the idea that the great apes are our close relatives and that we are part of the continuum. Around the time Darwin published *On the Origin of Species* in the 1860s, many scholarly research "findings" were still attempting to demonstrate that Native Americans were unrelated to people in the Old World and that blacks were more like gorillas than human beings. The reluctance to dethrone humanity's privileged position also led to strong emotional reactions to the scientific evidence for evolution.

The Eastern, more realistic view of humanity, in contrast, has always placed the value of humanity in a relative rather than absolute position. Following Eastern wisdom, the general principle to apply to the posthuman world is that the value of humankind, along with all natural and artificial sentient beings, should be considered in the context of its place in cosmic evolution. Treating humanity as an end in itself is still a valuable principle, but it is a local perspective and should be applied only to the political life of humanity.

In the higher perspective, the end is the human spirit rather than its flesh. With no species enjoying guaranteed long-term prosperity in history, there is little doubt that humanity as it is will become obsolete—there will be better physical organizations to carry on the human spirit.

We can face the transhuman future with the upbeat mood of scientist Harold J. Morowitz, who—in his enlightened mysticism—was in a state of *cosmic joy and local pain* (joyful understanding of our place in the universe, but continuing to face the usual moral dilemmas in daily life), and who treated the state of human nature as one of *cosmic pain and local joy* (existential despair and hedonic indulgence). It is one thing for us to make peace with our limited morality. It is quite another for us to shirk our God-given responsibility to contemplate and to advance the Cosmic Creation and what will come after humanity.

11.3. Moral Guidance: Religion, Science, Humanism—or a Transcendental Perspective?

Arguments about morality often boil down to differences between the partisians of secular science and traditional religion. In this section I will propose a third alternative. I believe what we need is not more debate along these lines but unifying moral guidance from the perspective of transcendental morality. The conventional wisdom is that religious beliefs provide (or have traditionally provided) moral guidance. But then what role should science play? Is science undermining the foundation of morality by discrediting specific statements in the scriptures? Should we restrict science to the investigation of facts and religion to the nurturing of the spirit?

The "experts" have sharply divided opinions on these questions, and the debate rages on today. However, the theory of perspectives suggests that these are not the right questions. *The best moral guidance is a higher human perspective,* which can be expressed in either religious or scientific language. *The worst moral guidance is a parochial perspective*—a rigid, unyielding faith in a personal divine being or, alternatively, dogmatic and reductionistic interpretations of science. Rather than asking the question "Religion or science?" we should ask, "What kind of religion and what kind of science?" These questions and their answers can be approached regardless of one's religious affiliation or belief, or lack of it.

There is also a common misconception that science and religion must be separated since they serve different purposes: science tells us *how* to do, and religion tells us *what* to do. As Galileo famously put it, "The intention of the Holy Ghost is to teach us how one goes to heaven, not how heaven goes." In reality, religion and science have been intricately linked in the development and application of both scientific theory and theology—they have always been more or less indivisible.

At their best, transcendental religions guide and promote scientific exploration, while science explains and stimulates transcendental "oceanic" religious feelings.[15] Rational philosophy and empirical science originated in Greece; but with the pioneering work of St. Basil (331–379 CE), St. Ambrose (339–397 CE), St. Augustine (354–430 CE), and many others, the tradition of rational thought was carried on by erudite Christian and Muslim theologians who sought to understand God's mind through Greek logic and natural philosophy.

The modern scientific spirit was born in the medieval monasteries. It was based on the scholastic theological notion that, to know the mind of God, natural laws *can* be and *should* be discovered through rational and empirical investigations. During that period, science and religion were not only compatible; they were inseparable.[16] Roger Bacon (1214–1294), considered the first modern scientist and the author of the empirical method, viewed science as a quintessentially religious enterprise.[17] Much of natural science in Newton's time was inspired by the idea of the glory of God, rather than seeming to contradict it; so were social sciences such as economics.[18] Einstein, while not an orthodox believer, was a man of profoundly religious sensibilities, which were only increased by his scientific investigations. The same can be said for many modern scientists who claim no affiliation with any organized religion.

The theoretical framework of an expanding universe (from the Big Bang) owes its origin not only to Hubble's empirical measurements, but also to the pioneering work of Belgian priest-scientist Georges Lemaître and the Soviet physicist Alexander Friedmann.

The scientific worldview can be seen as theological in its core.[19] The naturalistic revolution merely replaced the name God with "nature" and the "divine law" with "natural law," but left almost everything else unchanged in the Western theological framework. Even Darwinian evolutionary theory has roots deep in monotheistic theology. While Darwin did not believe in the personal God of Christianity, his work was influenced by his early, inherited belief in a theistic God who was the prime mover of the cosmos.

By adding a theological dimension to both the realm of history and the realm of nature, many higher religions have given us an understanding of a deity that manifests its power through natural laws, hence advancing the concept of a linear history hidden from nature's and civilization's cyclical patterns.

Modern technology itself can be seen as fulfilling a transcendent purpose. At the deepest level, today's technologies can be seen as going beyond satisfying basic human needs; some would argue that they have been aimed rather at the loftier goal of transcending our mundane concerns altogether.[20] The latest scientific research, especially in the fields of cosmology and evolutionary biology, solidly supports the religious vision of a purposeful universe. Again, one could say that the spirit of the Bible shines through.

Adding to the parallels between the scientific and biblical visions, we might note that humanity as it is depicted in the Bible (beginning with the stories of Genesis) gradually grew up from naughty childish behavior to become more mature and independent adults. In the end, humans do not want to only commune with God. They want to be *like* God.[21] This is exactly the spiritual vision we need to follow if we are to embrace open-ended transcendental morality—if we are to ultimately pursue God's objective for the Creation, giving up our self-centered view of humanity as the final end of all evolution.

The religious teachings that dominate the world today are indeed primarily concerned with moral values and personal wishes, but science is not limited to providing objective and value-free knowledge and finding ways to realize our values and wishes. Strictly speaking, no scientific inquiry can be value-free. With scientific explanations of why human moral instincts arose, how they could have evolved from primitive animal moral instincts, and what role they play in the functioning of human mind and human society, we can no longer agree with David Hume's dictum "Facts never prove a 'should.'" We can no longer treat our moral and religious sentiments as absolutes and black boxes.

Furthermore, modern science has not, as Leon Kass suggests, broken with its philosophic and religious forbears and abandoned "the large metaphysical-theological questions about the being or essence or causes of things, philosophical and religious concerns about the meaning of human life and how it should be lived."[22] On the contrary, through science, we are coming to an understanding that *ethics and morality must be grounded in an understanding of our evolutionary history*, and that human values and meaning must be devoted to the largest cause we can identify, which is not a humanistic but a cosmic one.

It is a common misconception that getting rid of the idea of a personal God will legitimize unethical behaviors. Both the scientific understanding of our moral instincts and the experience of many nonreligious cultures show that a personal God (a supernatural God that reads minds, answers prayers, performs miracles, and controls a person's fate after death) is unnecessary to achieve high moral standards. Many atheists have lived ethical lives and enthusiastically performed unselfish deeds. Buddhism, which has been called "Catholicism without God," contains some of the most powerful moral teachings known. On the other hand, in the name of God, religious people have sanctioned slavery, anti-Semitism, racism, homophobia, torture, rape, genocide, witch hunts, and ethnic cleansing.

We have to stand at a higher vantage point if we wish to address deep questions such as why we are born with innate moral sentiments. This is increasingly the task of science. Science gives us perspectives not only on the world but also, increasingly, on ourselves. It unlocks the black boxes of many natural phenomena, including the workings of our mind, and shows us how moral instincts evolved. Without science and the cosmic perspective, the big picture would be hard to see even by intelligent and thoughtful people. For example, we would not blame the writers of Genesis or any other scriptures for not mentioning species extinction, because it was not established as a scientific fact until late in the eighteenth century by the French anatomist Baron Georges Curvier when he demonstrated that mammoth bones are different from those of the elephants. Extinction has since become the central piece of the Creation story; we now know that 99 percent of the species that once existed on Earth are gone.

Faith and religious sentiments are too great a natural creative motivation to ignore and to not take advantage of. Like a powerfully drawn bow, faith stretches the soul, enabling us to aim at the furthest goals. Religious sentiments can awaken our commonality with the larger world. Our lives matter not just for other people (which is the concern of human morality), or even to all of humanity, but also for the entire universe (which is the concern of transcendental morality).

This sentiment that our lives matter for the universe can give us the "unnatural" emotion of embracing Cosmic Being (CoBe), which may share no genetic or physical likeness with us. Belief in our cosmic significance is needed to inject a spiritual motivation into the modern movement of secular humanism, which promotes "God-free ethics." Humanism is a noble and wonderful concept in that it sees life as "a great adventure." It also embraces the scientific principles of skepticism and freedom of mind, and endorses Julian Huxley's "most enduring" project: humanity's further evolution. But it has no answer for such questions as *What is our existence for?* or *What is the purpose of human evolution?* Humanism's answer is simply human fulfillment. The danger of humanism is that if "human fulfillment" is defined too narrowly, then it becomes "merely human"—or perhaps not even the best that we can be as humans.

Indeed, secular humanism defines "the supreme value" as humanity itself—and individual human beings in particular. This attitude is an understandable reaction to the supernatural, irrational, and arbitrary "leap-of-faith" elements that dominate premodern Western religions, but it risks treating our own egos and desires as an ultimate end. In an attempt to free itself from the authoritarian personal God, what humanism ends up with are self-limiting claims such as "We are to look for strength not outside ourselves but within," and reductionist notions such as "The meaning of life is that which we give to it." In hindsight, the Enlightenment awakened our sense of reason but went too far in believing reason alone is sufficient. Humanists have inadvertently severed the connection between humans and the "starry heavens" (or the cosmos), resulting (at least unconsciously) in a lack of at-homeness in the universe—and thus making the universe uncaring and terrifying. And their faith in humanity might be described as blind as well as anemic.

The open-minded humanist may well embrace the highest perspective—which states that the Cosmic Vision should be seen as a central part of the great adventure of humankind—and may even acknowledge that it might turn out to be the greatest part, or even the only one that really counts. Yet overall, secular humanism does not offer nearly as rich a repertoire of perspectives as do the Western monotheistic religions. Hence the latter are likely to survive and even thrive, provided that they continue to evolve. When interpreted literally, the biblical worldview differs fundamentally from the Cosmic View in terms of time-scale, dynamic nature, design of specific species, supernatural miracles, and humanity's place in the universe. Yet the essence of monotheistic faith, which calls for a surrender of the self to the Creator and for living as pioneers for the coming Kingdom of God, poetically matches the Cosmic Vision of humanity as a transitional species and the coming explosion of CoBe. Of course, this is scant comfort to biblical literalists with relatively one-dimensional views, but the deeper biblical meanings might be seen to resonate with CoBe.

Does the reality that billions of people around the world adhere to the monotheistic religions and the vast majority of them tend to interpret the scriptures literally hinder wide acceptance of transcendental morality? It does not have to be the case, since the scriptures are richly worded and have always provided a fertile ground for the prevailing theology to evolve over time. Take the notion of wealth, for example. Even though Jesus famously proclaimed, "It is easier for a camel to pass through the eye of a needle than for a rich man to enter the kingdom of God," modern capitalism was born in the Christian world as the Protestant ethic of hard work, frugality, and diligence took hold.

Those who argue that we are offending God by modifying our genetic makeup love to quote the Bible. Yet, if one considers the implications of Jesus's stern demand for a radical purification of human beings beyond mere good behavior, one might be forced into an interesting conclusion:

> You have heard that it was said, "Do not commit adultery." But I tell you that anyone who looks at a woman with lust has already committed adultery in his heart. If your right eye causes you to sin, gouge it out and throw it away. It is better for you to lose one part of your body than for your whole body to be thrown into hell. And if your right hand causes you to sin, cut it off and throw it away. It is better for you to lose one part of your body than for your whole body to go into hell." (Matthew 5: 27–30)

Realizing that the only practical way to follow exactly what Jesus asked for is to change our human nature (probably through genetic engineering), one could argue that transhumanism is full of biblical inspiration and endorsement!

Of course, the recognition of some religions' historical and future contributions to the Cosmic Vision and to higher morality is not an endorsement of these religions—the majority of believers will probably always hold on to the narrow concept of a personal God and the dogmatic notion of human morality. Furthermore, the Cosmic Vision can also be shared by those who think of themselves as atheists, agnostics, deists, or followers of other religions and cults.

Again, if we are to turn for support to James Fowler's Stages of Faith framework, what is ultimately important is one's understanding of the ultimate reality and humanity's place in the universe, not specific religious inclinations or a particular cultural background. It is tranquillity of mind that can give us transcendental bliss free of instinctual constraints, whether it is Lao Tzu heading out of the city gate on the back of an ox, Buddha sitting under the Bodhi tree, a Christian martyr calmly facing prosecution, or the quantum physicist Heinz Pagels making peace with the danger of falling into the abyss on a mountain climb. Pagels wrote shortly before his death in 1988:

> I often dream about falling. Such dreams are commonplace to the ambitious or those who climb mountains. Lately I dreamed I was clutching at the face of a rock, but it could not hold. Gravel gave way. I grasped for a shrub, but it pulled loose, and in cold terror I fell into the abyss. Suddenly I realized that my fall was relative; there was no bottom and no end. A feeling of pleasure overcame me. I realized that what I embody, the principle of life, cannot be destroyed. It is written into the cosmic code, the order of the universe. As I continued to fall in the dark void, embraced by the vault of the heavens, I sang to the beauty of the stars and made my peace with the darkness.[23]

11.4. The Pragmatic Nature of Transcendental Morality

Transcendental morality is high in the sky as it embraces the Cosmic Vision, but it is also on terra firma down to earth as it takes a pragmatic approach to understanding the human condition. Both our bodies and our minds (intuition, logic, spiritual feelings, etc.) can be taken not as inviolable ends but as limited means—in fact, often so limited as to inspire disgust, as we will see in this section. In this regard, transcendental morality's realism is bound to conflict with conventional wisdom and "political correctness."

- It can be dirty—or unflinching in its view of filth and waste, and hence violating our instinct for cleanness.
- It can be messy, violating our instinct for order.
- It can be wild, violating our instinct for civility.

- It transcends human morality's exclusive focus on human intentions and motivations.
- It encourages the mind shaped by human morality to open up to new possibilities even if this entails moral angst and ethical dilemmas.

While human culture and morality aim to cover up and control our ungainly desires and acts, transcendental morality aims to do the opposite, so that we can gather our courage for overcoming human limitations at the most fundamental level.

Enough then for abstract principles—let us discuss a few concrete examples. Take the fact that most human beings eat meat and enjoy the taste. The images of pigs, cows, and sheep are conspicuously absent in American supermarkets, where meats are cut and packaged into neat pieces so that they no longer bear any resemblance to the lovely animals we see on farms. And even if farm or ranch animals are "ethically" raised, they are nevertheless raised for our consumption, against their "will." No ethical treatment of animals can change the fact that we are *heterotrophs*, unable to use sunlight or chemical energy for our metabolism, and must ingest the flesh of life—animals and plants.

Humans are built to enjoy eating meat, the corpses of other animals, even as another part of our mind may remind us of the cruelty of the act. As William Ralph Inge put it bluntly, if the animals were able to formulate a religion, they would depict the Devil in human form. Transcendental morality shatters the sacred cow of human nature as the be-all and end-all, and it commands that we place ourselves in the shoes of other types of beings. Thus it will ask why we still cannot get rid of our taste for meat—or even, ultimately, vegetables and fruits.

As another example, take the fact that we must discharge solid, liquid, and gaseous waste. There is no way to avoid these disgusting substances, so we try our best to cover it up. As Milan Kundera put it, "Toilets in modern water closets rise up from the floor like white water lilies. The architect does all he can to make the body forget how paltry it is, and to make man ignore what happens to his intestinal wastes after the water from the tank flushes them down the drain. Even though the sewer pipelines reach far into our houses with their tentacles, they are carefully hidden from view, and we are happily ignorant of the invisible Venice of shit underlying our bathrooms, bedrooms, dance halls, and parliaments."[24] Out of sight usually means out of mind, a tendency reinforced by unconscious cultural taboos.

It is interesting what we do to either transform or cover up aspects of ingestion and elimination. Modern societies have managed to turn kitchens (at least those of the affluent) into culinary paradises. Kitchen appliances and cookware have become ever sleeker and shinier, and the food that is displayed is often gorgeous. Even some bathrooms have begun to look like spas, and toilets have undergone further disguises. In both cases, "Out of sight, out of mind" is the rule, though with a difference: although we don't want to know "how sausage is made," cooking and ingesting food has become an art form (we now have scores of "celebrity chefs" on TV and in high-end restaurants); on the other hand, elimination is mainly something we just don't think or talk about, except in medical terms. But perhaps we should.

In one of Jonathan Swift's poems, a young man suddenly loses his innocence as he faces the grotesque contradiction that is tearing him apart about his romantic lover:

> No wonder how I lost my Wits;
> Oh! Caelia, Caelia, Caelia shits!

The transcendental moral outrage is that we have a digestive system that is disgusting, inefficient, and vulnerable to virus attacks, and requires extensive maintenance and conscious attention.

Take the fact that we have an innate dislike of anything dirty, such as food waste (whether feces or garbage). For sound reasons, we want to avoid contamination and to eliminate filth at all costs, without realizing that purity is sterility and there is utility in rotting logs and municipal garbage. As Guy Murchie put it in a Zen-like state of cool reflection:

> Honestly now, if you were God, could you possibly dream up any more educational, contrasty, thrilling, beautiful, tantalizing world than Earth to develop spirit in? If you think you could, do you imagine you would be outdoing Earth if you designed a world free of germs, diseases, poisons, pain, malice, explosives and conflicts so its people could relax and enjoy it? . . . I know it

> seems almost blasphemous to associate pollution with spiritual beauty, yet overcoming such a prejudice is one of the first lessons of the Soul School, where decay is as much a part of life as growth and to be found in many of its loveliest features. . . . And by mystic law, no depth of earthly imperfection but harbors as great a potential height of perfection.[25]

Nothing perfectly neat can be truly alive in nature. Not only our society but our minds and our cells are rough, irregular, and unclean; even basic elements such as genes and proteins are "contaminated" with junk DNA and alien ions and molecules.

The perfect image of human morality is a white lotus flower: a spotless beauty of breathtaking elegance and balance. In contrast, the perfect image of transcendental morality is the rise of a white lotus flower out of filthy mud. The argument of human morality is that we shall not allow the filthy mud pond to exist, since it can be a fertile ground for germs and mosquitoes. The argument of transcendental morality is that the lotus flower is (in nature) impossible without the filthy mud—and this reality should motivate us to create artificial flowers that can thrive without mud. Leaving the filth out of the picture creates an illusion; we should know how plants grow, just as we should know the processes by which we ingest and eliminate energy.

Satisfaction with the human condition is what we all desire—moderating one's expectations is the secret to happiness—but moderation can also make us content and conservative. Human morality has always been about making our lives tolerable within the natural limits of our human nature and biology.

Yet humanity's drive for perfection and religious transcendence goes beyond the goal of mere happiness, as I have argued previously; it is fundamentally driven by dissatisfaction with, and deprecation of, the human condition. Given our condition as the only *self-conscious animal,* the ultimate hope and orientation of human beings has to be always beyond our bodies and our animal instincts. The truly healthy individual, the self-realized soul, the "real" human, is the one who has *transcended* himself or herself.

The critics argue that this drive for self-transcendence is equivalent to escapism. In *Science as Salvation,* the philosopher Mary Midgley suggests that the scientists-turned-prophets of the transhuman world are actually motivated by the "crude" fear of death and the lust for power. "They are really distressed by the contrast between the narrowness, meanness and brutality of much existing human life and the far better things of which humanity seems capable. . . . They want a nobler mental life . . . [and they] want science to provide salvation."[26]

Indeed, this is true. But what else would you expect from people with basically the same brain circuits as the religious fanatics? Wasn't the greatest nation on Earth today, the United States, founded by people seeking better living conditions and a freer spiritual life? Isn't the growth of urban centers in all great civilizations fueled by young men and women who wish to escape the material and spiritual poverty of their villages? Haven't some of the best technologies and scientific discoveries sprung from male adolescent fantasies?

What we often miss in the name of pursuing saintly, pure, and noble moral goals is the power of "dirty" motives and wild behavior.

It is time to loosen up.

Improving within the constraints of morality. The guaranteed method to prevent somebody from doing something considered antisocial or "evil" is to keep that person ignorant of the undesirable activity. Without that guarantee, perhaps some "sinful" sentiments could (in theory) be modified at the genetic level, so that committing those acts would be as disgusting as eating feces. On the other hand, desirable sentiments can be made instinctually automatic and pleasurable.

But any attempt to fiddle with humans in this manner, including the attempt to improve morality through such external manipulation, may be itself a violation of our moral principles. Take the case of language study. Child language study has exercised its fascination on rulers and scholars alike for over 2,000 years, especially in relation to such questions as the origins and evolution of language. In a famous historical event, the Holy Roman Emperor Frederick II of Hohenstaufen (1194–1250) carried out an experiment with children in order to find out what "natural language" humans speak without being taught. He ordered that a number of children be raised from birth in total silence and isolation. The "scientific" experiment failed, as all the children died young from severe psychological deprivation.[27] James IV of Scotland (1473–

1513) is said to have carried out a similar experiment. Few people can condone such inhumane treatment of fellow human beings, and rightly so.

Today we are learning much about language without resorting to such crude and inhumane research methods. Given our characteristic determination and ingenuity, many ethical dilemmas that current technologies confront us with may eventually be dissolved by unexpected technical advances. Eugenics practices, for example, are now possible without arranged marriage, forced sterilization, or genocide. In the mid-1920s, Joseph Stalin commissioned his top animal-breeding scientist, Ilya Ivanovich Ivanov, to create an ape-man chimera that would serve as super soldier and industrial worker, "a new invincible human being, insensitive to pain, resistant and indifferent about the quality of food they eat." But the attempt to impregnate chimpanzees in West Africa did not succeed; neither did the attempt to inject monkey sperm into human "volunteers." The whole idea was finally dropped after a report in the *New York Times* led to strong American protests over the research.[28]

Today, the superworkers have been created—there are numerous industrial robots in operation around the world. They are obedient workers, insensitive to pain, and have no complaints about the food (energy) they consume—exactly what Stalin wanted. The creation of these robots has caused no ethical outrage, but it would be quite a different story had the robots been born out of a human womb!

In addition to the development of autonomous intelligent agents without human flesh, improving existing humanity within ethical and commonsense boundaries still remains an attractive option. Although we cannot treat individual citizens as experimental subjects and violate their human rights, it is still ethically feasible to enhance existing human capacities if it is the individual's own desire. It is essential to retain the perspective of the intrinsic worth of each individual, rather than seeing them as mere cogs in a social machine. In fact, individual rights and liberties have to be emphasized even more in the process of human enhancement.

Individual liberty does not mean individual perspectives and motivation only. Actually, it means we should allow both lower and higher perspectives to be put in practice, and allow winners to prevail. In selective breeding, we purposely do not allow most individual animals and plants to reproduce, violating their most urgent natural desires. Farm animals and household pets are the fruits of such practices through hundreds of generations. We are "indifferent" because we understand too well that the whole "ruthless" process of selective breeding is good for the species—if the yield of wheat does not improve, for example, farmers may abandon it altogether to plant corn. In educating our children, we require them to endure repetitive practices and boring study against their desire to play. Our decisions are based on what we believe is good for the kids over the long haul.

The same higher-perspective principle can be applied to many of our instinctive values:

- We long for equality, but for natural selection to work, some individuals must be different from others and more suited to the environment. These favored individuals (and in modern society, favored ideas and institutions) will naturally be more successful and therefore able to reproduce (or continue to exist). Those populations, ideas, and institutions not so favored may cease to exist. From the standpoint of the Cosmic View, equality is a means to realize individual potential (especially for the disadvantaged), not an end in itself.

- We long for stability, but the point of evolution is not to make species that last forever. To be alive is to act on the cosmic potential and adapt to a world in constant flux. It is not a cosmic failure but rather a cosmic triumph that 99 percent of all species went extinct. The world today is much more interesting and much more valuable and meaningful—much more alive and closer to any imaginable vision for God's design than at any time in cosmic history. Stability is a means to set the stage for the next stage of evolution, not the end.

- We long for empowerment in order to satisfy our instinctive drives, but the higher-perspective principle reveals a different vision. Rather than becoming Faust in the German legend, who sold his soul to the devil in exchange for supernatural power to do what he wanted, our aim should be just the opposite: to use our seemingly supernatural power to create better souls in the future from the cosmic perspective.

Water and house. Almost all bold attempts to push beyond humanity have been criticized on moral and ethical grounds. So far we have laid out many lines of counterargument. But still, the best approach to this ethical debate is to relax and have no debate.

The Taoist Chuang Tzu said the need for debate arises from lack of understanding. Great wisdom has the appearance of dumbness and ignorance. A sage holds everything in his mind, and his silence is the most powerful eloquence. As Shakespeare wrote, "Smooth runs the water where the brook is deep" (*Henry VI, Part II*). And water, said Lao Tzu in the *Tao Te Ching* (chapter 8), never fights the stones in a stream: it just runs around them (in the short run) and erodes them gradually (over the long run).

Henry Ford once complained that "most people spend more time and energy going around problems than in trying to solve them." But going around problems is not necessarily a bad thing; once the problem is behind you, it is no longer a problem. Often obstacles on the path are much more easily circumvented than bulldozed. As the legendary UCLA basketball coach John Wooden put it, "Do not let what you cannot do interfere with what you can do."

"Water going around the stone" is a great metaphor, since the world is fundamentally conservative, but fundamental changes can occur quietly in a conservative world. With many individuals poking around and chipping away, breakthroughs usually occur at the most unexpected, and thus unguarded, places. If selection at the personal level is morally and culturally unacceptable (which it is), then progress must proceed through selection and refinement of genes. So far there has been no conservative argument for the "dignity" of individual genes. Similarly, very few of us are against medical treatment, but many are concerned about genetic enhancements. We may be able to overcome restrictive barriers of enhancement in places where there is no sharp distinction between treatment and enhancement.

The road to immortality and the coming era of the transhuman will be a continuum as we live ever longer, stronger, smarter, and healthier lives with both biological and inorganic enhancements. The Christian evangelical preacher Rick Warren uses the house as a metaphor for life in the human body. Human life, he says, is like a tent—it is set up temporarily on this earth—but our eternal life in Heaven is like a permanent house. Interestingly, the genetic-engineering pioneer Aubrey de Grey also uses the "house" metaphor to illustrate how we could keep living forever by doing a good job of regular maintenance (such as diet and exercise) and by continuous replacement of individual parts and components that have worn out. Human immortality might be an exaggeration, but increased longevity and enhanced functioning even in our present bodies is increasingly becoming a reality.

A wood house is still a house if some metal pieces are inserted to strengthen it. With knee replacement and pacemaker implants, we have already started primitive nonbiological replacement projects on failing bodies. Thus, de Grey's metaphor of a house can be extended further to illustrate the transhuman future. In addition to maintenance, some homeowners may decide to add solar panels to the roof, to convert a spare bedroom into a home office, or to turn the entire house into a bed-and-breakfast inn—the structure or even the purpose of a house can continue to evolve well beyond the design/intent of the original builder/owner.

Finally, it should be noted that for "water going around the stone" to work, a flexible social and regulatory environment must be in place. Morality requires liberty. A suffocating environment can kill life and stop evolution. Water cannot go anywhere if it is frozen. While it is difficult to build a dam that completely blocks rising water flow, overly strict zoning laws can prevent homeowners from doing some creative things with their houses—just as overly strict laws and regulations can block progress to a posthuman future.

11.5. Human Dignity: The Pseudo-Spiritual Argument

Perhaps the starkest contrast between human morality and transcendental morality is the allowance for going above what we are born with—our body, our environment, our mentality.

Manipulation of human life has in general met with resistance from the religious establishment. The Western religious tradition regards humans as created in God's image. The human genome and identity are considered sacred, perfect, and untouchable. For example, the Vatican says, "The freezing of embryos, even when carried out in order to preserve the life of an embryo—cryopreservation—constitutes an offense against the respect due to human beings."[29] It also says cloning humans denies "the dignity of human procreation and of the conjugal union." A Muslim cleric agrees and demands that "science must be regulated by firm laws to preserve humanity and its dignity."[30]

The religious objections are echoed by many bioethicists and academics. The bioethicist George Annas is firmly against cloning because it "threatens human dignity and potentially devalues human life."[31] He has called genetic engineering a "crime against humanity" and argued for a UN treaty prohibiting it. "Genetic engineering," writes Michael J. Sandel, a professor of political philosophy at Harvard, is "the ultimate expression of our resolve to see ourselves astride the world, the masters of our nature. But the promise of mastery is flawed. It threatens to banish our appreciation of life as a gift, and to leave us with nothing to affirm or behold outside our own will." Richard Heinberg captured professional bioethicists' conception of biotechnology's seeming disrespect for life's dignity with this warning, which he himself no longer endorses: "Biotech develops and reinforces a certain way of looking at the world—a mechanistic, reductionist, utilitarian view not only of genes but of life itself. This is part of the source of both our 'yuck' and 'wow' responses to it. For many people, the act of treating an animal such as a cow or a sheep as merely a living machine to be genetically manipulated violates a sense of organic wholeness and of compassionate connection with another conscious being."[32] The Center for Bioethics and Human Dignity, a leading Christian bioethics think tank, devotes itself to the defense of the God-given "innate and irreducible" human dignity against all new biotechnologies.

What do all these arguments amount to? Appeals to "human dignity" have the seemingly noble intention of protecting humanity's worth and honor, but as we have seen, these conceptions have the consequence of suffocating the human urge for self-actualization and self-transcendence. The argument limits our vision to humanity alone, as if it were some eternal or static entity, and ignores humanity's connection and responsibility to the broad environment that created it. An inward-looking humanity is ultimately a species living without larger meaning, which can only be achieved if we are aligned with the larger purposes of cosmic evolution. Yes, we are all for individual flourishing and happiness, but again, what is that *for*?

While "human dignity" has been the rallying cry of some on the Christian Right's fight against transhumanism, that has not always been the case in the past. In the fourteenth and fifteenth centuries there was a Catholic humanistic movement that condemned the theology of original sin and argued that man should become more like God. In the Italian humanist philosopher Giovanni Pico della Mirandola's 1486 book, *Oration on the Dignity of Man,* God speaks to man, saying, "To you is granted the power, contained in your intellect and judgment, to be reborn into the higher forms, the divine."[33]

Paradoxically, the statement that humanity is created in the "image of God" (*imago Dei*) has been a strong rallying cry for those who hold that humanity is the noble and untouchable apex of creation, and that human self-worth is innate and irreducible. In other words, believing we are created in the image of God actually *prevents* us from fulfilling divine potential. In the 2004 study "Communion and Stewardship," the Vatican's International Theological Commission (ITC) held that *imago Dei* defines man in his totality and "the being created in [God's] image cannot be the object of arbitrary human action."[34] But what exactly is the *image* of God? Interpreted literally, humanity has not only God's shape but also God's capabilities and His view of the world. The ITC agrees that it is certainly not just the physical flesh (*sarx*) that is in contention, since we are not talking about the human entity or human identity. What defines humankind most saliently is its spirit and capability.

Modern Catholic theology presumes that humanity takes on God's image through conversion from sin and supernatural salvation, but it also holds that humanity has a responsibility to gain scientific knowledge about the universe and take on God's image through godlike creative acts in this world—the privilege of sharing in the divine governance and "visible creation." Humanity imitates the divine rule (the natural law) but cannot displace it. As Gregory Stock points out, "When we imagine Prometheus stealing fire from the gods, we are not incredulous or shocked by his act. It is too characteristically human."[35]

What the Catholic theologians should realize, as I have argued before in a different context, is that human history is a history of pushing the limits of what can be considered divine. Nothing that is a physical manifestation of natural laws has turned out to be sacred and eternal. There is no "ontological discontinuity" between humanity and other animals. What is really special is not our "biological integrity," but our potential in an evolving universe. Like our genes, most of our behavioral and psychological characteristics, including our moral instincts, are shared with other animals, especially the primates. What fundamentally differentiates humankind from all the other animals is that we are a species with the capacity for *conscious self-creation*, without which we would not be living in this affluent postindustrial high-tech information society, a huge jump from the society of farming and herding that existed in most of the world in recent time (and certainly in biblical times). Humanity's freedom from sin is ultimately the freedom from its genetic bondage.

Furthermore, regardless of whether humanity was created in God's image, we might ask the theologians what we know about the Supreme Being. Certainly the conclusions should be drawn based on the evidence around us. Something of God's spirit and capability have been depicted poetically in the creation myths from ancient civilizations, such as Genesis, but we also must look at modern scientific cosmology. God has created the world, including humanity, while humankind has built upon that creation with new concepts, ideas, tools, social organizations, and visions of further creation. Human dignity should be all about respect for and protection of that precious spirit and capability. Conscious evolution is impossible without it.

Francis Fukuyama argues in *Our Posthuman Future* that there is a universal human essence and we are close to breaking it in the posthuman world. But in an earlier book he also agrees with Hegel that "it is human nature to have no fixed nature, not to *be* but to *become* something other than it once was."[36] Yes, humans must never be treated as merely a means but always as ends; but human ends, endowed by the Creator, have never been static and never will be. Yes, human societies often stagnate and sometimes even digress, but the overall pattern of human history is that we humans always want more and are always capable of more—often much more than the previous generations could imagine. Yes, self-preservation (including the preservation of pride) has always been, and will forever be, the paramount concern of humanity, but the understanding of human essence has continuously been pushed deeper with the aid of science, and the knowledge of how to best preserve that essence has been rapidly increasing.

Human dignity—our duties and rights—is not about freezing the human condition, but rather about free will, about taking the fate of humanity and the universe in our own hands! This Promethean aspiration certainly serves our purposes and satisfies our desires, but again, our purposes and desires are a means, not the end, as critics such as Michael Sandel suggest.[37] The end is Cosmic Creation, not "human hubris." In the New Testament, St. Paul's epistle to the Philippians (3:21) provides an expansive take on God's creating humanity in His image when he states that God (through Jesus Christ) "will transform the body of our humble state into conformity with the body of His glory."

The Vatican's International Theological Commission decreed in 2004, "Given that man was also created in God's image in his bodiliness, he has no right of full disposal of his own biological nature."[38] But treating our own genetic makeup as sacred is pseudo-spiritual: at best, it merely reflects our self-preservation instincts; at worst, it severs life's deep historical bond with the universe and ignores the deep understanding of life and mind as magnificent expressions of natural laws and potential. Leon Kass's idea of retreating to the most primitive value system, for example, is not rooted in Christian theology, but in the concept of a Golden Age, a concept that is in direct contradiction to the core Christian value of progress.

Human rights fundamentally boil down to the right to create. What we are is best defined by what we have done. The great creators are usually at peace with the idea that the products they create are greater than themselves. Shortly before his death at age 95, Peter Drucker, the greatest management thinker of the twentieth century and author of thirty-eight books, had this self-assessment: "I'm totally uninteresting. I'm a writer, and writers don't have interesting lives. My books, my work, yes. That's different."[39] The famous French-born Columbia University professor Jacques Barzun expressed the same sentiment—his own personal life is not "a subject I am interested in."[40] I am deeply moved by Drucker's remark, and I am sure he would have enjoyed reading Nietzsche's comment that "an author must become silent when his work begins to speak," as well as Jorge Luis Borges's comment that "Shakespeare resembled all men. In himself he was nothing, but he was everything that all others are, or what they can be."

In everyday life, we are all proud of our work when we have tried our best, even though we know someone else might have done it better. A parent's dignity is by no means hurt by the superior performance of their child, nor a teacher by a pupil's star performance.

Making the human body and human nature "off limits" is also at odds with the fundamental democratic principles of individual rights and freedom. If you ask people if they would choose to live today or in any historical human society, including the primitive tribes before animals and plants were domesticated, most will certainly say today, even though they are still dissatisfied with today's society or technology. We can imagine that people who lived in the past would love to jump into today's world if they had a chance.

The ultimate demonstration of freedom and human rights is the faith that collectively, people will make a better choice regarding what to do with their body and how to live in the future. The freedom for everybody to pursue selfish interests is, relatively speaking, better than having anybody setting limits, regardless of how visionary or benevolent that person is. Yes, people are shortsighted, but the minds that survive over the long term will always tend to be those that put a premium on long-term survival, whether it is a conscious effort or not.

For those who argue that the new practices are "unnatural," we have been crossing that threshold since the start of civilization when we created new kinds of plants, animals, and even laws and religious beliefs to organize and regulate ourselves. (Just consider how "unnatural" the mule is!) Over the course of history we have also gradually retreated from various taboos about our inviolable body. The study of anatomy was once regarded as a violation of the "temple of the body" in Europe.

There are people who are against altering human nature on the ground that the struggle of the conscious mind against animal instincts is the true purpose of human endeavors. There are also people who believe we are put on Earth to live in a "natural" state. We should allow these people to live out that choice. With growing evidence that a peaceful and nurturing "mother nature" is a modern delusion,[41] most people will not choose that option. As John Stuart Mill pointed out, there is no torture that a human being has inflicted on his worst enemy that nature does not inflict on thousands of diseased human beings every day. But genetic advances make it possible for such struggles to be increasingly unnecessary, and, indeed, that is what the practice of medicine is all about.

Let us sing. Let us conclude by briefly discussing and summarizing some of the issues that people often ask me about on this topic—such as *whether the transhuman enterprise is moral, whether we are offending God through these efforts,* and *how transcendental morality differs from human morality.*

Suffering has always been part of the evolutionary process since life emerged billions of years ago. If we want to imagine that a conscious being (God) is the Creator of the universe, then this God must be amoral in His or Her creative acts. The transhuman enterprise will involve suffering, but the process will not be any more "immoral" than the creation process in history. We cannot offend God by taking on the responsibility of conscious evolution. Yet, this act does not mean human morality should be discarded—in fact, it should continue to guide our efforts to minimize human suffering with our ingenuity—but human morality should be considered as a special case in the posthuman future. Its validity and value should be carefully assessed if we want to apply it to nonhuman relations. Humankind is not the ultimate end, but what is good for humankind has been good for furthering cosmic evolution. We just have to be aware that this may not always be true.

Since Immanuel Kant expressed his wonder, admiration, and awe at "the starry heavens above and the moral law within me," we have gained much knowledge of both and are ready to make connections between the two[42] to reach a transcendental morality. Ultimately, what is right and what is wrong must be judged on the ground of what is good for cosmic evolution. Humanity's longing for the heavens will be answered someday, but only when the inner "moral law" is expanded to embody the transcendental spirit. This morality, not the human morality of so-called human dignity, is truly sacred because it is dedicated to the creative process itself rather than to a fleeting, albeit extremely important, transitional species within the process.

What outrages us morally depends on our point of view. Humans are morally outraged that some conscious human beings could commit possibly evil acts toward fellow human beings. Those people who have embraced the transcendent perspective will be morally outraged that a reflective conscious mind with a Cosmic View could continue to allow human nature to be left alone and let humans define humanity and human welfare as the ultimate end.

Beyond the demand for being selfless, the mandate to foster the emergence of a transcendental morality that will expand human-centered morality is the most difficult to realize but also the most honorable. To play our role in cosmic evolution is the ultimate moral imperative for humankind. Its success is not assured, but the effort will be honorable and the possible failure memorable.

In the face of adversity, let us sing with the sort of human spirit expressed by the Tang Dynasty poet Li Bai (李白):

> *The apes' loud cries from the river banks cannot stop it;*
>
> *The light boat has already passed thousands of mountains.*
>
> (两岸猿声啼不住, 轻舟已过万重山.)

Part Five

THE COSMIC FUTURE

CHAPTER TWELVE

THE COSMIC BEING

The question that will decide our destiny is not whether we shall expand into space. It is: shall we be one species or a million? A million species will not exhaust the ecological niches that are awaiting the arrival of intelligence.

—Freeman Dyson

IN THESE TWO final chapters we will attempt to determine in a concentrated way what the future might bring. Of course, the future by its nature is inherently unknowable and unpredictable. Even technological development, completely under human control, is only foreseeable in the near term. Nevertheless, as we will presently see, there is much that we can learn about future probabilities from what we already know.

12.1. What Science and History Tell Us about the Future

What prevents this forward-looking chapter from being pure speculation are broad patterns we can discern from the past—many of which we have examined in previous chapters. Patterns that have been repeating during astronomical, biological, and cultural evolution have a good chance of repeating in the future. One of the most salient patterns is *emergence*—in other words, novelty and complexity arising from organized interactions of simpler elements. We have seen inanimate molecules forming biological organisms, mindless neurons forming brains, and selfish individuals forming markets that serve public interests. The only thing we know with a great degree of confidence is that the future will be richer and more interesting than the past or present, because this phenomenon of emergence will continue to produce ever more novelty and complexity. This understanding is the essence of the optimistic Cosmic View.

To have anything concrete to say about the future requires examining everything we know about emergence as well as all other known evolutionary patterns. The best forecasters tend to have deep knowledge about both science and human nature. Culture is increasingly shaped by technological developments, but humanity is in charge of technological evolution, and what motivates us determines its future direction. As the saying goes, "The pill didn't invent the sexual revolution"—the sexual revolution was a result of both human motivation and technological breakthrough.

Our intuition and knowledge are perhaps the best guides for predicting the future of our own lives, but not much beyond. Traditional religious, mystical, and political ideas cannot provide much guidance for looking beyond humanity since their subject is human experience and longings.

The writers of the Bible and the *Tao Te Ching* did not know or reveal the wonders of the universe that even laypersons are aware of today. They knew nothing of spiraling galaxies, black holes, subatomic particles, trilobites, or dinosaurs. Science fiction is the freest literary form that does not necessarily hold human nature as given, but it still mainly serves as metaphor or commentary on contemporary human concerns and longings. It can provoke our thoughts and inspire our actions, but as a genre, science fiction is fascinated by the alien but practically wedded to the familiar.[1]

Perhaps forecasting technological development is best left to scientists and technical experts. But even the experts have a poor track record of forecasting the technological future: research found that less than one in four of their forecasts turned out to be correct, as the forecasts tended to reflect more about the spirit of the forecasters' time than about the future.[2] But Arthur C. Clarke's famous First Law of Prediction has so far stood the test of time: "When a distinguished but elderly scientist states that something is possible, he is almost certainly right. When he states that something is impossible, he is very probably wrong." In 1984, for example, the biologists James McGrath and David Solter stated in the prestigious magazine *Science* that the cloning of mammals by simple nuclear transfer is biologically impossible. Just thirteen years later, a cloned sheep named Dolly was born in Scotland.

Beyond everything technological and human we also have to look at broader aspects of nature and biological evolution to enlarge our vision and discern broad evolutionary patterns. We had long thought, for example, that intelligence in an animal was a result of complex social life (to survive in a society, one has to read intentions of others, etc.), long life span (so one can take advantage of learned skills and knowledge), and a long and protected childhood (so one has the time and opportunity to learn from parents and others). As it turns out, our bias toward viewing all intelligence through the lens of human intelligence obscured our view. Recent research on shell-less mollusks such as the octopus and cuttlefish has proved our deep-seated notions wrong. These very intelligent invertebrates, whose ancestors split from ours some 1.2 billion years ago, live a short life (one to four years), do not develop complex social relations, and have no long and protected childhood like us. Their cleverness, however, is absolutely critical for their survival: with a soft body and without a killer weapon, they have to be versatile and efficient foragers and use cunning to evade predators.

Again, to make long-term predictions, we have to utilize long-term patterns in our cosmic knowledge bank. If mindless natural evolution can generate a conscious mind and Einstein-like geniuses, accomplishing as least as much is certainly technically possible with artificial intelligence. If the extremely intricate and complex protein nano-machine called the cell somehow came into existence, then duplicating the same degree of "miraculous" transformation in the nanotech lab should not be impossible. On the other hand, there are obvious limits to looking at existing nature. We have only one life-form (DNA-based) and one conscious species on which to base our flights of fancy—a hard limit for now. And it is not impossible that our assumptions about emergence are wrong; we cannot absolutely rule out the possibility that a divine mind far superior to ours is the ancestor of biological evolution, as the Intelligent Design theorists argue.

With the latest scientific knowledge, we already know a lot about what is possible *in principle*. For example, it is possible to cram nano-computers with the equivalent computing power of a billion human brains into a sugar-cube-sized space! It is also possible for downloaded computer minds to blaze off to other galaxies on beams of controlled energy. And just as with the path of biological evolution, the domain of possible scenarios is enormously large but not infinite. Overall, known natural laws make it possible to draw broad outlines of the future and to make educated guesses about some specific future developments. Going forward, we are not in total darkness. The trick is to identify the leading edge. For example, moving back 1,000 years to a time when the Asian economies and technologies were still far ahead of Europe's, the trick was to identify the emerging power of science and rational thought that eventually led to the accelerating growth spurt of the Industrial Revolution in Europe while the rest of the world remained stuck in the traditional linear growth pattern.[3] Today, the trick is to identify the social developments and technologies that will lead the world to the posthuman future and to the emergence of Cosmic Beings (CoBe). The privileged and courageous few who develop these institutions, mind-sets, and technologies will be at the forefront of conscious evolution.

12.2. Social and Technical Challenges

Rather than offering rosy fantasies of the posthuman future, where all our dreams come true with the aid of technology, it is far more important to lay bare the social and technical challenges ahead, in order to underscore the amount of time and energy needed as well as the potential setbacks and losses that we should be mentally prepared for. The path from bacteria-like primitive life to the advent of *Homo sapiens* involved passages through periods of bitter struggle, long-time stagnation, and numerous massive failures. We emerged in the end not by predetermined design but through overcoming seemingly overwhelming odds. The path from humankind to CoBe is unlikely to be much different, even as we transition into conscious evolution rather than natural evolution. Both top-down (involving the improvement of existing life and brain forms) and bottom-up (involving the engineering of artificial life and intelligence) approaches (separately or in combination) will present daunting technical and social challenges.

We should be wary of all the talk about smooth sailing to the inevitable Singularity,[4] where technological progress is said to accelerate so fast that it can be represented by a rising curve eventually going straight up to infinity. According to this view, all we have to do now is to live long enough for that day to arrive when we can upload ourselves into the so-called techno-heaven.

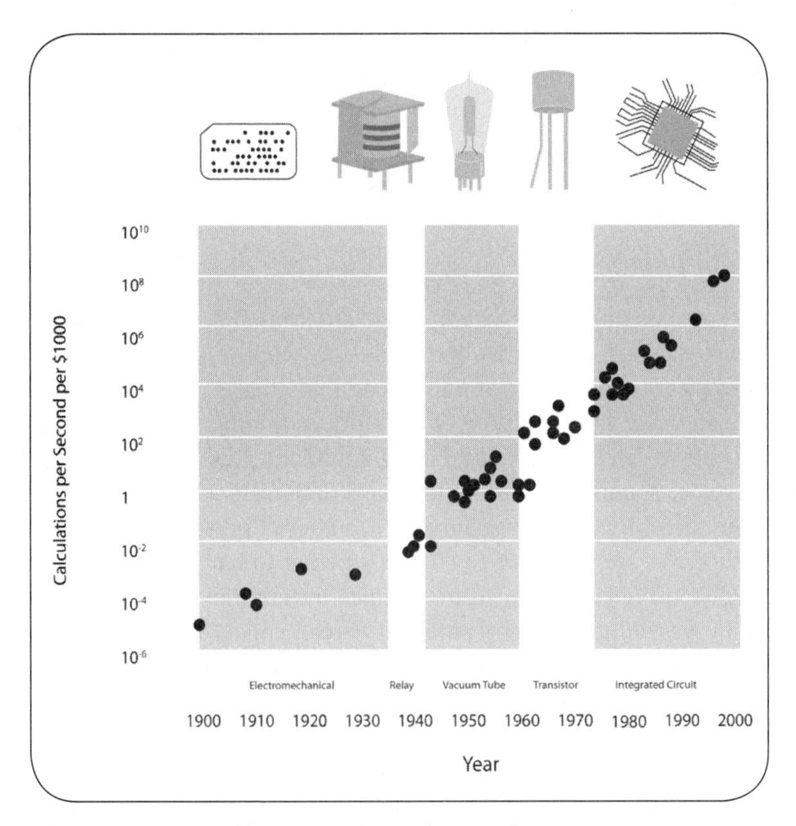

Figure 9: Kurzweil's Extension of Moore's Law

This graph depicts Ray Kurzweil's extension of Moore's law to a succession of technologies—from integrated circuits to earlier transistors, vacuum tubes, relays, and electromechanical computers. (Moore's law states that the number of transistors on a chip doubles about every two years; it has given rise to many similar "laws" of logarithmic increase involving processing speed, memory capacity, and the like.) Kurzweil and others believe such exponential improvements will ultimately lead to a period where progress in technology occurs almost instantly. However, for reasons we outline in this chapter, the future is likely to be far more difficult than the techno-optimists are predicting or hoping for.

Social barriers. As already discussed, social barriers have historically slowed or blocked technological innovations on many fronts and will only become dramatically higher and tougher as we proceed to transform humans. Realistically, breaking the dam of our natural tendencies and emotions against transhumanism can only be a gradual process and can only be successful if many pressure points are working simultaneously.

The greatest source of pressure to break social barriers is likely to be economic and political. Global competition is adding urgency to the need to upgrade the quality of human resources. Investment in human resources (education, health care, and various social services) has become the biggest category of investment in developed countries. Yet the marginal returns are diminishing. From the macroeconomic perspective, "brainpower" is slow and expensive to build up, with finite peak years of production and high maintenance costs during the lengthy stage of terminal decay. (Of course, quick disposal is not an option due to wholly legitimate moral concerns.[5]) From animal-pulled carts to digital computers, the development of civilization is a history of finding ways to replace manpower with cheaper and more abundant substitutes.

The potential pressure for transhumanism can also appear from the individual perspective as a matter of personal freedom of choice. We are now able to choose whether to have a child or not, but we do not have much of a choice in terms of what kind of a child we will get. By comparison, we can collect lots of information and carefully choose before we purchase a product, a service, a house, or an investment. We choose which college or job to apply for. We do not randomly pick a mate to marry or friends to hang out with. With a newborn baby, however, parents are making a lifetime wager on and commitment to a very unknown quantity bearing a high risk of serious problems. To be fair, the child too has no control over what kind of parents he or she will inherit genes from or how these unknowns will affect his or her quality of life.

These basic needs for social efficiency and personal choice, combined with the perennial personal longing for happiness and immortality, will be the "water" (forces of change) pushing the "dam" (barriers of resistance) to upgrade human genetics, enhance control over reproductive processes, and reform personal relationships in the near future. It is almost inevitable that the "water" will overcome the two greatest forces of resistance—fear and moral concerns—which we discussed in the last two chapters.

Still, the fallout from clashes between the "water" and the "dam" is likely to be highly uneven. If history is any indication, initially the dam is likely to hold back the rising tide in all places. The reader might have noticed that the discussion of diminishing returns of human-resource investment is not exactly politically correct. But one or two breakouts could happen by accident or occur during extraordinary circumstances—metaphorically, water breaches the dam during a big storm, although nobody can predict where and when.

In free societies, the dam is unlikely to be "rebuilt" once it is broken, since people can see the results. When smarter, nicer, happier, healthier, and more productive enhanced humans or androids emerge, popular demand for democratization of the new technologies will transform the special into the norm as cost, quality, and safety improve dramatically. This process will pave the way for further breakthroughs despite the fact that new dams of social resistance will be constructed farther "inland."

Technical barriers. Because of strong social resistance based on both safety and ethical grounds, the practical path to the posthuman world is likely to be one of gradual change rather than a sudden, dramatic shift on a global basis. That is in line with the way natural and cultural evolution work. The likely gradualness of the change may also make building nonbiological intelligent beings (robots) a useful strategy despite its own challenges. The difficulty of developing fully autonomous intelligent robots is that we have to more or less recapitulate the evolution of animal intelligence on Earth. The technical hurdles involved in both the "remodeling" of existing human biology and the creation of human-like autonomous robots with advanced engineering design are formidable.

Nature is not a designer that develops each species from scratch but a tinkerer with existing structures. Evolutionary novelty is often only skin-deep. *Conscious* evolution will do better. For example, we will borrow insights gained from studying the human brain in our design of robots. Yet even with our best efforts, the most likely scenario is still likely to be one with sluggish improvements most of the time and only the occasional great leap forward (as well as the occasional catastrophic collapse).

One reason for this is that "remodeling" can be very difficult given how complex the human organism is. In general, once the complexity is built to a certain level, further improvements face diminishing returns, and major "redesigns" are practically impossible. As Karl Ernst von Baer first discovered in his study of embryological development (von Baer's Law), gains in complexity almost always come at the cost of surrendering flexibility and potential for further improvement.

Evolution seldom takes place quickly, but it is very slow for complex organisms because beneficial mutations become increasingly improbable when a single change could trigger a cascade of consequences. The evolution of artificial intelligence has not fared much better. As a structure becomes more hierarchical and its components become more interdependent, the problem of legacy issues and unintended side effects can quickly overwhelm even extremely fast processors. In her book *Close to the Machine*, Ellen Ullman tells the story of a sixteen-year-old computer program that, having been worked on by ninety-seven programmers, became so unwieldy that nobody has dared to touch it; the program is being kept alive only because it still works.[6]

The difficulties of scaling up artificial intelligence give some hint at the challenge of improving the human machine. Our bodies are like computer "legacy" systems—mainframe dinosaurs that contain a tangled web of poorly understood and documented codes from previous generations. The hypothalamus, for example, is so compact that its basic structure and function have remained intact since the amphibians first cobbled it together, and its status in the entire neural system remains central and complex.

In general, it is too simplistic to say we can insert or remove a certain gene to achieve a certain functional objective. Even the assumption that there is a single structural unit called a gene is invalid in many cases. The links between structures and functions are messy since existing genes and organs are often recruited to serve new purposes.

Only the adaptation of simple organisms such as viruses and bacteria has been lightning-fast. Germs can adapt quickly to new drugs but cannot quickly evolve into something much more complex, like an earthworm. It takes time for complex organisms to complete a life cycle. A bacterium can reproduce itself many times within a day, a small animal in about a year, a cultural system in many decades.

It is nearly impossible for complex organisms to take an evolutionary jump. None of the thirty-five or so basic body types that emerged from the Cambrian explosion changed in the subsequent 500 million years. For civilizations, the more specialized and adapted a cultural system is in a given evolutionary stage, the smaller is its potential for progressing to the next stage within any given period of time.[7] Startup companies have consistently shown a greater capacity to innovate than large and more established ones. The route that evolution has often traveled is to kill an organization or an organism and start over rather than to change it.

Given the compressed and convoluted nature of the human genotype, it is tempting to conclude that it is better to make a new start with robots and take advantage of the positive feedbacks of exponential growth during the early stages of the S-curve—a period of the product's life cycle when the pace of progress can easily accelerate.

There are several challenges in climbing the learning curve. Autonomous robots initially can probably only survive in tightly controlled artificial environments. Beyond a certain point, autonomous robots will have to evolve on their own. That requires a minimal level of structural complexity for the new intelligence to be autonomous and adaptable. To use an analogy in nature, scientists have come to realize how complicated a simple animal such as the cockroach is, even though it seems to be completely driven by hard-wired instincts. An autonomous living organism must be able to obtain energy and materials for sustaining itself, discharging waste, defending itself when necessary, and replicating itself (since fatal "accidents" happen). On the other hand, a robot, if structurally simple, can never exhibit the richness of human behavior regardless of how much computing power it has. The brain is the most complicated organ in the universe, not just because it has 100 billion neurons interconnected by 100 trillion synapses, but also because it has an odd collection of competing motives and interests to function as an individualistic social animal.

An advanced autonomous robot that can truly stand on its own will be so complex that its behaviors will be unpredictable—it may eventually have free will similar to that of humans, with all the awesome consequences that implies for the future of the planet and for the universe. We can do much testing before releasing such robots, but testing with computer simulation has its limits: reality is forever richer than virtual reality. Only the real world provides the ultimate testing ground with (among other things) all its political implications, something we will discuss in a moment.

Penetrating the barriers with tinkering. While radical innovations are easier in robotic engineering than in biological engineering, tinkering with the human genome should be the most rewarding strategy in the near term since there are many low-hanging fruits that are easier to grab first. Great benefits can be had without inventing anything truly new. It is conceivable that many desirable physical and mental traits and exemplary personalities that are currently at the tail end of the population distribution can become the norm through genetic tinkering. Perhaps we will someday have the Iron Man, Einstein, Mozart, and Buddha all rolled into one "perfect person."

Reducing (but not eradicating) undesirable behavioral traits may not be so difficult, either (at least technically). Take nature's tinkering with the primate's tendency toward violence. As an example, the sexual and competitive behaviors of bonobos are very different from those of the other chimpanzees, even though they share a common ancestor from as recently as three million years ago. All it took for bonobos to drastically alter their social behavior seems to have been a restructuring of female grouping behavior through homosexual activities, which proved to be enough to alter the behavior of the males, along with other subtle genetic differences in the species. We have also discussed how humanity seems to have "tamed" itself in recent history. With only little tweaks of our genome, we managed to vastly strengthen our moral instincts and greatly cut back antisocial behavior. And only minor genomic changes enabled us to speak and use languages.

In the foreseeable future, moving (or, more precisely, creeping) beyond humanity can be more about expanding than remodeling the existing genome. From the informational and cybernetic perspectives, the boundary of a person is not defined by our epidermal surfaces. A blind person's cane or a professional programmer's computer can be considered as part of the person, except that these things can be separated from the body and can be replaced or upgraded more easily than our natural body parts.

More advanced technologies will make a cane (or a computer) more of an integral apparatus than a simple tool. We cannot live solely in cyberspace—Hubert Dreyfus's critique is correct to the extent that if our body goes, so does relevance, skill, reality, and meaning—but this does not mean that our limited bodily equipment cannot be augmented or even replaced by more sensitive, powerful, and flexible alternatives. We can gradually mutate into a hybrid, and then, eventually, a cyborg.

Although improving the existing human genome has an inherent speed limit—it will take at least eighteen years to see the impact of genetic change, given the length of the human life cycle—it is not an outlandish prediction that before long, enhanced humans may look back at us with more astonishment than that with which we look back at the lives of the hunter-gatherers today. In their eyes, we may become those ignorant beings who lived only seventy to eighty years; were often killed by senescence, awful diseases, mental errors, and crude technologies (automobiles, airplanes, etc.); spent most of our childhood studying in school and most of our adulthood working (all the while longing to be doing something else); and conceived children by a *reproductive roulette*—the random, unpredictable meeting of sperm and egg.

That said, we have to be aware of the limits of the add-on/enhancement approach: stiffer social resistance, drastically reduced flexibility, and unavoidable compromises. This has almost always been the case in history, and I doubt the future will be different. As Freeman Dyson points out, the contrasting histories of motorcycle and nuclear reactor development are pertinent. The small, simple, and low-tech motorcycle was able to succeed because thousands of designs were tried until a few types of motorcycles that were efficient and reliable emerged and were adopted for common use; whereas the large, complex, and expensive nuclear reactors for power generation largely failed to catch on because fewer than a hundred different types of reactors have been operated. For economic and environmental reasons, society simply cannot tolerate the failure of a nuclear power plant to the same degree as a malfunctioning motorcycle.

Going beyond humans. Looking further into the future, some fundamental changes may not be feasible with the human species. There is a good reason why basic instincts related to survival and reproduction are so hard to suppress. Technically speaking, we might be likened to a car powered by an internal combustion engine (ICE). In the early history of automotive power-train development, the ICE was the winning technology, practically eliminating electric and coal-fired steam engines in the early twentieth century. But once the ICE dominated the vehicle market, its drawbacks—polluting tailpipe emissions and low energy efficiency—became obvious. In the last several decades, tinkering with the ICE has largely worked. Harmful emissions have been drastically reduced with the catalytic converter and other technical improvements to the power-train. But further improvements are offering diminishing returns. The ultimate solution to tailpipe emissions is not further refinement of the ICE, or even hybrid power-trains, but fuel cells or electric vehicles. The adoption of radically new power-trains will not be a panacea—with zero-emission vehicles, we still have to deal with emissions in car manufacturing and electric power generation. For wide consumer acceptance of the new power-trains, lots of technical breakthroughs are needed to drastically lower the costs and improve the reliability and flexibility of the batteries and fuel cells.

The function we must continue to serve when we replace an ICE with a fuel cell or electric motor is clear—to supply power to four wheels—but the situation is far more complicated when we are dealing with the complete overhaul of the biological cell and the brain. To create an artificial being that is truly alive, we may have to replicate some yet-to-be-determined essential functions of biological life. To create an intelligent being, we may have to replicate those aspects of "no brainer" common sense that are actually building blocks for intelligence. In the case of an electric vehicle, when we solve one aspect of technical issues, other problems pop up elsewhere in the automotive ecosystem. Likewise, I am sure that artificial life-forms and artificial brains will have numerous problems and shortcomings that we have to tackle.

From the social perspective, there will also be transitional costs and adjustment pains, just as existed when the appearance of automobiles put thousands of blacksmiths and horsemen out of work at the beginning of the twentieth century. Luckily, the transitional pains, however intense, will soon be forgotten. Dealing with human beings, however, does involve emotional and moral issues that other technologies do not, and these can have long-lasting repercussions, as we saw in the last chapter.

To summarize, technical and social challenges may dictate the pace of progress in both the tinkering (enhanced-human) and ground-up (autonomous robot) approaches, since they have their own pros and cons. While the overall trend is toward more complexity, technologies that are too big, too expensive, and too challenging to our moral sensibilities do not have much hope of success.

At some point, these two approaches—leading to enhanced humans on the one hand and autonomous robots on the other—are likely to start to reinforce each other. The near-term progress in human enhancements could greatly upgrade human resources to tackle the progress in robots and nanomachines—imagine increasing the percentage of students capable of completing a science and engineering Ph.D. from the current 20 percent to 80 percent, or prolonging the peak productive years of scientific research from ten to forty years! In turn, breakthroughs in computing, genetics, neuroscience, and nanotech will greatly facilitate our efforts in tinkering with the human genome.

Eventually the line between the two approaches will blur, since they end up using the same technologies and achieving the same ends—intelligent beings with unheard-of structural flexibilities and capabilities. In principle, all the limits of the current human condition listed in chapter 7 can be overcome. This will result in a proliferation and competitive selection of new life-forms and organizations beyond our imagination.

As I write these words, I notice that my three-year-old son, Megene, is fascinated with the opening and closing of the tray door in our old cassette tape player—he pushes a button here and the door opens there, so he shuts the door and tries it again, wondering how can it happen. We adults are seldom intrigued by that particular mechanical operation. We treat it as just a necessary step so we can listen to the music. And we do not miss that step at all when we transition to an MP3 player that has no tape or door but offers high-quality digital music that does not wear out like tape.

Just as hearing the music—rather than being concentrated on the means for playing it—is the goal in the above example, our deeper essence (intellectually, morally, transcendentally) is the proper goal of our lives, rather than the particular life-forms (or "players"). Therefore, let us focus on the "music" and create better ways to play it and preserve it.

12.3. CoBe: The Cosmic Being

David Brooks has called America's zeal for future vision "a Paradise Spell: the capacity to see the *present* from the vantage point of the *future*," and he notes that "future-mindedness is a trait that repeats in the biographies of inventors, entrepreneurs, and political leaders, and it is a prominent feature in the literature of the pioneers."[8] Indeed, articulation of the goal is essential for any goal-directed actions.

Out of a wide range of possible new life-forms, what are the essential characteristics of CoBe—those who will be pushing the frontier of cosmic evolution? In considering this, we are like an alien visiting Earth and trying to identify which species, out of millions thriving on this planet, will lead the expansion into space. The following discussions are speculative, but I hope they can be useful in stimulating further thought.

No death. Death is bad for the individual, but it leaves necessary room for successors, room for change. With new capabilities to self-transform in continuous evolution, however, the successor can be oneself—as I will explain here.

Once the body and brain can be made plastic throughout life, death and reproduction become unnecessary. With no biological death, a whole set of concepts is pushed aside—including generation, infancy, parents, education, retirement, etc. New concepts that we have never heard of will emerge to describe the self-renewal of immortal beings.

Immortality is an eternal human longing, but its wider implications are still underappreciated. Coupled with other changes, immortality can be qualitatively different from a limited life span. Our sense of a lifetime does not change much as long as life has a visible end. Just witness how little our religious faith and lifetime perspective have changed as our life spans have doubled from those in the Middle Ages (from 40 to 80 years). Doubling that again to 160 years is unlikely to make a fundamental difference, either.

With unprecedented physical and economic satisfaction and security, some of us are probably even more saddened and frightened by the prospect of personal oblivion than we were when life was a constant hard struggle. Death is said to be the greatest equalizer and demotivator. Life's finiteness is the greatest obstacle to embracing the cosmic perspective—what's the point of making an effort if I have no chance to live to see the outcome? It is the biggest reason ancient religious beliefs in a personal God remain attractive—we still seek salvation, or fulfillment in the promise of an eternal life, after death.

Without death, whole generational dynamics will be different. Aging and fading scientists, artists, and everybody that we care about will be rejuvenated rather than replaced at great economic and emotional costs. Traditional education will be obsolete. For CoBe, all existing knowledge is either built-in or easily accessible.

As we transcend death, the emphasis of education will turn to providing the widest evolutionary perspective on knowledge that is imparted. Teaching science and math without showing students how these theories and theorems were formulated (as an evolutionary process) is not teaching much. The emphasis on building new individuals will shift from instilling the existing stock of knowledge to understanding and even experiencing natural and human history.

In addition, the elimination of biological death (and the sexual drive, which we will discuss shortly) lays the foundation for the transcendental morality that is so hard for mortal humans to grasp. For example, for Jews who move beyond death, the ancient Jewish vision can be realized: every member of the community will have relived the entire history of the community and will share a collective memory. For the Jews, reliving history means soaking in the rich tribal and community experience such as the exile from Egypt and the covenant with God. The Passover dinner includes bitter herbs or horseradish to symbolize the tribe's traumatized history.

For CoBe, then, the goal of education is reliving the entirety of cosmic history. Each individual CoBe should have the whole thirteen billion years of cosmic experience imbedded at "birth"—this will ensure that every CoBe identifies itself with the Cosmic Creation process, whereas we naturally attribute our birth only to our parents and associate our life experience with our social and earthly environment. The God that most of us naturally worship is primarily a personal one that appears to be cosmic; the God that CoBe will naturally worship is a cosmic one that appears to be personal.

Since the cosmic experience can only be a simulated and selective representation of history, each CoBe may still have different perspectives on the same process. There may still be uncertainties about which aspects of "cosmic history" better reflect the fundamental reality and thus serve as the best guide for the future. The search will go on forever for any finite being; and it will be part of the continuing growth experience of CoBe.

To be on the frontier of Cosmic Evolution means that every CoBe will have a childlike open and curious mind such as we see in some of today's scientists and thinkers; individuals such as Stewart Brand, Freeman Dyson, and Hans Maravec come to mind as examples for me. The generational renewal will be replaced by continuous improvements in knowledge, skills, and perspectives in a perpetual fashion. Idle retirement, a peculiar product of the Industrial Age, will not be part of CoBe's life. For CoBe, purposefully purging what is obsolete and retaining what is desirable will be something like a human's starting a new life.

The elimination of natural death does not mean the end of death itself. The life-and-death struggles against adversaries have always stimulated breakthrough evolutionary developments. In the near future, new forms of deadly "viruses" will certainly play the role of determining which technologies and organisms will become obsolete and possibly be forced to die. To overcome the technical challenges of diminishing returns on certain organizational platforms, we may make the conscious decision to abandon or completely restructure certain designs or organizations, just as we discard fully functional or repairable electronic equipment and appliances today. The upshot is that we need not worry about a Malthusian doom of unsustainable exponential population growth with limited resources.

For the distant future, we can speculate about whether something else will replace forced death as the main mechanism of selective pressure. What is certain is that selective pressures must be in place to prevent degeneration and spur further evolution—as this is a fundamental feature of the universe.

"All Artificial." Despite accelerating technological developments, our creations thus far have barely begun to match the sophistication of nature. A tree is more versatile and economical than any of our artificial devices.[9] The label "All Natural" on food products is still a trusted sign of the safest and purest quality. The words "artificial" and "synthetic" still carry a negative connotation for most consumer products. Artificial intelligence still cannot hope to mimic human intelligence because it lacks the "common sense" that each of us is born with.

On the other hand, we also realize that nature is imperfect. The belief that "natural substances" are in every case inherently safer than synthetic chemicals is a myth—organically produced plant foods may contain natural toxic or carcinogenic chemicals. We have already created some artificial fabrics for clothing that are superior to cotton or animal hair in quality and durability. And this trend of the artificial beating the best that nature can offer will only accelerate. As this technical evolution continues, the creations of CoBe are likely to be unimaginably more efficient, flexible, powerful, and versatile than anything we find in nature. And it will be "All Artificial."

For the artificial to beat the natural, the artificial will have to be more complex than the natural. It may have simpler structures but still have more "depth" than the natural—depth is a measure of the information accumulated *and* discarded during the process of evolution.

All current technologies are primitive piecemeal structures designed to meet a specific need in human society. There is much hope that computers will exceed human intelligence soon, even during our lifetime. But such machines, however fast and powerful in performing computational tasks, are still extremely limited tools and utterly dependent on humans. To be better than nature, the artificial must be able to do *everything* that the natural does, plus much more.

Specifically, CoBe cannot simply be a super-fast digital computer, which is a vast oversimplification of the brain at the structural level. For example, the brain has at least two modes of electrochemical transmission of signals: the wiring transmission through synapses, which the computer mimics, and the volume transmission of electrical signals and various chemical compounds through the bloodstream over both short and long distances in a broadcast-like fashion, which the computer totally lacks. The chemical and electrical dynamics of the brain resemble the sound and light patterns and the movement and growth patterns of a jungle more than they do the activities of an electric company, which is what a digital computer most resembles.

Within the brain, thought and feeling have a deep unity that is unbreakable. Accumulation of emotional experiences helps a person to form judgments and make decisions. Even a seemingly trivial task such as a child learning when to say hello is based on the mastery of subtle, complex cues.[10] It is those emotional contexts that form the basis of our decisions, often unconsciously. We have a long way to go before matching that in robots.

Emotions are so central to us that there is every reason to believe they will also be a critical part of CoBe, but at a far more complicated level. The Dr. Frankenstein image of intellect run amok is a vivid warning about the danger of mating primitive emotions with raw physical or intellectual power. It also highlights one area that may see astonishing innovations in our efforts to make the artificial better than the natural.

No sex. In terms of our technological future, many people (especially young males) may be looking forward to virtual or enhanced physical sex that could give them more intense and lasting orgasms. Virtually all of us as well may want to seek out all kinds of other extraordinary virtual or physical sensory experiences, such as feasting on rich gourmet food without gaining weight. The dreams of enhanced senses and desires will continue to fuel technological development, but when it becomes possible to turbocharge our hedonic indulgences with genetic engineering, it also becomes possible to pull the plug on the instinctive drives altogether!

Likewise, the androids, for whom sexual reproduction would not be necessary for their perpetuation, might evolve into supersexed machines, or they might have no sex at all. Sex and sexual reproduction may persist for some species, new and old, but sex could disappear from the frontier of evolution once its reproductive function is obsolete and its social and relational functions are taken over by other types of mechanisms.

Sex (and the entire sexual reproduction process) is so critical to humans' very survival that it is by far the biggest burner of human energy and attention—think of how much time and money we spend on satisfying and managing our sexual urges, making ourselves attractive, gossiping, dating, forming families, raising kids, etc. "Remodeling" our sexual drive, and the limbic system in general, will be a necessary step for CoBe, because its drawbacks (sex forces one to focus on self-interests—it is at bottom about passing on *your* genes) clearly outweigh its advantages (such as providing the basis for romance and pair-bonding). Love is one of the noblest human emotions, but sex is also closely associated with our instincts of aggression toward fellow human beings and status seeking.

Civilization only has managed to suppress our primitive motives, not to eliminate them. But with sex and other sources of "sin" removed rather than simply suppressed by our consciousness, most of the traditional moralistic preaching on sexual habits becomes unnecessary. We will have no "confessions" to make and no sordid scandals to fill the pages of our tabloids. Without the sexual drive, future minds will not miss the fun of sex, just as we do not miss whatever forms of "fun" are had by other species.

But how can we live without the animal instincts? Although we have already managed to have sex without having babies, and to have babies without sexual intercourse, the sexual drive still has absolute control over us, just as it does with any other species that reproduces sexually. Technically, eliminating the sexual drive in humans could be very difficult, since it involves remodeling the hypothalamus, the seat of most human motivations and the "control room" for many vital bodily functions, such as body temperature, daily rhythms, sex, appetite, and hormone making. Touching the sex drive means touching almost every function of the brain.

True environmentalist and artist. Going way beyond today's environmental protection efforts, CoBe is likely to have fundamentally different relationships with the natural environment on Earth. For them, most natural requirements for sustaining biological life can be removed. Both the skeletal and metabolic functions of the human body are inefficient and inflexible even compared with those of many animals, not to mention with what is theoretically possible; these will likely be replaced by much more complex sensory and motor systems. In particular, CoBe will not have to consume any organic life; the act of eating organic materials such as dead animals, fresh vegetables, and fruits will be made redundant and probably even emotionally repellent.

Still, some human populations may be preserved—there will be debates over how to keep humans "natural" while minimizing their suffering from things like diseases and psychological agony over moral issues—but much of the Earth could be allowed to revert to a prehuman state. Further into the future, when the ever-more-powerful CoBe has expanded into the cosmos, preserving the Earth by shielding it from natural shocks in the Milky Way will not be difficult at all; it may be as routine as preserving historical artifacts within a natural history museum in a bustling city of skyscrapers.

As CoBe moves into outer space, its aesthetic values will move far beyond those associated with the earthly environment (such as snow-capped mountains and pristine lakes and beaches). Biophilia, our innate love of the earthly environment, will be a trivial sentiment in the cosmos.

Some may wonder, "How can CoBe still find life worth living given that they are no longer human? How interesting can the dark emptiness of outer space be?" It is true that the cosmos (as we would actually see it) is nothing like the beautifully colored images that science promoters have put on the Internet, but we just simply cannot even imagine what will keep our fundamentally transformed progeny occupied and fascinated in outer space. Forget about the frivolity of the lives of the gods depicted in Greek myths. The intensity of CoBe's new motivations and the enhanced responsiveness to a wider range of stimuli would, it seems certain, be tremendously empowering and enjoyable. The creation of earthly environments in space for humans has been envisioned by many futurists.[11] While this vision may be psychologically satisfying to us, it is highly unlikely to be achieved, since that would be a very inefficient and restricted way for intelligent beings to fill the universe.

Cosmic science. It will be up to CoBe to break the hard limits of human science. Both the theory of evolution (by Darwin) and the law of conservation of energy (by Julius Robert von Mayer in 1840) were discovered following a long sea voyage. There is no doubt that far more exciting and surprising things will be discovered in long *space* voyages.

CoBe's sense of time is likely to be altered fundamentally as well. Relativity, for example, may become intuitive. They will incorporate multiple senses on different frequencies that encompass cosmological, geographical, ecological (a thousand years may feel like a minute), and biological time scales, and all the way down to biochemical time (the scale at which our brain operates). Both the slowest geological time and the most rapid vibrations of physical phenomena will be directly discernable. In terms of spatial perspective, CoBe will be able to move quite effortlessly not only between distant locations in the universe but between perspectives, such as were vividly depicted in the classic book *Cosmic View: The Universe in 40 Jumps* and in the short film *The Power of Ten*.

CoBe will obtain new insights of the fundamental nature of the universe, a new "cosmic science." Human science is grounded in the modes of human thought and full of convenient simplifications that are excessive and misleading. The enlarged mind of CoBe will find it unnecessary to break down complex issues to analyze each aspect or component separately. They will develop totally new and growing bodies of "common sense" or tacit knowledge on which to base their understanding of the universe. Compared with CoBe's cosmic science, our mere human science, which Michael Polanyi called a "token of reality," may turn out to be superficial and the natural laws we identify today only skin-deep. Ultimately, CoBe's thinking about the universe as a whole with more depth and details will approach a scale of cosmic reality unimaginable to us.

New motivational drivers. The desire for everything from sex to food "from the earth" may be eliminated, but not all animal instincts will disappear, at least not for the first generations of CoBe. On the contrary, some our instincts could be enhanced and carried to an extreme from our vantage point, somewhat in the same way that we humans exaggerated the rudimentary culture of our mammalian ancestors (as we discussed in chapter 6).

Normally, humans are grateful to be alive. We don't like to be alone. We have expectations for the future and want to know many things that will help us to guide our choices and actions. These are just some of the fundamental characteristics that are likely to persist in CoBe. Hope is both the earliest and the most indispensable virtue inherent in the state of being alive.[12] Closely related to it are confidence and trust, which are expressions of faith, the enduring belief in the attainability of fervent wishes. Just as the pleasure of sex and love is essential for our reproduction, the urge to survive, to search for new things, and to join with others may be essential motivators for future travel and colonization of the cosmos.

The urge to embrace the universe, to be part of the creative whole, may become as strong as the urge to embrace a lover for a young man or woman. However much we may strive to answer life's larger callings, we are always limited by our selfish and competing animal instincts. With increasing flexibility, however, CoBe may regard itself as part of the whole in a way that far transcends even the most devoted humans.

The instinct of self-preservation will still be strong, but there will be a clear sense of what preserving the self is *for*. The overarching devotion to the whole means that there will be little fear of one's own death if tone feels that such an event, however it comes about, is in the interest of the whole (of which we are all, in fact, a part). What has been achieved with great effort for a few exceptional human beings—sacrificing one's own life for a greater cause—is likely to become a natural part of CoBe's existence. Love and care are likely to remain great virtues, but they will be expanded beyond the boundaries of the human and the familiar to the cosmos itself—just as love for many of us has expanded beyond intimates, tribe, and country.

When any of us are asked to "sacrifice for the greater good," the first reaction may be to question the motives of the requester, rather than to consider our lives as intrinsically a part of Cosmic Creation. I am not assuming that we can be like CoBe—we cannot, since we cannot change our nature. In fact, the vast majority of future intelligent beings will *not* be part of the pathbreaking CoBe. But we would all do well to think about what is possible beyond the limitations of human nature. And by simply considering this in depth, we may naturally become more identified with the greater good.

New emotions. Entirely new emotions are also likely to be a growing and continuously evolving part of the "tool-kit genes" for CoBe. We hardly know anything about how to build a new emotional system because little research has been done in this area. But evolutionary history offers some hints. Technically, human motivation functions mainly through specialized chemicals transmitted through the bloodstream rather than neural synapses.

Sexual arousal, alertness, and attention are also regulated by brain chemicals.

During the course of evolution, the designers of CoBe's mind are likely to tinker with many different emotional or mood states and transmission mechanisms. New motivations will surely emerge once the dam of fixed human nature is broken, and under continuous selective pressure, the successful motivations will be those that are oriented toward higher perspectives.

It is often said that animals are driven by instincts while man is driven by a rational mind, but actually man has more instincts than any other animal. Instincts are computational shortcuts. CoBe will likely have an even larger repertoire of instincts, most of which are unknown among earthlings. And correspondingly, they will be likely to evolve more complex coordination systems to handle the large collection of instincts. The system may have many hierarchical layers. If the evolution of the human mind is any guide, the problem of uncontrollable unconscious instincts and yearnings will appear, and it has to be effectively addressed.

Beyond intelligence. There is little doubt that intelligence is the winning formula in the near future and that CoBe will be superintelligent. The literature on future beings is almost entirely about the future of intelligence and has little to say about the motivational and social developments that must be in place.[13] That is unfortunate because CoBe may eventually move beyond a preoccupation with intelligence. Even as we humans become more intelligent, the limits of intelligence are already visible.

From the informational and computational perspective, intelligence is about emergence and compactness. As Ray Kurzweil points out, "Intelligence is: (a) the most complex phenomenon in the Universe; or (b) a profoundly simple process. The answer, of course, is (c) both of the above. . . . Understanding intelligence is a bit like peeling an onion—penetrating each layer reveals yet another onion. At the end of the process, we have a lot of onion peels, but no onion."[14] In general, the more intelligent a mind (or a "program") is, the more compactness (simplicity) it is able to achieve, with its modules enjoying great reusability in different contexts. The reuse of basic "codes" in our language, for example, is extensive as it takes advantage of metaphors in our understanding and reasoning. As an example, the Chinese language has only fifty thousand to eighty thousand characters, of which only about five thousand are sufficient for a person to use to communicate in a wide variety of subjects and create beautiful novels or poems.

The challenge in creating intelligent machines may not be speed but structure. As the processing speed rises exponentially according to Moore's Law, so does the scale of problems that have to be solved. Designing flexible and robust organizational structures that can combine and coordinate numerous relatively simple computational modules may become more and more difficult and costly. We all know humans are highly intelligent—the human genetic code, though far from optimal, possesses a "best in a million" degree of efficiency as compared with randomly generated DNA genetic codes.[15] We also understand that natural evolution took a very long time and a great deal of energy to generate the structure we possess. Even with conscious guided evolution, to push the envelope of intelligence with even more complex organizational structures will be slow and unpredictable. So the question may become: Will there be a fundamentally different alternative to human intelligence for a vastly more complex artificial organism?

Evolution often makes unexpected turns. In both biological and technological evolution, an optimal growth strategy in one moment may suddenly be superseded by a very different strategy. If such abrupt switches (examples of which are given below) are any indication, our current preoccupation with the expansion of human intelligence (adding more and more modules to the existing ones to become a "super-mammal" as we described in chapter 7) as representing the frontier of future evolution may also prove a local phenomenon (to be superseded).

The growth of biological complexity used to be closely associated with the growth of genetic complexity. For example, fungi continued with the bacteria's earlier winning strategy of freely exchanging genes. They are successful adapters, but their potential for further "progress" turned out to be fairly limited. From brachiopods to echinoderms, fish, and finally to amphibians, the evolution of animals showed a generally rising tide of genome complexity.[16] However, the genomes of reptiles, birds, and mammals actually became significantly smaller than those of amphibians and most cartilaginous fish as they took different growth strategies, the most significant of which is the larger and more complex brain. In fact, the human genome is smaller than that of the lungfish!

We can see a similar shift of development direction in recent military history (to cite an instance of technological history). The technological frontier during World War I was developments in the chemistry of explosives and poison gas and the creation of fixed-wing aircraft. Larger and larger bombs were made in the arms races that followed. But the bigger bombs became so destructive that they were effectively impossible to use. The deployment of the atomic bomb marked the end of that line of "progress" in weaponry. The trend toward ever more destructive bombs, a characteristic of the Industrial Age, quickly gave way to the new trend of "smart bombs" in the Information Age. Likewise, totally unexpected frontiers are likely to be discovered in many areas in the future, expanding the potential for the existence and creativity of future mind.

Finally, one thing is certain: individual happiness cannot be the ultimate aim for CoBe. Surely, these advances, like all the advances in human and prehuman history, will also bring into play new and sharper paradoxes, new tensions, and more painful disharmonies on a grand scale that we cannot even begin to imagine.

12.4. New Political Concepts and Relationships

Societies have always evolved to keep pace with new technologies. This dynamic will be the same for the transhuman world. The questions that concern us the most during the transitional period are: how will autonomous androids and cyborgs become part of human society; and how will genetically modified humans be dealt with politically and socially?

If history is any indication, some robots will be more than welcome because they can become a useful and enjoyable part of our life. Think about our love affair with dogs and cats. It started with utility—dogs as hunting and herding aids and cats as mouse catchers—but our lasting reason to keep them turns out to have been as playmates and companions. Aside from their childlike size, they also have childlike characters—not too smart and unpredictable, but also not too dumb and boring. One critical premise of our love is control. We love dogs, even those that are old, weak, or sick, but never feral dogs. We love cats but find it necessary to trim their claws so that they won't damage our furniture. Initially, androids will evoke a similar set of issues as we begin to live and work with them in the near future.

Cyborg rights. Once robots are welcomed into our homes, offices, and playgrounds as aids and pets, however, they may rapidly gain ground in terms of capabilities and versatility. Soon, some of them could become our colleagues and playmates. At some point, once truly autonomous androids or robots are designed, we can assume that they will "work" for pay as we do, and that they will even invest monetarily in their own upgrade. We may also assume that the most emotionally complicated and sensitive ones can easily become our "soul mates." Perhaps the autonomous robots will try to mimic various human lifestyles. Of course, many of them will not succeed; only those that have been designed with a strong will to survive and evolve could last in a competitive environment.

Will robots come to resent their social status as our slaves or pets? Not if they don't have human psychology. Their values and motivations may well be drastically different from ours, depending on our design choices and on the upgrades they perform on themselves. The ethical treatment of autonomous robots must be consistent with their nature, not necessarily with *our* nature. Nevertheless, as the sophistication of autonomous robots approaches that of humans, their needs and desires will have to be addressed, since they may no longer like to be treated like servants or pets. Some may eventually demand equal or even higher political status than humans.

The more extreme aspects of the animal rights movement failed to gain wide social acceptance because of the perception that animals do not deserve equal treatment or the rights we believe are deserved only by humans. Animals may share with us a common ancestry and a wide range of emotions,[17] but their rationality, intelligence, communication skills, and moral sentiments are rudimentary by human standards. No society in history has ever granted equal political rights to children, animals, or those with severe mental deficiencies. Reflective consciousness, intelligence, and moral sentiments are going to be some of the key criteria for judging whether artificial lives qualify for citizenship in our societies. Following the Western liberal democratic tradition, Chris Hables Gray has drafted the "Cyborg Bill of Rights," which advocates the following freedoms: (1) freedom of travel; (2) freedom of electronic speech; (3) the right of electronic privacy; (4) freedom of consciousness (rights to modify the conscious mind); (5) right to life; (6) right to death; (7) right to political equality; (8) freedom of information; (9) freedom of family, sexuality, and gender; and (10) freedom from violent harm.[18]

Many of these anthropomorphic rights are probably acceptable political arrangements for androids/cyborgs that are not too different from humans in terms of motivational and perspective-based capabilities. However, as they continue to diverge—some may remain at a dog's level, some may reach human capacities, and some may go way beyond those of humans—political principles such as equal political rights and responsibilities will surely be challenged by the fact that they cannot be treated as a single group.

Changing minds and new democracy. Our legal and criminal justice systems have evolved over millennia under one unchanging condition: nobody can alter human nature. As the ability to manipulate the human brain increases with advanced genetics, law and order may undergo profound revolutions. For example, the emphasis of the legal and police system may shift from retribution after a crime has been committed to deterrence and prevention through monitoring and early warning systems. A convicted criminal may not be thrown into prison, which is a horrendous waste of public and private resources, but instead may undergo treatments that will enhance his or her functionality much more effectively, and with fewer downsides, than anything we have to offer today. In the past, society has had no choice but to imprison or kill antisocial individuals; in the future, there could be alternative solutions less damaging to both the individual and society.

But these possibilities raise the issue of how to prevent the abuse of mind-control and mind-altering technologies, as depicted in Anthony Burgess's dystopian *A Clockwork Orange* and in numerous other science-fiction movies and books before and since. In the case of an autonomous robot, should its creator *also* be held responsible for breaking the law? Or will robots themselves be policed, arrested, and brought to civil and criminal courts? There will be interesting debates, but the issues may have to be resolved in practice as different kinds of legal frameworks are tried out in different political jurisdictions.

Once technical issues of how to define social capability and how to account for motivation are addressed, liberal democratic societies will face pressure to transform themselves through redefining political rights. Constitutional systems that stress equal political rights may need to be modified since they were developed for societies made up of homogeneous members—though this homogeneity has always been something that the slaveholders disagreed with. Creative innovation will be as necessary for political structures as for advanced technologies. Different structures may be developed, tried, and refined for the new posthuman reality.

However, the core political wisdom of liberal democracy is likely to remain and even be reinforced. The ideal of political equality arose from the Enlightenment insistence that since no one has access to absolute truth, no one has a moral right to impose their values and beliefs on others. All citizens are responsible moral agents who can distinguish right from wrong, and therefore all deserve equal consideration before the law and a respected place in our political community.[19] Human beings will retain as a basic political right the freedom to pursue their happiness in the transhuman world because their human nature has not changed. Every member of the posthuman society, humans as well as transhumans, will enjoy the same political freedom of being protected from harm by others.

The fundamental moral principle we have enunciated, "Put yourself in another's shoes," applies to all in terms of rights and responsibilities. The critical difference, however, is that a transhuman/robot's moral endowment may well become adjustable, whereas "natural" persons have no control over the genetic endowment and cultural environment that shape their morality and personality.

The adjustability of moral instincts (if we can call it that) will pose tremendous legal and political challenges to the current systems and eventually lead to fundamental changes. For the individual transhuman or robot, political competition will exert pressure to upgrade one's moral standards. If a societal consensus develops on how to determine if one is free of evil motives, the pressure to upgrade one's morality will certainly increase—just as, with today's general consensus about the need for higher education as a prerequisite for certain kinds of jobs, the pressure to go to college increases.

Threat of evil. How about the downside of adjustability? If the history of human moral evolution is any indication, the ability to create transhumans who perpetrate inhumane evils worse than Hitler's will become a reality, just as human beings have managed to commit far more evil acts than any other animal. But there is a strong counterargument: the selective pressure to be "good" will be stronger than ever as new technologies enable easier and more intense social interactions. On the one hand, there is no guarantee that a new species will be less selfish than its ancestors—it pays to be selfish—but the overall trend of biological and cultural evolution shows that the winners tend to be those who take advantage of positive-sum cooperative gains. As we have noted before, the species that emerged as the winner from the evolution of hominids is one with much greater compassion toward others (outside of one's kin and even one's species) than any other animal.

How things will play out is unpredictable, but in principle, it is unlikely to be different from the gene-meme coevolution that shaped human moral sentiments and "tamed" humans. Without our being able to "tame" ourselves, it would have been impossible for civilization to take hold, and moral teachings would have fallen on deaf ears.

In short, the battle between good and evil may (once again) become unavoidable during the transitional stage. Which side is likely to prevail? Again, let's look back at the common pattern in history. As in previous human endeavors, we can expect massive failures and casualties during the formative growing stage, and we can anticipate that most new developments will turn out to be dead ends; but the likelihood is that Cosmic Beings with transcendental morality will emerge victorious and spread to the cosmos.

New political organizations. In the posthuman world, there is no doubt that fundamentally different political structures will emerge and flourish. Living as we have through some of the greatest ideological battles in history in the twentieth century (and, in a different form, in the twenty-first), we tend to think rival political systems are as different as night and day. In a sense this is true. Yet, despite their obvious overt differences, human political structures such as aristocracies, oligarchies, plutocracies, tyrannies, despotisms, and democracies all work according to the same "primate political biogram," with ritualized competition for hierarchical dominance and concentrated political attention on leaders as its core characteristics.[20]

The ongoing concentration of political power in the hands of powerful individuals or "faceless" institutions has been a long-term trend, but it always faces an inherent limit. Given human nature, no government, regardless of its sources of legitimacy and mechanism of control, has been all-powerful over the individual. Bureaucracy, which was invented to cope with the large-scale organizational demands of urban civilization, has always been disliked for its inhumanity and impersonality. Its lofty ideal of efficiency and purity has always been impossible to maintain. Yet bureaucracy survives and thrives, simply because our natural emotions adapted for small-scale personal relationships cannot handle large-scale organizational tasks.

When the constraint of fixed human nature is relaxed, waves of unprecedented political innovation will begin. Fundamentally different organizing structures (as a part of the overall artificial universe) will be created, tested, duplicated, or discarded. The selective pressure for the emergence of social emotions that are flexible and focused on higher perspectives will surely generate novel organizations and results, if history is any indication.

Traditional government, the territorial monopolistic agency of legitimatized coercion, may eventually disappear altogether. Although rule-based political control dominates the landscape today, legitimacy and willingness to submit to an order have always been determined by a variety of psychological motives—legality is only one conceivable mechanism.[21] Democracy and popular sovereignty may very well prove to be critically dependent on human nature. Human nature is very complicated, and future experience will show which parts of human nature matter the most. For example, is it true that the most popular and successful political system today depends on our mixed conformist (follow the majority) and nonconformist (be an innovator or follow the innovators) tendencies in social interactions? Along the same lines, we can also think about another concept that the Greeks invented—the *citizen*. Is that concept also critically dependent on particular aspects of human nature?

The nation-state has been a very effective but costly institution. Throughout the history of government, warfare and related expenditures have been—with some possible exceptions, such as the Old Empire in ancient Egypt—the single most extravagant and continuous drain on the fiscal and economic resources of the state.[22] Hence selective pressure for eliminating intraspecies aggression from its roots—the selfish human instincts—will be immense as soon as selection becomes technically possible.

It is impossible to predict which type of organizational structures will prevail and how networks and inter-organizational relationships will evolve. The field will be wide open, and conflicts and clashes could be fierce. Furthermore, the mechanism of the organizational selection process itself is also likely to evolve. Although it has been done over and over, it has never in human history proved *necessary* to impose one's ideas on others at gunpoint. In the future, wars between political organizations could become not only improbable but also inconceivable as the standard of moral sentiments and the power of destruction both grow exponentially. But new forms of selective pressure will surely emerge.

12.5. The Second Axial Age

Having learned how to travel across vast space (by harvesting stars as heat engines, for example), CoBe will enter and adapt to environments totally alien to humans, and in doing so become incredibly sophisticated, converting matter and energy into what Ray Kurzweil calls "computronium." Relatively isolated CoBe tribes may form in distant galaxies, possibly marking the beginning of a second Axial Age.

With new knowledge and technologies, CoBe may manage to travel at ever faster speed, maybe even faster than the speed of light. But unless CoBe can discover or invent instantaneous communication and information processing techniques such as so-called wormholes, space will be "space"—an absolute barrier to communication and interaction among distant stars and galaxies. In other words, we only know the *history*—not the current reality—of others living at great distances in space. The future might then be a return to a "tribal environment" in the sense that the instant communications we have established on Earth might no longer be possible. CoBe would be isolated by space, each "tribe" taking time to learn and to evolve.

In our introduction to cosmic evolution, we noted the two most significant explosions on Earth: that of life and that of human culture. The proliferation of species and cultures had its bedrock in the cell and human nature, respectively: the march toward more complex organisms and technologies experienced numerous crashes in mass extinctions and collapses of civilizations, but cells and people persisted, always ready for another march in a new direction. What will be the bedrock for the coming CoBe explosion? I believe it will be the conscious understanding of the universe (for us with our limited mind, the understanding is neatly summarized as "natural laws"). While different technological and organizational approaches may come crashing down, the understanding that CoBe processes will not perish, and it will always be ready to serve as the basis for another campaign. Over time we will increasingly approach the state of being "one with the universe."

The development will not be even. It is very likely that whatever the pathfinder "nation" turns out to be—with its ability to articulate the indivisible Cosmic Vision and its extraordinary capacity for creativity—will be the next superpower in the galaxies and will spread its influence, just like Venice in the fifteenth century, Portugal and Spain in the sixteenth, Holland in the seventeenth, Britain in the eighteenth and nineteenth, and the United States in the twentieth.

Just as our African ancestors could speculate that their children would fill every corner of Earth, yet they could not imagine the emergence of civilizations and technological developments, it is beyond our mental power to imagine any specific development of CoBe. Beyond certain points, technologies will be powerful enough to alter natural laws. We simply do not know whether natural laws can be modified or whether the laws we perceive are specific manifestations of deeper and immutable laws. The speed of light and other hard constraints may eventually be broken, as they may well have already been broken by extraterrestrial civilizations. Looking back at human history, we see that industrialization carried us into the Newtonian universe, where ships and wheels broke the space barrier. The invention of the telegraph and radio brought us into the Einsteinian universe, rendering Newtonian space and time relative. Communication between any two individuals on Earth has now become almost instantaneous. Will CoBe's history follow the same pattern?

If the barriers to instantaneous communication and information processing are broken in space, all of a sudden the isolated "tribes" of the universe will experience an "urbanization" and political centralization process, and the "second great Axial Age" (speaking in terms of Earth's past) will lift cosmic evolution to a new level of technical sophistication. And with that, we might well expect spiritual enlightenment about the universe, its origin, and its future. Just as our enlightened political ideal is "All persons are created equal," the cosmic ideal may be "All beings in the universe are created equal"; and as this ideal is realized, the enlightenment that comes from experiential contact with the Cosmic Spirit begins to manifest. Even the Cosmic View articulated here will become a historical relic at that point.

Our universe will be fundamentally transformed, new dimensions of existence will be discovered, and just as we have created artificial universes, new universes will be created as the "mind-children" of CoBe. Just as today we mere humans contemplate *life as it could be* versus *life as we know it,* the future CoBe will ponder *the universe as it could be,*[23] as opposed to the universe as it is given.

CHAPTER THIRTEEN

REFLECTION AND EXPECTATION

> *We appear to be on the eve of having a hand in the development of our bodies and even of our brains. . . . [And] we may well one day be capable of producing what the earth, left to itself, seems no longer able to produce: a new wave of organisms, an artificially provoked neo-life. . . . Thought might artificially perfect the thinking instrument itself. . . . The dream upon which human research obscurely feeds is fundamentally that of mastering . . . the ultimate energy of which all other energies are merely servants; and thus, by grasping the very mainspring of evolution, seizing the tiller of the world.*
>
> *I salute those who have the courage to admit that their hopes extend that far.*
>
> —Pierre Teilhard de Chardin,
> from *The Phenomenon of Man*—written circa 1937–1940

IN THIS BOOK we have been exploring the cosmic perspective on humanity: its origin, its growth, and its coming elevation. We as a species have existed for about a quarter of a million years, which is just an inch along the mile of cosmic evolution, but our next frontier—the posthuman frontier—is already visible on the horizon. And when we cross that frontier, we will understand that humanity is enduring but not eternal. *Homo sapiens* plays a pivotal role in growing and spreading life and intelligence into the cosmos, but it is not the final or highest expression that evolution can achieve.

13.1. A New Perspective on Humanity

The cosmic perspective, or Cosmic View, that I have proposed represents a paradigm shift comparable to the Einsteinian revolution in physics. Recall that in the Newtonian model of the universe, space and time were absolute givens, an unmoving backdrop against which everything played itself out. Einstein's conceptual revolution turned that upside down: space and time became relative, while the speed of light became the absolute given.

A similarly radical shift in perspective will result from the new paradigm I have proposed. Until recently, humanity has been treated as an absolute given in all major schools of thought. Even the great monotheistic religious traditions consider man as God's central concern. Likewise, secular humanists treat the well-being of humans as their ultimate concern, as do technology visionaries such as James Albus, who considers human benefit as the ultimate measure of goodness for any technology and social/economic system. Just as Einstein relativized space-time, the Cosmic View puts humanity in a transitional position in history. I have argued that the new absolute goal is the process of cosmic evolution itself, which moves from a single starting point (life on Earth) toward greater complexity, beauty, and transformative power in the cosmos.

From the cosmic perspective, the potential reality beyond humanity must suddenly become our central concern. With human beings as the absolute be-all and end-all of the Cosmic Creation, discussions about new technologies and new social organizations are always centered on potential benefits and risks to human beings. With cosmic evolution as the absolute given, human prosperity and happiness can still be our preoccupation, but it must be pursued along with a deeper sense of purpose—the realization of our cosmic potential.

When I speak of "our cosmic potential," I mean that humanity has the ability to complete a critical step in this process of cosmic evolution. In order to do so, we must realize that our greatest responsibility is not to keep ourselves living happily forever, but to create new beings and pass the baton to them. Nor does it mean that humanity will be discarded, but only that in the future our spiritual descendants will take over our cosmic responsibilities. As we discussed in chapter 9, this creative process must rely on our strongest motivations—and those motivations certainly include whatever can make us happy, including developing capabilities to modify ourselves so that we can be free of the genetic bondages and live a longer, healthier, and happier life.

There is nothing extraordinary about this transition, considering the fact that we once took over that very same responsibility for being at the vanguard of cosmic evolution from our ancestral species—although without the conscious intention that will likely be the case this time. After *Homo sapiens* emerged, *Homo habilis* and *Homo erectus* were not the leaders of cultural and technological evolution any more. It is unfortunate that our direct ancestral species have ceased to exist, but if they had survived today, they would have been well protected, nurtured, and respected—it would surely be our desire for them to lead a good life, but whatever they might do would have been irrelevant to the further evolution of a conscious humanity—that is, they would not be able to contribute anything to all of the social, cultural, political, technological, and economic revolutions that have marked our ascent to primacy in the known universe.

The cosmic understanding of humanity I have presented here is complex but can be boiled down to three main points: (1) the human is is a transitional being in the long-term sweep of cosmic evolution; (2) the evolution of humanity and its culture was made possible by our sense of freedom (free will) and understanding of nature; and (3) the potential exists for our species to create new beings that will fulfill our destiny and that of the Creation.

The coming elevation of humanity will be realized by passing on its central insight and judgment (or human wisdom) to higher beings in order to amplify the wisdom. What is central for humanity is not to be found in our biology; rather, it is to be found in our transformative potential for the universe and possibly beyond.

Pointing out the cosmic value of humanity does not devalue the attributes that we normally consider human or humane. We have developed a dazzling array of wonderful lifestyles, institutions, technologies, art, science, faith, and so on. The noble goals we have for ourselves—freedom, liberty, morality, and happiness—are still worth pursuing. But rather than treating these goals as ends in themselves, I have shown that we must ask what a prosperous and happy humanity is *for*.

When Immanuel Kant said we should treat each human being as an end, not merely as a means to some further end, he was concerned about the proper relationship between human beings, not about the direction of human evolution as a whole. It is unfortunate that even the more enlightened prophets usually let their imagination stop at some static Utopia on Earth or paradise in heaven after death in spite of all evidence pointing to an ever-increasing acceleration of change.[1] It is time to step outside of ourselves and place the meaning of our existence in a cosmic context.

From the cosmic perspective, we cannot be the ultimate being—nothing ever was and nothing will ever be. As with any other species, the human being is not a perfectly designed shell to house the soul. A person, as the shape of its Chinese character (人) suggests, does not stand still but continues to move forward.

For those immersed in the Western monotheistic religious traditions, there is no need to fight Darwin's idea that we are "merely" the outcome of a trial-and-error process. It is nothing short of a miracle that "mindless" natural evolution generated a wonderful mind capable of reflecting on this creative process and shifting it into a higher gear. Likewise, for secular humanists, there is no need to regard ourselves as nothing but a freak accident in a random universe. We are not the first but simply the latest manifestation of cosmic evolution, which for billions of years seems to be heading toward more order and more interconnections—in other words, more meaning. We are fortunate to live in this universe; but that fortune means little if we don't take advantage of it.

Self-evaluation is always trickier than sizing up others. If the first cell had a thinking mind, surely it would have regarded itself as the world's ultimate being; the same would be true for the first multicellular organism, the first animal species, the first animal species with a brain, the first land animal species, and so on. We know they would have been mistaken, and future minds will know that those of us who hold the same self-regard are also wrong. By the same token, if the first generations of cells or the primitive animals had believed that the evolutionary process was so random that it could not have had any discernable trend, we are here to prove them wrong. And again, the future Cosmic Beings will show that the process has much further to run beyond the stage of humanity.

From our human perspective, it is natural to believe that evolution is all about creating a world for us to live in, or that God cares about us and loves us. We have no choice but to use our intuitive minds to understand everything that is related to us. But at the same time, we have the rational mind to ask why this is the case. The source of creation and order (often conceived as the cosmic God) rewards us for understanding the universe. Let us ask ourselves: how can we express gratitude toward the Creation? Reciprocal altruism is embedded in us by natural selection as a manifestation of the nonzero principle. Combined with science, altruistic love is probably the best emotion we can use to understand the cosmic perspective.

Humanity is a critical transitional being, and there is nothing wrong or sad about our transitory status, given what we know about human nature. We know that, structurally, we are a peculiar mix of our ancient animal heritage and a newfound rationality and higher consciousness. The realistic view of this super-mammal, with its narrow limits of physiology and psychology shaped by a particular evolutionary path on Earth, can be a source of both existential despair and eternal hope. Humanity is neither a perfect, eternal form designed by some supernatural being, nor just a "naked ape." The human being is the most advanced known result of Cosmic Creation, although that status is temporary.

In a sense it is true that the human is created in the Creator's (God's) image, because of its ability to create. Humanity has created an incredible array of artifacts and cultures like no other animal. Consciously or unconsciously, we have been learning how nature creates—a process best characterized as "creative destruction"—and in the process we have been building up the capacity for the ultimate creation, which is *creating something better than the creator (humanity) itself*.

Only when we fully accept and utilize our unique ability to create can we become truly significant. In a sense, we are like the orca, or "killer whale," which, as described in the prologue to this book, sits at the top of the food chain in the ocean, but it occasionally jumps out of the water and sees magnificent landforms. Just as the orca cannot return to the habitat that we might imagine it longs for, humanity cannot fulfill its cosmic potential by itself—but it can participate in its fulfillment by creating CoBe as the being that inherits humanity's spirit and continues its quest, leading ever closer to fulfillment of the Cosmic Vision itself.

The appearance of human intelligence and higher consciousness is perhaps comparable to the appearance of the first cell on Earth. Both of these were truly momentous events. The emergence of cell life involved the first complex biological organism with metabolic pathways and self-replication. But that was not the end. Much more exciting things had yet to materialize—life would spread to the entire surface of the Earth and develop millions of different species, with many organisms consisting of billions of specialized cells. Through countless long and uneven paths, with several mass extinctions as well as many quantum leaps along the way, suddenly human beings emerged to reflect on the past and to initiate the next act of Cosmic Creation.

Can anything be more grand and thrilling than this?

13.2. Do Not Settle for Too Little

We have accomplished a lot, but we should have even greater expectations for the future. To paraphrase Thomas Merton, the biggest temptation we must resist is to settle for too little. If there is anything that the Cosmic Vision is *against*, it is against setting the ultimate goal of our lives as the maximization of human happiness.

The Cosmic Vision is not a wishful-thinking philosophy; rather, it is a pragmatic perspective grounded in science and cosmic history. Life has shown a tendency to fill every niche on Earth. With conscious evolution, mind is likely to do the same in the universe. The awareness of progress itself made conscious progress possible.

It is neither necessary nor possible for everybody to embrace the transcendental perspective, but those who do will live a more meaningful life and become an energetic and fearless vanguard. Such pioneering efforts matter, because results themselves bear eloquent testimony during revolutionary times—whether it is science against religious dogma, or a liberal democracy against a top-down system of excessive controls and oppressive limits on personal and economic freedom.

Depending on how it is interpreted, the Western idea of humanity can be either a big obstacle or a motivating force for setting in motion the cosmic perspective described here. In our attempt to cross the river by feeling the stones, we must be careful about two conceptual traps: the human-centric God and existential nihilism. We must find meaning in a universe that is not created *for* human beings. This requires both reason and faith.

Not knowing the exact path forward is what gives us freedom. However, freedom is useless, even dangerous, without responsibility and purpose. Humanity can very well be a vital part of the answer to the question "What is the fate of the universe?" Mao's declaration that "man's determination can overcome his fate" (人定胜天) is meaningful in the long run.

Since the Axial Age, we have asked the questions only we can ask, but we have not found definitive answers. Why does the universe exist? By chance or by design? Are other universes possible? We have now found a better way to answer unanswerable metaphysical questions, which is to create more capable beings. If we can know far more than do animals with simpler minds, then more complex superminds will know far more than we do.

We will move forward without definitive answers to many questions. Instead, the feedback from our actions will tell us if we are on the right or the wrong path. This faith in the test of reality is both stubbornly restrictive and inspiringly open. Contrary to popular belief, science and scientific discoveries—which, as we pointed out in chapter 4, have become (along with their rigorous methodologies) a major source of feedback in our lives—are highly religious, filled with a transcendental spirit, because they provide the best available guesses about the fundamental questions.

So far, there is no scientific evidence that contradicts the feasibility of implementing a higher purpose for our lives in this lawful universe: if we possess the correct perspective on our place in the cosmos and if we have the courage to unravel natural laws, we can fill the universe with our successors and eventually embrace God, the Tao, or whatever symbol you use for the supreme and the ultimate.

We sense that providence is on our side, but we will always have to act on incomplete information. Faith in a larger purpose is necessarily only our best guess based on experience, but it has to be there to give us the extra motivation in addition to whatever natural motivation we are equipped with. One of the most important lessons from history is that the individual (or organization) with a larger purpose (or higher inspiration) tends to prevail in the end. That is something very concrete we can (and must) bank on.

In the eternal cosmic dance, we will often be wrong, but never in doubt. To paraphrase Lincoln at Gettysburg, "This species under God shall have a new birth of freedom, and the intelligent consciousness with a Cosmic Vision and an awareness of its mission shall not perish from the universe."

Buddhism and Taoism both claim that the human spirit existed from the very beginning. I had previously thought this flatly contradicted modern science until I read Freeman Dyson's remark that the universe must have expected us from the beginning. The highest human inspiration is, then, to find out what else the universe is expecting—what can be realized through our conscious efforts. The only way to find out for sure is to make it a reality.

13.3. Leaders, Pioneers, and Favorable Environments

To be a leader of cosmic evolution, we have to be a player rather than a watcher. Contemplation of our destiny is just the starting point. We inherit Greek science and logic but denounce their assumption that the detached theoretical viewpoint is superior to the involved practical viewpoint.

To be a leader of cosmic evolution, we have to identify ourselves with the cosmos. This means we do not allow ourselves to be controlled by our animal instincts. This means letting higher consciousness and the higher perspectives guide our destiny. In Victor Frankl's terms, an "authentic" human being is one not driven but responsible—responsible for what the conscious mind recognizes as our mission.

It is difficult to be "authentic" because responsibility can be both a source of achievement and a burden. As Fyodor Dostoyevsky wrote in 1864, "Man is pre-eminently a creative animal, predestined to strive consciously for an object and to engage in engineering—that is, incessantly and eternally to make new roads, *wherever they may lead*. . . . [P]erhaps the only goal on earth to which mankind is striving lies in this incessant process of attaining . . . but to succeed, really to find it, he dreads that when he has found it there will be nothing for him to look for. . . . [C]onsciousness is the greatest misfortune for man, yet I know man prizes it and would not give it up for any satisfaction."[2]

Pioneers are those who take on our evolutionary responsibilities despite knowing the possible consequences. We have always had pioneers in history, but "cosmic consciousness" has never been experienced by more than a tiny minority within human society. There have been many dark hours and dark ages when few people were concerned about more than mere survival, although cosmic consciousness would eventually be rediscovered by reflective individuals. This situation is unlikely to change much in the future.

Still, we should not despair about the human condition. Given positive cultural influences and a degree of social order, we can rely on significant numbers of people to provide diverse ideas and innovations that together can help lay the groundwork for cosmic evolution. Extending Adam Smith's "invisible hand" metaphor, the Cosmic Vision can be furthered even through the creative efforts of relatively self-absorbed individuals. Creativity often transcends any "selfish" motivations behind the creative act. And, if historical experience is any indication, some people can be brought to the point of transcending personal dreams of human happiness by identifying with the greater goal of cosmic evolution.

What about dogmatic religious faiths that so many people hold dear to their hearts? The clash between science and religion is sometimes unavoidable, but it is important to recognize that at the highest level, the faith in a supernatural God and the faith in reason are not that different—both are an optimistic bet that the forces that contribute to positive outcomes—however we may conceive them—will survive and thrive, as they always have. What the Cosmic Vision does is shatter the illusion of a "happily-ever-after" Golden Age/Heaven; rather, it sets our sights on another Axial Age, with its splendid flowering of new minds into the universe.

Embracing the evolutionary perspective wholeheartedly means that each generation must do its best. History is a product of both necessity and contingency. We can never predict who or what will produce a breakthrough; we can only say that certain individuals and particular events can be significant in blazing the path of cosmic evolution.

The fact that people have adapted well to abstract and impersonal laws and social objectives offers much hope that the seemingly impersonal Cosmic Vision can stick and spread. Certainly, personal perspectives will continue to dominate people's lives, and animal instincts will continue to play a part in human motivation. But just as a tree can pull water upward against gravity, we are capable of embracing spiritual unity even in the face of animal instincts. To our human-centered morality we can add transcendent morality.

The fact that the Cosmic Vision is unlikely to be embraced at once by the population at large is no cause for despair. In natural history, evolutionary changes usually proceed more rapidly in isolated small groups than in large, widespread groups since the novel genes can have an easier chance to spread to the population rather than getting inundated by the gene pool of the masses.[3] Cultural breakthroughs usually happen in the same fashion.

We should despair, however, if personal freedom and the freedom to experiment with a variety of social organizations are taken away by a global political entity. While global integration in many areas continues to deepen, it is a blessing that no centralized "world government" is on the horizon. By conscious will or by chance, there will be a few people who can "dream of things that never were" (in the words of John F. Kennedy),[4] people who will seek to realize their dreams in some nurturing social environment with a high degree of enthusiasm and risk tolerance.

Our venture will certainly take many baby steps and many collective efforts; but at each step, once initial breakthroughs occur, the snowball and trickle-down effects will take over. Concerns over risks and ethics will have to be addressed again and again—but as practical issues, not as grounds to doubt the overall direction. We do not know if conscious evolution will proceed in a more humane fashion than natural evolution, but we will certainly adopt the least taxing and most efficient strategies if they prove to be practical.

Witnessing just one case on Earth, we have no way of knowing whether natural evolution could have had a shorter and less bumpy route from the first cell to humanity; but under the most advanced social environment, we have already managed to elevate the unit of elimination from person to organization. Individuals won't get their head cut off if they make a mistake in the trial-and-error process. Just like the turbulent and uneven transition from agricultural society to industrial and informational societies, the Cosmic Vision will be realized, first and foremost, by offering unprecedented rewards to people who are devoted to it.

13.4. "What can I do? What's in it for me?"

So at last, we get down to the personal: "What can I do?"

The short answer is: embrace the highest perspective and realize your potential.

The highest perspective reflects truths that have become self-evident: the universe evolved from a simple beginning; life and humankind emerged from that process; and the universe will continue to evolve. This perspective is the deepest meaning of your life because it connects the realization of your personal potential and satisfaction to that of the cosmic potential by answering two questions: "What is this world all about?" and "What is my place in it?"

You can feel really happy and pain-free as an individual. But without a connection to the outside, it is just a particular molecular arrangement of your brain—perhaps a rush of dopamine to your pleasure centers—that gives you the subjective feeling and nothing more. The same is true about the collective happiness of your family, your country, and the entire human race.

Why do we have to care about larger meanings? It is both a curse and a blessing that we have evolved to be a meaning-seeking species with our reflective consciousness. While many have argued otherwise, I firmly believe it is a blessing that we are not merely an animal.

But how can the connection to the "outside" be made as we seek larger meanings in our lives? For most people, there is little direct connection between their lives and the broad flow of human history, let alone cosmic evolution. Most people cannot "know themselves" from the broader perspectives I have suggested in this book. This has always been the case in history and is likely to be true in the future. As Gregory Paul and Earl Cox pointed out in their book *Beyond Humanity*:

> The American sitcom *Seinfeld* is popular because the show is explicitly about—nothing. The gang of four young, hip, post-modern New Yorkers really has nothing to do with themselves; they have no great causes to fight for, no need to struggle for existence, except lead moderately amusing, but rather harmlessly hedonistic and aimless, lives in the Big Apple. As will humans in the cyberfuture, Jerry, Elaine, George, and Kramer do not have critical jobs that produce goods others

> must have. . . . After the robots come, all that remains of humanity may be living one oversized episode of *Seinfeld*.[5]

In his Pulitzer Prize–winning book *The Denial of Death*, Ernest Becker describes the "automatic cultural man,"[6]—a modern-day version of Søren Kierkegaard's "philistine," one whose identity is limited by mass culture and who does not (unless forced by extraordinary circumstances) look beyond the immediate environment he or she lives in. For such an individual, the *Seinfeld* life is good enough. And from the conventional social perspective, this kind of life is nothing to sneeze at. Most of these people are law-abiding, taxpaying, working-and-spending citizens—sometimes described, not incorrectly, as the "building blocks" of society—with daily challenges in their careers and social lives. Even with their obvious limitations, they contribute to the stability, diversity, and collective "common sense" of society. And although they as individuals are probably not aware of it, they are all connected to a much bigger picture than they can imagine.

Those who *are* aware of this bigger picture hold a higher perspective than that of the "automatic man" (who is roughly at Stage 2 or 3 in terms of faith that we discussed in chapter 8). And such awareness is necessary to become part of the vanguard of cosmic evolution and the posthuman future. Although such a future has long been a product of our creative imaginations, now we can also say that this future can be real.

Timing is everything for a finite being. One can conceive of great technological advances, but for the conception to be realized, certain external conditions must be met. In the mid-nineteenth century, Charles Babbage and Ada Lovelace had no chance to realize their dream of a digital computer because their brilliant ideas were way ahead of the availability of supporting technologies. Leonardo da Vinci designed an airplane—a flying machine that could "reach the heavens"—but he did not have the means of creating what he had designed. Today, we are fortunate that the dreams of manipulating our own genome and creating new intelligent beings are no longer pure fantasy.

Granted, only a lucky few are directly working on the technological frontier. For the rest of us, how can we prioritize our actions? Actually, there are many things we can do to make a difference.

- **Spread the Cosmic Vision.** Those of us who accept the arguments I have laid out in this book should have the confidence that the Cosmic Vision is as "self-evident" as, for example, the U.S. Founding Fathers' view on human rights and freedom. We should confidently assert that the Cosmic Vision boasts far more scientific evidence than any competing worldviews about our place in the universe.
- **Support political reform.** We can help to lift the benchmark of the "bioethics" debate from human happiness to cosmic significance. We can fight excessive regulations or outright bans on certain scientific research based on outdated dogma. We can provide support for public funding in scientific research and development.
- **Isolate the neo-Luddites.** Their opinions, coming from both the right and left sides of the political spectrum, will never die out because they have strong appeal to our instincts of nostalgia and longings for an illusionary bygone Shangri-la. We should pay our respects to the instinctive beauty of their vision and acknowledge their right to express their opinions—and then move on to the Cosmic Vision, which is *more* beautiful, profound, and useful. Our best chance to refute the neo-Luddites is not rhetoric, but results that are attractive and beneficial. That is how the test-tube-baby controversy was settled yesterday, and this is how today's controversial technological issues will be settled tomorrow.
- **Keep cool close to home.** Be ready for temporary setbacks. Keep a higher perspective when we are caught at the losing end of the posthuman world. If our kids—or our grandkids—come home and complain about unfair competition from genetically enhanced classmates, does it make sense to lobby for a complete ban on genetic technologies? No. The appropriate policy response should be to actively mitigate the adjustment pain and share the benefits with those who are less fortunate, but not stop the change. This is similar to policy adjustments associated with free trade, but at a deeper structural level.

- **Move to the right place.** Locality still matters even as we get more comfortable roaming the physical as well as virtual world. Nation-states still set the clearest boundaries for distinctive political and social environments. If fighting against entrenched cultures and interest groups is too hard, we can vote with our feet to the country or region that offers the best regulatory environment for the posthuman world and put ourselves right in the thick of the most exciting actions. There are risks, but also rich rewards—both spiritual and material.
- **Be willing to try new technologies.** There is a fine line between courageous and reckless behavior when it comes to adopting cutting-edge technologies. We should do the necessary research and use good judgment based on our individual risk tolerance, but we should not avoid new technologies just because they are new—and if they show promise, we should embrace them. The fields of medicine and pharmaceuticals provide an especially instructive example. Only when individuals step forward and take part in scientific trials are advances made in these areas. By doing this, we are not only adopting technologies but advancing them, creating a ripe environment for cosmic evolution. Again, one should do this based only on an intelligent assessment of one's own circumstance. Most of us will benefit by researching results of trials and testing that have already occurred.
- **Join a "SETI" project for the human genome.** Compared with the star-scanning SETI, the inward-looking human "SETI" has a far better chance of success. We can contribute directly to the enhancement and improvement of the human genome by tinkering on our computer or in our garage lab. Overcoming the daunting technical challenges that we discussed in the last chapter needs massive grassroots support. Breakthroughs often occur unexpectedly in unexpected places by nonprofessionals.
- **Become a "libertarian" philanthropist.** Those of us who enjoy financial wealth might consider making this our motto: "It's easy to make a buck, but hard to make a

difference." There are many good causes that are worthy of our attention, but our investment in, sponsorship of, or patronage of high-risk, cutting-edge research projects as well as funding for social projects aimed at mitigating the negative consequences of new technologies are probably the biggest "bang" we can have in terms of an impact on cosmic evolution.

- **Organize or join support groups.** As in any social movement, organization is critical for building the Cosmic Vision's political influence and community support. We all need a place, physical or virtual, for mutual support, collaboration, information exchange, leadership, and, most important, that spirit expressed in Shakespeare's immortal words "we few, we happy few, we band of brothers." There are already some pioneer organizations that promote and support the posthuman vision, and I expect that more will emerge.
- **Create new stories.** Those of us who have artistic or literary talents should use them to promote the Cosmic Vision. Most science-fiction stories and movies are merely human dramas set in the galaxies—*Star Wars* is typical. We can create new stories about CoBe, which require true imagination in terms of their motivations and aspirations. Our works will serve as tangible images against those conveyed in the well-known dystopian stories such as *Brave New World* and *Frankenstein*.

These are but some examples. Each of us must think and decide what *we* can do to make our lives meaningful in the ultimate cosmic sense. Overall, we should be optimistic and energetic about what we do, knowing that "90 percent of life is just showing up."[7]

An interesting life is always an eventful one with unexpected ups and downs. The ultimate satisfaction comes from the ability to live with ease and purpose. Human motivation is what we naturally have, so we should use it to the fullest, whether our quest is for status, wealth, justice, or peace; but we must also remember that our personal potential can gain an extra gear when it is connected to the expression of cosmic potential. To paraphrase Ronald

Reagan,[8] there is no such thing as left or right, only up or down: up to humanity's ultimate dream, realized in individual freedom consistent with law and order; or down to the ant heap of totalitarianism and stagnation.

As I mentioned in the opening chapter, some people may ask, "What if I see humanity as the most desirable state of existence (or I see that fundamental change is too risky) and have no wish to change it?" The answer is that there is still room for Becker's "automatic man." As long as we are alive, we all knowingly or unknowingly have an impact on the broader society and on historical development. What is more, the civilized modern world is no longer "red in tooth and claw" in its treatment of citizens as long as they are law-abiding, and we expect this more compassionate state of affairs to continue. Personal freedom and dignity are now viewed as inviolable individual rights in an increasing number of countries.

That said, the most successful individuals, organizations, and states join competitions in which winners are richly rewarded, both spiritually and materially. We still live in a highly competitive world that is necessary for progress, for continuous improvements in products and services, and for better utilization of precious natural and human resources. In a sense, the world still treats dropouts and "downshifters" ruthlessly, even though bare necessities for living are usually provided by a social safety net. One could be left behind and made largely irrelevant in the same way that those subsistence farmers who insist on using Stone Age tools have been made irrelevant in the modern industrial and information society—and therefore have gradually changed their way of making a living or have gone out of business.

The faster the evolution, the faster the existing practices and attitudes will become irrelevant. Before we suggest "I want to change the ruthless way the world operates and make everybody's life happy and fulfilled," we need to think about the fact that our entire modern lifestyle is a result of continuous rejection of those things we deem less desirable (ranging from our Internet or TV choices to the clothes we wear) and fearless creation of the new.

So the last important question we must address is: "Even if we all agree with the Cosmic Vision, should we pay attention to something that is so far in the future while half of the global population is suffering from poverty and oppression?" This is the question raised by Langdon Winner in his essay *Are Humans Obsolete?* And I am sure it is on many people's minds. Invoking Condorcet's Enlightenment dream of social engineering, Winner argues that "the real challenge lies in realizing the potential of all humans regardless of their prior condition of poverty and oppression. Until that hope is fulfilled, post-humanist ambitions will seem irrelevant or patently obscene." This is an issue of what we prioritize based on our limited resources. It is a political and economic decision no different from the question of whether we should be spending money on R&D while people are starving—or how to allocate our limited spending on basic science, which has no immediate impact on social welfare, and on applied research.

There can be political and academic debates about how much weight should be given to distant benefits and high perspectives. A global consensus could be nearly impossible to reach—but such a consensus is unnecessary. Even within our own brain, different motivators and perspectives imply that we have different priorities competing for our conscious attention. Rather than imposing a global priority, we should not restrict the choices; instead, we should allow countries, organizations, and individuals the freedom to focus more on one versus the other, and let results guide their further actions and decisions.

More important, we must recognize that the tasks of helping the disadvantaged and supporting the pioneers do not necessarily conflict. Actually, they tend to reinforce each other over time. On the one hand, we must try to secure adequate nutrition, sanitation, housing, health care, and education for the three billion people who are still in desperate need. Improving health and education for the impoverished increases the world's human resources; securing political stability and property rights increases people's motivation to work when their efforts are properly rewarded. On the other hand, products ranging from vaccines to mobile phones to inexpensive food created through high-tech procedures benefit those same populations and end up lifting the well-being of all.

We must also realize that while the cosmic future outlined here seems fanciful, it is much more realistic and satisfying than the dream of earthly utopias of eternal peace, absolute equality, no suffering, and effortless eternal bliss. Yes, we have to seriously commit ourselves to making the posthuman future happen, but the unbounded potential of the universe is out there waiting for us to realize.

A thousand-mile journey begins with one step: ours. (千里之行, 始于足下.)

NOTES

Please see the Bibliography for titles and full bibliographic information.

Author's Preface

1. Nietzsche, 1976, 172. Emphasis mine.
2. Nietzsche, 1976, 129.

Chapter 1

1. Bostrom, 2005b.
2. "Jean Rostand and others say that *Homo sapiens* is becoming *Homo biologicus.* But not so. It is precisely because men are sapient that they can control their biology. If we like word play it would be better to speak of *Homo autofabricus*" (Fletcher, 1974, 4).
3. Joy, 2000.
4. Kass, 2003, 9.
5. Kass, 1999.
6. http://www.salvomag.com.
7. Fukuyama, 2002, 183.
8. Peter Lawler, "Pursing Happiness," *National Review,* December 22, 2003.
9. Kurzweil, 1999, 185.
10. In a recent World Transhumanist Association survey, less than a quarter of transhumanists claim to be "religious or spiritual" (Hughes, 2007).
11. The arguments made by apologists and proponents of human cloning include replacing a beloved spouse or child; providing a child for infertile couples; avoiding genetic disease; securing a supply of transplant organs and tissues; replicating exemplary individuals; aiding medical research; or special missions that require a large number of genetically identical humans (Kass and Wilson, 1998, 16).
12. Singer, 1993.
13. For the powerful American business elite, the top three nonprofit objectives are social equality (32%), education (17%), and environment/energy (17%), based on the Skoll Foundation's survey of 109 social entrepreneurs (*Financial Times*, April 5, 2007). The billionaire Chuck Feeney gave away his entire fortune to four causes: disadvantaged

children, the care and treatment of the elderly, global health problems, and human rights (O'Clery, 2007).

14. When Lincoln died, Secretary of War Edwin M. Stanton is reported to have uttered his famous remark, "Now he belongs to the ages."

Chapter 2

1. Dennett, 1995, 21.
2. Mayr, 2001, 9.
3. The story of evolution has been told many times; see, for example, Bowler (1984), Dennett (1995), Mayr (2001), and Larson (2004). Natural selection was not Darwin's original idea. At least two others—William Wells and Patrick Matthew—came upon it years before Darwin did. But their brief descriptions did not receive public attention, a situation changed by Darwin's publication.
4. See works of, for example, Henri Bergson, Samuel Alexander, Pierre Teilhard de Chardin, J. D. Bernal, Freeman Dyson, and Frank Tipler.
5. Gareth B. Matthews, "Aristotle on Life." In Boden (ed.), 1996, 303–13.
6. Mark A. Bedau, "The Nature of Life." In Boden (ed.), 1996, 338.
7. "In the Beginning: The debate over creation and evolution, once most conspicuous in America, is fast going global." *Economist*, April 19, 2007.
8. Spetner, 1997, 177.
9. Drucker, 1993, 275.
10. Darwin spent two chapters in The *Origin of Species* appearing to apologize for the lack of support from the fossil record for his new theory. At the time, his arguments were based almost exclusively on the evidence from living organisms. But soon fossils of transitional species were discovered. Today, the amazing diversity of fossils unearthed provides a powerful testimony to natural evolution (Prothero, 2007, xix).
11. However, the idea of social progress is not equivalent to the idea of evolution, although they are historically closely related (Ruse, 1997). As Peter Bowler pointed out, Darwin didn't simply apply Adam Smith's economic principle to biology, because this would ignore "the fact that for Adam Smith and his followers, the principle of *laissez-faire* itself was intended to allow the natural harmony of human interactions to flourish for the benefit of all. Political economists of the period did not see individualism as a license for unlimited competition" to eliminate the weakest members of society, but as an encouragement for everybody to contribute his or her best. In fact, "Malthus himself only used the crucial phrase 'struggle for existence' in his discussion of primitive tribes" (Bowler, 1984, 96–97).
12. To name just a few: Thomas Hobbes, Thomas Robert Malthus, Georges Cuvier, James Hutton, Sir Charles Lyell, William Paley, and Darwin's grandfather Erasmus Darwin.
13. Darwin might have been influenced by a wider social context than Marx recognized: for example, there are the problems of the urban poor, commercial breeding, and Victorian sex roles. See http://nti.educa.rcanaria.es/penelope/uk_confcartwright.htm.

14. Dawkins, 1986, 316.
15. Dennett boiled the process down to this in modern information language: "Life on Earth has been generated over billions of years in a single branching tree—the Tree of life—by one algorithmic process or another" (Dennett, 1995, 51).
16. Gerald Edelman called this process "neural Darwinism" in his book of the same title (Edelman, 1987). William Calvin has proposed a similar idea.
17. Darwin stated, "The formation of different languages and of distinct species, and the proofs that both have been developed through a gradual process, are curiously the same" (Dennett, 1995, 135). However, Darwin's idea also inspired a misguided search for the inviolable logic and laws of social development. Karl Marx professed to be a "sincere admirer" of Darwin.
18. The natural laws in our universe could be the result of natural selection. Theoretical physicists such as Lee Smolin and Alan Guth have speculated that black holes could give birth to universes, each with slightly different characteristics. And only those with the ability to develop black holes and spawn baby universes proliferate. Smolin has argued that this is a testable idea based on properties of elementary particles. (See Brockman, *The Third Culture,* 1995.)
19. Cziko, 1995. Alexander Vilenkin developed a variation of an inflationary model of the expanding universe that accounts for the birth of the universe by quantum tunneling from nothing—a state with no classical space-time, the realm of unrestrained quantum gravity. (See also William Carroll, "Thomas Aquinas and Big Bang Cosmology," at Jacques Maritain Center, Thomistic Institute site. http://www2.nd.edu/Departments//Maritain/ti/carroll.htm.)
20. Ridley, 1997, 367.
21. Quoted in Broderick, 2001, 11.
22. Nietzsche, 1976, 129.
23. Alfred Wallace, 1907. "How Life Became Possible on the Earth." http://www.wku.edu/charles.smith/wallace/S646.htm.

Chapter 3

1. The word "Utopia" was coined by Thomas More in 1516 as a pun on a Greek root meaning both "nowhere" and "good place," although the concept can be traced back to at least Plato's *Republic*.
2. Of course, many people care for much more than their own personal lives. They live for their children, their friends, their community, their country, and so on.
3. Quoted in Angell, 2000, 126.
4. Mumford, 1934, 6.
5. The social functions of eating meals together could be replaced by new emotions and sentiments.
6. In some sense, this is the same as what Kenneth Boulding calls "spiritual species."

7. The *Mahabharata,* quoted from Lokamanya Bal Gangadhar Tilak, *The Arctic Home in the Vedas* (Poona City, India: Tilak Bros. Gaikwar Wada, 1903).
8. http://en.wikipedia.org/wiki/Golden_Age.

Chapter 4

1. Jaspers, 1953, 2.
2. Jaspers, 1953, 2.
3. Quoted in Backman, 1991, 161.
4. Bergson, 1932. Closed morality is concerned about the survival of the particular, demands strict obedience, and wages war to eliminate the antagonists; while open morality is concerned about universal good, demands creative solutions, and embraces peace and tolerance.
5. Gellner, 1988, 80.
6. Quoted in Eisenstadt (ed.), 1986, 63.
7. The distinction of power versus order was made by Jean Pierre Vernant, in Eisenstadt (ed.), 1986, 43.
8. Eric Voegelin made this point in his multivolume *Order and History.* See Eisenstadt (ed.), 1986, 46.
9. Peter Machinist, "On Self-consciousness in Mesopotamia." Eisenstadt (ed.), 1986, 200–201.
10. Murray, 2003.
11. In Hart's list, over 70 percent of the most influential persons are Europeans/Westerners (Hart, 1978).
12. Crosby, 1986; Diamond, 1997.
13. This longing to be a child in front of larger-than-life figures is a basic human psychological need (Bloland, 1999).
14. According to Dudley Young, "The storm is an appropriate vehicle for divinity on several counts: it arises from nowhere and prevails everywhere, its source is invisible and yet its designated effects are palpable, and it manifests a quite astonishing power to move and disrupt the pattern of things as they are into a new arrangement—all in all, a plausible origin for the high winds of *pneuma*. One might further note that the storm's power is both real and symbolic—a mixture of theatrical thunder and lightning and real wind and tree snapping power" (Young, 1991, 120–21).
15. There are over 9,000 religions in the world today, not including many failed or marginal cults and ideologies.
16. Guthrie, 1993, 32.
17. Stark, 2001, 32.
18. Murray, 2003 ("For God's Eye"), 22–26.
19. Blainey, 2002, 118.
20. Quoted in Bloom, 1997, 304.
21. Quoted in Nisbet, 1980, 55. Brague (2003, 212) noted that later Western thinkers (Buffon and Herder) reinterpreted that as a sign of human domination over all other creatures on Earth. And for Marx and others,

"the cosmological gaze toward the above is associated with a historicist and progressivist gaze toward the future."

22. Drucker, 1993, 12. (Originally published in *Perspectives*, 1953.) Italics mine.
23. Mark Elvin, "Was There a Transcendental Breakthrough in China?" in Eisenstadt (ed.), 1986, 326.
24. Moltmann, 1975, 20.
25. For a long time, each generation of Christians grew up believing that theirs would be the last generation in the world. This tradition is still alive in the modern era, as the book of Revelation has stirred some dangerous people to act out their own private apocalypses (Daniels, 1999; Kirsch, 2006).
26. Novak, 2002.
27. The plague, the breakdown in family bonds, and the disarray in Roman social structures provided the favorable backdrop for Christianity's rapid growth in its fledgling stage (Stark, 1996).
28. The fact that science was invented once does not in itself make science special. The same can be said for wheeled vehicles and many other inventions. Also, many basic inventions that appear to be easy eluded most primitive peoples around the world: nails, needles, scissors, and saws are a few examples. (See Lowie, 1929, 119, 88.)
29. Kitto, 1951, 1.
30. Brunschwig and Lloyd, 2003, 287.
31. Wolpert, 1993. An example is the Taoist alchemists.
32. Quoted in von Mises, 1996, 37.
33. Bloom, 1987, 296.
34. Gopnik et al., 1999.
35. Turner, 1996.
36. Feyerabend, 1975, 9.
37. Ferris, 1988, 386.
38. Grinnell, 1992.
39. For the tip of the iceberg, see Brooke, 1991; Barbour, 1997; Witham, 2005.
40. Ernst Mayr (1998, 536) identified several significant stages in the modification of Darwinism: Weismann's influence (1883–86), Mendelism (1900), Fisherism (1918–33), evolutionary synthesis (1936–47), post-synthesis (1947–70), punctuated equilibria (1954–72), rediscovery of sexual selection (1970s–1980s).
41. Lewis, 2002; Aslan, 2005.
42. Whitehead, 1967, 182–89.
43. The isolation was never absolute, as was the case for ancient civilizations in the Americas. China maintained contact with other Eurasian civilizations throughout its history, but unlike Christianity and especially Islam, its civilization is largely a local one, geographically, racially, and culturally.
44. Merson, 1990, 14.

45. The Gospel of Thomas, with its strong Eastern flavor, was denounced as heresy following the creation of the Nicene Creed in 325 CE.
46. Tagore, 1970, 24.
47. Lakoff and Johnson, 1999, 566. See also the clinical discussion of the Philosophic Physician from the neurological perspective in Sacks, 1990, 279–88.
48. Brothers, 2001; Damasio, 2003.
49. Nisbett, 2003, 140–41.
50. McNeill, 1963, 377. Tantrism (found in both Hinduism and Buddhism) and secular Taoism believe that man can absorb vital energy from woman through controlled sexual intercourse, *maithuna*, and "the art of bed chambers."
51. Wang, 1987, 248.
52. Chuang Tzu, *Great and Small*, quoted in Jantsch, 1980. Chuang thus anticipated recent scientific research that emphasized the emergence of order out of chaos; see Prigogine and Stengers, 1984.
53. Raymond M. Smullyan, "Is God a Taoist?" In Hofstader and Dennett (eds.), 1981, 339.
54. An offshoot of Buddhism, the word *Zen* ultimately derives from the Sanskrit *dhyana*, meaning concentration or meditative absorption.
55. Aristotle believed that man's highest capacity is not *logos* (speech or reason), but *nous*, the capacity of contemplation, whose chief characteristic is that its content cannot be rendered in speech. (See his *Nicomachean Ethics 1142a25 and 1178a6*, quoted in Arendt, 1958, 27.)
56. Campbell, 1996, 33–34.
57. Wright, 2000, 171.
58. Arendt, 1958, 15.
59. Christians often say everything is done for a purpose—that God always has a positive motivation in mind for every event that mere humans cannot understand. That is *Inshallah* (God's will) in Islamic sayings.
60. The early Vedas said the origin of the universe was "not even nothing." However, the early Hindu cosmogony, like most of the cosmogonies of the ancient world, understood the cosmic creation through the more immediate understanding of human procreation. The *Rg Veda*, for example, casts the universe itself as a cosmic immortal person.
61. Hopkins, 1971, 44.
62. Quoted in Mark Elvin, "Was There a Transcendental Breakthrough in China?" Eisenstadt (ed.), 1986, 345.
63. Quoted in Zubko, 2004, 410.
64. Quoted in Norretranders, 1991, 326.
65. Quoted in Austin, 1998, 16.
66. See Duh Bau-ruei (杜保瑞), "The construction of the Confucianism by Cheng-Yi through the approach of the metaphysics and practical theory and the I-Ging study," 2004. http://ccms.ntu.edu.tw/~duhbauruei/4pap/1con/35.htm.
67. See Ronan, 1978. This work is a condensed version of Joseph Needham's classic book on science and technology in Chinese civilization.

68. Landes, 1998, 57.
69. Quoted in Capra, 2000, 289.
70. Jaspers, 1955, 414–15.
71. Turchin, 1977, 102.
72. Lin, 1942, 33. The most diverse intellectual activities in Greece were around 500–300 BCE. But not all schools are equal: within a community where intellectual argument flourishes, the number of major active schools that can be passed on from generation to generation is on the order of three to six. Most minor schools die out quickly. (See Collins, 1998, 81.)
73. Fair, 1969, 92.
74. This is confirmed in modern information theory, which holds that perception or mental simulation of absolute reality is impossible.
75. Stevenson and Haberman, 1998, 52.
76. According to R. Eisler, the initial idea of the world as a totality of things goes back to Babylonian political theory (Popper, 1945, 204).
77. Brague, 2003, 11–12.
78. Johann Arnason, *The Axial Age and Its Interpretations.* In Arnason et al. (eds.), 2004, 33.
79. Brague, 2003, 23.
80. Goldhill, 2004; Kaplan, 2002.
81. Whitehead, 1979, 39.
82. An Egyptian scientist named Hero wrote a paper titled "*Spiritaliaseu Pneumatica*" that included a sketch of steam from a boiling cauldron used to open a temple door. (See Kessler, 2005, 13.)
83. The antikythera mechanism (150–100 BCE) has been called the world's first (analog) computer, and there are similar devices mentioned in ancient literature. See http://en.wikipedia.org/wiki/Antikythera_mechanism.
84. In his majestic survey titled *Ideas That Changed the World,* Felipe Fernández-Armesto suggested that there are probably no more than a dozen ideas emerging after the Axial Age that are comparable to those of that age (Fernández-Armesto, 2003, 103). But the real number could be zero. The "new" ideas could be mere extensions of earlier ideas, and most original Axial ideas are unknown to us. Many of the sages were indifferent or hostile to written teachings, and there were episodes of "mass extinction" of ideas, when books were burned and people were killed.
85. Kwok et al., 1994, 17.
86. Chaitin's Theorem says that no program can calculate a number more complex than itself (Casti, 1995, 145).
87. James, 1997, 288.
88. As Freud himself put it, "Poets and philosophers before me discovered the unconscious. What I discovered was the scientific method by which the unconscious can be studied" (quoted in Lionel Trilling's 1940 essay "Freud and Literature").
89. Heraclitus of Ephesus (late sixth century BCE) conceived a universe of flux animated by a conflict of cosmic forces, in which "harmony consists of opposite tensions like the bow and the lyre. . . . We must know that

war is common to all, strife is justice, and all things come into being and pass away through strife" (Harrison, 1985, 53).

90. However, one can make the case that education and social environment have lifted the perspectives of today's masses beyond those in ancient times.

Chapter 5

1. Backman, 1991, 31.
2. Stanley E. Hyman, in Barlow (ed.), 1994, 65.
3. Lovejoy et al., 1935.
4. Gellner, 1988, 118–26.
5. Barrett, 1986, 23.
6. The term "noble savage" may have originated in John Dryden's 1670 stage play *The Conquest of Granada* (part 1, act 1, scene 1): "I am free as nature first made man, / Ere the base laws of servitude began, / When wild in woods the noble savage ran."
7. Among major scientific theories, Darwin's may have endured more ridicule and abuse than any other in the early years. "Rotten fabric of speculation . . . Utterly false . . . Deep in the mire of folly . . . I laughed till my sides were sore . . ."—these were some of the early reactions to Darwin's work (Jastrow, 1977, 19).
8. Frankl, 1992, 85.
9. Garreau, 2005, 196.
10. There are four types of explanation used in science. Evolutionary theory utilizes genetic explanation, which depends crucially on the actual occurrence of a historical sequence of events. There is no central deductive or probabilistic law constraining these occurrences; hence it is impossible to predict, or even explain, particular events in the future. In this light, Karl Popper was mistaken in his challenge about the scientific credentials of Darwin's theory, which has stood the test of numerous scientific findings and proved highly successful in providing a diversity of historical explanations. (See Horst Hendriks-Jansen, "In Praise of Interactive Emergence, or Why Explanations Don't Have to Wait for Implementations," in Brooks and Maes [eds.], 1994, 79.)
11. Primack and Abrams, 2006.
12. Dobzhanksy, 1973. Quoted in http://www.pbs.org/wgbh/evolution/library/10/2/text_pop/l_102_01.html.
13. The global historian William H. McNeill claimed his research had no particular method, but when pressed he described it as follows: "I get curious about a problem and start reading up on it. What I read causes me to redefine the problem. Redefining the problem causes me to shift the direction of what I'm reading. That in turn further reshapes the problem, which further redirects the reading. I go back and forth like this until it feels right, then I write it up and ship it off to the publisher" (Gaddis, 2002, 48–49).
14. Teilhard de Chardin, 1965, 219.

15. When we ask "What does it mean?" we are asking a connection of "it" to something we are familiar with. At the most elementary level, meaning is generated by the assimilation of neuron patterns with external stimuli.
16. Creationism largely relies on revelation for knowledge. Materialism is also called reductionism, which denies the hierarchical complexities of nature and life. For detailed discussion, see Boulding (1978), Dennett (1995), Shermer (2000).
17. J. L. Monod, "On the Molecular Theory of Evolution." In Harré, 1975.
18. The word "human" is from the proto-Indo-European root *dhghem*, meaning simply *earth*, which suggests that we've had an etymological hunch about our origin since the start of language (Thomas, 1993, 19).
19. Pearson, 1997, 166.
20. By "big picture" I mean rough outlines. Many book-length and multivolume video treatments of the Cosmic Story have been produced in recent decades. A couple of recent works are Chaisson's *Epic of Evolution* (2006) and Primack and Abrams's *The View from the Center of the Universe* (2006). See also *The Cyclic Universe,* by Paul Steinhardt, discussed in Brockman, 2003.
21. Collins, 1998, xvii.
22. Smoot and Davidson, 1994.
23. See Dennis Overbye, "Knowing the Universe in Detail (Except for That Pesky 96 Percent of It)," in the *New York Times*, October 24, 2006. (Article with appended correction appears at http://www.nytimes.com/2006/10/24/science/space/24essa.html?ref=science.)
24. Examples include human fossils found in the same rock as the dinosaurs, human stone tools found in rocks that are more than 100 million years old, the appearance of a new species within a few generations, or the discovery of a species whose DNA is totally unrelated to that of any other species.
25. In 1917, Einstein was still advised that the Milky Way was the universe, which led him down a legendary path of error and regret (Kirshner, 2002, 51).
26. Estimate based on 2003 Wilkinson microwave anisotropy probe satellite data. The confidence band is about 200 million years.
27. Thus the metaphor of the egg as the origin of an organism is also inappropriate. Although the egg contains all the information that the organism needs to grow, its growth depends on interactions with a preexisting external environment, whereas the universe's growth seems to be self-contained.
28. The expanding universe can be likened to a rubber sheet being stretched uniformly in all directions.
29. As the French poet Paul Valéry once quipped, "God made everything out of nothing. But the nothingness shows through."
30. At the base level, shape is destiny. Life is about complicated organic chemistry, about fitting molecules together in lock-and-key mechanisms like sticky Lego: different shapes and sizes, but with standardized studs and corresponding holes to allow them to be assembled. The basic set of molecular building blocks is fairly simple—20 amino acids, a few

nucleotides, a dozen or so lipid molecules, and two dozen sugars. Life is not magic; rather, it is very complex chemistry in action, so complex that life can only come from life through a self-replication process. Another vital characteristic of life is metabolism, which is always a separate function in the cell, suggesting a different origin from the nucleic acid replication function (Dyson, 1985).

31. See, for example, Gabrielle Walker's somewhat speculative study, *Snowball Earth: The Story of the Great Global Catastrophe That Spawned Life as We Know It* (2003).
32. De Duve, 1995, 60.
33. Delbrück, 1986, 83. Individual nerve cells can exhibit irritability (simple reflex), while a nerve net can show a complex reflex and an associating (conditioned) reflex (Turchin, 1977, 98).
34. Watson and Berry, 2003, 242.
35. As Samuel Butler said, "To do great work, a man must be very idle as well as very industrious."
36. Chandler, 1977.
37. Angus Maddison, "When and Why Did the West Get Richer Than the Rest?" (2004). http://www.ggdc.net/conf/paper-maddison.pdf.
38. The affluence is reflected in how "wasteful" we have become. Take clothing, for example: in agricultural societies, most people repaired their cloth until it was no longer repairable; in industrial societies, worn cloth is thrown away; in information societies, cloth is disposed of simply because the style is out of fashion.
39. The unreliability of recurring patterns is the essence of risk (Bernstein, 1996).
40. Minsky, 1986, 20.
41. Pico, 2002.
42. Pierre Leon once wrote, "Historians have usually remained indifferent to the long term." Ernest Labrousse wrote, "I have abandoned any attempt at an explanation of long-term movements." (Quoted in Braudel, 1979, 83.)
43. Complexity is a fuzzy concept that is easy to describe but difficult to define. Scientists have defined it as a system's information content (the amount of meaningful information content that cannot be compressed), the length of its underlying grammar, or its mathematical intractability. It attempts to describe that interesting area between order and chaos. Complexity is something we can feel is neither trivial nor dull. (See Gribbin, 2004; Kurzweil, 2005, 36–38.)
44. Kauffman, 2000, 85.
45. Robert Wright has called it the logic of nonzero (games), and Paul Seabright has identified risk-sharing, specialization, and knowledge accumulation as key advantages of a large complex society.
46. Named after Richard Gregory (1981), who first pointed out that cultural artifacts not only require intelligence to produce but also enhance their creator's intelligence.
47. Termites, for example, add chemical traces to their mudball deposits; these traces attract more mudball deposits, in time resulting in impressive

arched columns—but without any individual termite designing or coordinating the action (Clark, 1997, 75).

48. The modern scientific vision of cosmic hierarchy, however, shares only superficial resemblance to Aristotle's "Great Chain of Being."
49. Human groups often behave differently from separated individuals. This is true of packs of teenagers, and it is true of political parties. The group dynamics of conformity and cascade are emergent properties of human society, guided by what Cass Sunstein (2003) called "the law of group polarization."
50. Wright, 1988, 72. Fredkin's colleague Tom Toffoli shared his view and quipped that of course the universe is a computer—the only trouble is that somebody else is using it (Davies, 1992, 123). The recent "upgrade" of the universe computer to a quantum machine makes it inherently "random" based on the current interpretation of quantum mechanics (Lloyd, 2006).
51. Barbour, 1997, 240.
52. See Murchie, 1978. Also see Wolfhart Pannenberg's lecture "Modern Cosmology: God and the Resurrection of the Dead." http://www.math.tulane.edu/~tipler/tipler/tipler3.html.
53. Deutsch, 1997, 187.
54. Gould, 1989, 289.
55. As David Pearce pointed out, even if a humanistic heaven could be built on Earth that banishes pain and suffering, we would still need their functional analogues, or gradients of well-being rather than uniform bliss, to provide motivational satisfaction. http://www.life-enhancement.com/neofiles/default.asp?id=17.
56. Jaynes, 1976, 225.
57. Benjamin Uffenheimer, "Myth and Reality in Ancient Israel." In Eisenstadt (ed.), 1986, 152, 163.
58. Friedman, 1995, 17–82.
59. Saint Anselm (1033–1109) concluded that for God's existence to be beyond doubt, it must include its own existence; for if it were not the case, there would be a more perfect being. Thomas Bradwardine (1290–1349) argued that God and infinite space are one—that God "is infinitely extended without extension and dimension" (Grant, 1996, 123).
60. Quoted in Barrow and Tipler, 1988, 156.
61. Bird, 2003, 21.
62. Templeton and Giniger (eds.), 1998.
63. By the same token, we celebrate December 25 as Jesus's birthday (Christmas), although his true birthdate is unknown. In 440 CE, the Church fathers decreed that his birthday be celebrated at the time of the winter solstice, which was traditionally the most important festival. The actual date of Christ's birth had always been debated (Giovannoli, 1999, 165).
64. This is not apparent since they are so small. The bacterial genus *Prochlorococcus*, probably the most abundant species on earth, was only discovered in 1988 (Watson, 2003, 214).

Chapter 6

1. Fair, 1969, 29.
2. Stephen J. Dubner and Steven D. Levitt, "Monkey Business: Can Capuchins Understand Money?" *New York Times Magazine*, June 5, 2005.
3. Pinker, 1994. Language is an instinct because even if children are not taught any language, they will have an urge to speak and will invent language to communicate.
4. Deacon, 1997, 434.
5. Mumford, 1970, 417.
6. Clark, 1997, 53 and 68–69.
7. Richard Wrangham, "The Evolution of Cooking." In Brockman (ed.), 2003.
8. Dunbar, 2004. For example, we understand this sentence: Shakespeare intended his audience to realize that Othello believed that Iago knew that Desdemona loved Cassio.
9. This is based on the game theory that von Neumann developed in the early twentieth century.
10. Wright, 2000, 9.
11. Life as we know it is impossible without water. For example, water is an excellent solvent for all kinds of molecules; it also has a high surface tension (important for plants' water transport) and viscosity (important for animals' blood circulation) when compared with other normal liquids (Denton, 1998).
12. Day, 1978, 80.
13. Silicon possesses some advantages over carbon, but the advantages are more than offset by the drawbacks (Conway Morris, 2003, 24–31). The number of known organic (carbon-based) compounds is enormously larger than the inorganic ones (Laidler, 2004, 232).
14. Mallove, 1987, 142.
15. Paul and Cox, 1996, 459.
16. Ward and Brownlee, 1999, 35.
17. Water, for example, has the "wrong" color in its crystallized form, snow, because it exacerbates winter chill by deflecting solar radiation into space. The extreme transparency of water also provides no shield against solar radiation and forces life to develop elaborate defense mechanisms against ultraviolet and visible light. (See George Williams, "Gaia, Nature Worship, and Biocentric Fallacies," *Quarterly Review of Biology* 67 [December 1992]. Also J. L. Monod, "On the Molecular Theory of Evolution," in Harré, 1975.)
18. Abbott, 1952.
19. Preface to *The Complete Science Fiction Treasury of H. G. Wells* (New York: Avenel Books, 1979), quoted in Yeffeth (ed.), 2005, 236.
20. Quoted in Fenton Johnson, "Beyond Belief," *Harper's*, September 1998.
21. As a result, she suggested, we are not mere earthbound creatures. Arendt, 1958, 10–11.
22. Moltmann, 1975, 32.

Chapter 7

1. Tipler, 1994.
2. Department of Education, 2000. Neisser et al., 1998.
3. The maximum number of languages any person has acquired, for example, is just under sixty. Sir William Jones (1746–94) is said to have known over forty (Whitrow, 1989, 13).
4. If a well-educated person gets stranded in a primitive jungle, cut off from modern technologies and unaided by primitive people living in that environment, he or she would not know what to eat, where to sleep, or how to shield himself or herself from predators (Sowell, 1980, 7).
5. Miller, 1956. The only way to expand our mind is through chunking or grouping information, so that we can consider more things with less "resolution."
6. The advantage of a featherweight consciousness is its ability to jump from one area to another easily—you could be thinking about the taste of the hamburger one second and an investment decision in the next moment during a business lunch. The light and floating nature of consciousness also makes telling a lie possible.
7. Descartes in 1632, quoted in Davies, 2003, 3.
8. Quoted in Ferris, 1988, 205.
9. Kohlberg, 1981.
10. Asked to memorize a list of words including "pillow, bed, pajamas," people would recall words such as "sleep" that were not on the list.
11. DNA evidence indicated that 90 percent of eyewitness identifications of perpetrators were wrong (Schacter, 2001).
12. De Duve, 1995, 8.
13. Pinker, 2002, 294.
14. There is a grain of truth in Schopenhauer's words "Thoughts die the moment they are embodied by words."
15. Tom Wolfe, "Stalking the Billion-footed Beast." *Harper's*, November 1989.
16. Boyer, 2001, 57.
17. In the early twentieth century, most companies had a "vice president of electricity," since each firm had to have its own generators. The situation is similar to today's information technologies, with most large firms having a "CIO" or "CTO." Eventually the IT work will be farmed out, as was the electricity-generation work (Carr, 2004).
18. Camus, 1991, 17.
19. Quoted in Csikszentmihalyi, 1993, 61.
20. Quoted in Kosko, 1993, 267.
21. Edelman, 1992, 162–63.
22. Fleck, 1935, 90–92.
23. Guthrie, 1993, 166–76.
24. Davies, 2003, Preface. The emphasis is his.
25. Davies and Hersh, 1986.
26. Dreyfus, 1992, 189.
27. Gaddis, 2002, 60.

28. Davies, 2003, 236. Emphasis mine.
29. Lloyd, 2006. When Lloyd says, "The universe *is* a quantum computer," he is *not* intending this metaphorically.
30. Seife, 2006. Information is powerful because it is physical, and laws of information theory literally govern everything in the universe.
31. See the Ideonomy website (http://ideonomy.mit.edu/) and the review of the book *A New Kind of Science* on http://www.amazon.com.
32. Montaigne, 1993. (Book One of his *Essays*, Chapter 50: "On Democritus and Heraclitus.")
33. Harth, 1990, 115.
34. See also Gould, 1997, 224–25 for the suggestion that the great excitements of scientific and artistic discovery are largely over.
35. Kindleberger, 1978, 32.
36. *The Rambler*, no. 60. Quoted in Pinker, 2002, 142.
37. Mead wrote this in her 1928 book *Coming of Age in Samoa: A Psychological Study of Primitive Youth for Western Civilization*. Quoted in Pinker, 2002, 25.
38. Konner, 2002, 148–50.
39. Brown, 1991.
40. Konner, 2002, 131.
41. Boyer, 2001, 249.
42. Wright, 1988, 190.
43. Strauss and Howe (1991) suggested that America exhibits a four-part generational cycle, which consists of *idealist, reactive, civic,* and *adaptive* generations. Similarly, Philip Jenkins (2000, 13) identified a loose four-stage cycle of American cults in the twentieth century: *emergence, reaction, speculation,* and *second peak*.
44. Weiner and Brown, 1997. The tearing of our existing social fabric, however, is costly and damaging (Jacobs, 2004).
45. Quoted in Brockman, 1995, 18.
46. Negroponte, 1995, 204–5; Johnson, 2005.
47. The extreme nonviolence sentiment was instilled in children with the adults' nonaggressive behavior: not only were the children never punished, they were also never given a chance to show aggression or rebellion. In a practice called *bood,* a child can refuse to do anything if "I don't feel like doing it" (Watson, 1996, 154–56). Furthermore, "their general level of interpersonal emotionality is low, whether indexed by the absence of aggression, by the muting of mourning, or by the lack of any strong exhibitions of anger." Instead, their chief emotional outlet is fear—not fear of people, but of storms, strangers, supernatural beings, and animals (Mandler, 1984, 215). Thus, when ordered by the British to kill in the early 1950s, they killed people remorselessly like robots, in a kind of insanity they called "blood drunkenness."
48. Helena Cronin, "Getting Human Nature Right," in Brockman, 2003, 56.
49. See also FBI reports of police arrests and serious crime by age distribution (Lykken, 1995, 30).
50. Watson, 1996, 196.

51. Bloom, 1997, 259.
52. Lykken, 1995, 6 (quoted in Jane M. Murphy, "Psychiatric Labeling in Cross-cultural Perspective," *Science* 191, no. 4231 [1976]: 1019–28).
53. Thomas Jefferson, secretary of state, letter to Francis Willis Jr., April 18, 1790.
54. Reiss, 2000, 20–21.
55. Quoted in Edelman, 1992, 165.
56. Thayer, 1976.
57. Dunbar, 1997.
58. MacHale, 2003, 157.
59. Crick, 1994, 3.
60. Quoted in LeDoux, 1996, 267.
61. Pinker, 2002, 10.
62. Quoted in Barrow, 1996, 220.
63. Pinker, 2002, 50.
64. Turner, 1999; Ramachandran and Hirstein, 1999.
65. Barrow, 1996.
66. Hofstadter, 1979, 326.

Chapter 8

1. Hayles, 1999, 111.
2. Durant, 2002, 2.
3. This metaphor has been used by Thomas Dewar.
4. Hofstadter, 1979, 250–51.
5. Brockman, 1995, 91.
6. A person can be a mother, a wife, a sister, a daughter, a granddaughter, an aunt at home; a boss, a subordinate, a colleague, a team member, a consultant at work; etc.
7. Edelman, 1992, 109. This is the same principle we discussed earlier about language as symbolic patterns and maps.
8. Wernicke's area is responsible for understanding, while Broca's area is for producing speech. Wernicke's area matures earlier than Broca's area, creating a frustrating period for toddlers (referred to as "the terrible twos"). In addition, the supralaryngeal vocal tract needs to be developed for speech.
9. Goldberg, 2001.
10. There are several parallel stages of legal development (Kohlberg, 1981, 235–36): In stage 1, disputes are settled in an "eye-for-an-eye" fashion; authority derives from physical strength. In stage 2, disputes are settled through negotiations, bargaining, and reciprocal exchange. In stage 3, disputes are settled in conformity with the expectations of an impartial mediator or with conventional social expectations or standards. In stage 4, disputes are formally adjudicated in accordance with substantive and procedural rules uniformly enforced by a judge or court.

In stage 5, procedures for dispute settlement embody many of the structural features of stage 4 procedures; the judicial system is institutionally differentiated from the authority of the state.

11. Kohlberg, 1981, 409–12.
12. Martin Luther King, "Letter from Birmingham Jail"—an open letter written April 16, 1963, and published in many sources, including Kohlberg, 1981, 319.
13. The advances in the stage of moral reasoning are not irreversible. Those new to stage 5, for example, can slip back toward conventional reasoning, in order to remove the confusion posed by the moral dilemma.
14. Kohlberg, 1981, 340–43.
15. Rizzuto, 1979, xiii.
16. George Santayana once said, "We cannot know who first discovered water. But we can be sure," he continued, "that it was not the fish."
17. Turnbull, 1972.
18. Even then, there are clear internal limits to how high people can go. Kohlberg considered abandoning the concept of stage 6 after failing to find a single example in American and Turkish longitudinal data (Bergling, 1981, 19).
19. Quoted in Brooks, 2004, 243.
20. Strauss and Howe, 1991.
21. Charles Gray, for example, saw Classical history as a series of superimposed cycles, each going through Formative, Developed, Florescent, and Degenerate stages (Tainter, 1988, 81).
22. Colonel Jessup's words in *A Few Good Men*: "You can't handle the truth! Son, we live in a world that has walls, and those walls have to be guarded by men with guns. Who's gonna do it? You? You, Lieutenant Weinberg? I have a greater responsibility than you can possibly fathom. You weep for Santiago and you curse the Marines. You don't want the truth because deep down in places you don't talk about at parties, you want me on that wall, you *need* me on that wall."
23. Quoted in Shattuck, 1996, 307.
24. Even to the agnostic, "atheism is like telling somebody, 'the very thing you hinge your life on, I totally dismiss'" (Wolf, 2006, 187).
25. Quoted in Kohlberg, 1981, 356.
26. Barbour, 1997, 129.
27. Be careful, however, to avoid the misconception that morality cannot exist without religion. Rather, it is the other way around: our intuitive moral sense made religion plausible and effective.
28. Some German biblical scholars discerned four different sources in the first five books of the Bible. These were later collated into the final text as Deuteronomy during the fifth century BCE (Armstrong, 1993, 12).
29. As a typical "sticky" supernatural belief, the concept of the Trinity has both commonsensical and counterfactual aspects—it is intuitively compelling yet fantastic, eminently recognizable yet surprising.
30. The same is true for an organization. For example, the answer to the question of whether a religion is really tolerant or intolerant cannot be

found in its doctrine. Contrary to today's situation, Christianity was strikingly intolerant during much of its history; Islam, on the other hand, was remarkably tolerant during its period of political and cultural supremacy.

31. Tolstoy's autobiography, *A Confession,* has been published in multiple editions and translations.
32. Wilson, 1956, 272.
33. Fowler, 1995, xi.
34. Harrison, 1985, 101–2.
35. A big danger is that history often becomes a reinterpretation of past events through the lens of today's perspectives and values.
36. Wright, 2000, 244.
37. The arguments concerning the disappearance of and therefore the need to conserve certain natural resources have been made for fossil fuels, fresh water, topsoil, various forms of wildlife, and so on.
38. A good example is that the emissions of certain harmful chemical compounds in cars have been cut by 98 percent since the early 1970s, with near-zero-emission vehicles being developed today.
39. Seabright, 2004, 31.
40. See William Broad, "Deadly and Yet Necessary, Quakes Renew the Planet," *New York Times*, January 11, 2005.
41. Schumpeter, 1943, 83.
42. Quoted in Greenspan, 1997, 117.
43. These are what Kierkegaard called the "immediate" men and the "Philistines." They "tranquilize themselves with the trivial"—and as a result they can lead normal lives (Becker, 1973, 74 and 178).
44. Murray, 2003, 458.
45. Herman, 1997, 13.
46. The *Katha Upanishad*—from Lin (ed.), 1942, 45–46.
47. George Loewenstein and David Schkade, 1999, "Wouldn't It Be Nice? Predicting Future Feelings," in Kahneman et al. (ed.), 1999, 85–105.
48. Frankl, 1997, 142.
49. Myers, 1992; Lykken, 2000. From studies of identical twins, Dean Hamer estimated that purely genetically determined "subjective well-being" is about 40–50 percent. A study conducted by Bruce Headey and Alexander Wearing of the University of Melbourne found that neuroticism, extroversion, and other personality factors account for the bulk of happiness variations, dominating other factors such as gender, age, income, social class, and marital status (Nettle, 2005, 106–11).
50. Money seems to be a necessary but not sufficient condition for happiness. People who have trouble meeting their basic needs become happier as their standard of living improves. However, enjoying a better life beyond a certain point adds little to happiness.
51. While the percentage of Americans who say they are happy has stayed at 60 percent, the share of "very happy" Americans has declined from 7.5 percent in 1950 to 6 percent today (Lane, 2000).
52. Bernstein, 2004, 326, figure 10–11.

53. This is driven by our tribal instincts of group competition, even though the circumstances have changed drastically. Today's professional sports teams, for example, consist of coaches and players from out of town, are owned by someone extremely wealthy who is seldom a member of the local community, and are managed by "faceless" big organizations for the purpose of making a profit. All of these facts do not matter for the avid fans, and the entire community still gets excited when the home team is winning big.
54. Life satisfaction shows no decline with age, despite the fact that certain resources such as marriage and income that correlate with well-being do decline with age. See http://psycnet.apa.org/psycinfo/1997-36657-011.
55. Interestingly, there seems to be no equivalent saying in Western cultures, although contentment with one's lot is an attitude that can be found in all societies.
56. Gintis, 2003.
57. In artificial intelligence, neural inhibition also turns out to be central to the computational and representational powers of backpropagating neural nets.
58. Benedict Carey, "Why Revenge Tastes So Sweet." *New York Times,* July 27, 2004.
59. As Lily Tomlin once quipped, "The trouble with the rat race is that even if you win, you are still a rat."
60. Bailey, 2005, 16–17.
61. Friedman, 1995, 78.
62. Regis, 1990, 175.
63. Goertzel and Bugaj, 2006, 399–402.
64. See, for example, Gore, 2006, 296.
65. Midgley, 1978, 101.
66. http://www.brainyquote.com/quotes/authors/r/reinhold_niebuhr.html.
67. Barlow, 1994, 251.

Chapter 9

1. Erich Harth further elaborated on this point: "Modern physicists reject the notion of an action-at-a-distance and recognize only local events. . . . Present events result from present conditions, that is, conditions that are contiguous in space and time. The past has dropped out of existence . . . [but] the human mind is the joiner, fitting together the disparate elements of the world . . . I remember or reconstruct what no longer exists and call it the past. I project or guess at what has not yet happened and call it the future. I connect the past with the present and invent purpose, a kind of nonlocal causality. I do the same with present and future, and create intentionality, also hope and fear. All of these are constructs of the mind, because neither past nor future exists in the world of objects" (Harth, 1993, 8–9).
2. McKibben, 2003, 109–15. Italics his.
3. Kass, 2001.

4. Callahan defined the "research imperative" as doing research for its own sake, often as a cover for selfish profit motives and at the cost of compromising important moral and social values. He suggested that it can also be a motive to achieve worthy practical ends for human good; these do not include immortality or other "unnatural" human desires (Callahan, 2003, 3–4, 80–84).
5. Obituary: Francis Crick. *Economist,* August 5, 2004.
6. Grand, 2001, 104. Nature's "laziness" is actually the economic necessity of survival in a competitive world—never waste energy, whether it is physical or computational. See also Clark, 1997, 25–29.
7. Silver, 2005.
8. Ridley, 2003, 36.
9. Stark, 1996.
10. Douglas Hofstadter imaged a setup in which at each turn, the player would move the piece and change the rules! It is a "tangled hierarchy" in which "the moves change the rules, the rules determine the moves, round and round the mulberry bush." But this is not a game free of conventions. In fact, this setup can be changed continuously in a higher hierarchical system (Hofstadter, 1979, 688).
11. Rosner, 2002.
12. Berry, 2000, 43–44.
13. Winner, 1986, 168. I don't think Mr. Winner knew that the ancestors of the whale did not live in the sea.
14. Quinn, 1999, 20.
15. Silver, 2006, 202.
16. Lovejoy et al., 1935, 447–51.
17. Watson, 1996, 42–43. Many other factors, especially climate change, may have contributed to the changing landscape in Africa.
18. Sharon Begley, "Darwin Revisited: Females Don't Always Go for Hottest Mate." *Wall Street Journal,* May 5, 2006.
19. Bagemihl, 1999.
20. Nick Bloom et al., "Management Practices across Firms and Nations" (June 2005). http://cep.lse.ac.uk/management/management.pdf.
21. Heraclitus, Fragment 124. Quoted in Brague, 2003, 20.
22. John Wheeler, quoted in Barlow (ed.), 1994, 141.
23. Root-Bernstein, 1989, 249.
24. Mintzberg, 1994, 288.
25. Witham, 2005, 265.
26. As quoted in Bergman, 137.
27. This mimics the release of the true creative, imaginary power of the unconscious mind by relaxing the mental control of the conscious.
28. It is in this sense that Eric Hoffer asserted that "man is most peculiarly human when he cannot have his way" (Hoffer, 1963, 126).
29. A heart beating at an orderly pace is likely to be on the verge of cardiac death (Rothschild, 1990, 261–62).
30. Kaplan, 2002, 34.

31. Ackoff, 1978, 53. In a "fishy story," the problem of how to keep densely packed fish moving was unexpectedly solved by a naturalist, who suggested throwing a few predator fish in the tank. The moral: one is often moved deeply by another's hunger.
32. However, local setbacks can be severe. The island of Tasmania, 130 miles off the coast of southeastern Australia, lost many technologies and artifacts (such as fishing, awls, needles, and other bone tools) after the hunter-gatherers there were separated from the mainland. When they were discovered by the Europeans in 1642, they had the simplest material culture in the world (Diamond, 1997, 312).
33. Crosby, 1997, 3.
34. For example, in many freshwater lakes, there are "species flocks" where species specialization is made possible by differentiated lake basins, lake-bottom substrates, and spawning sites (Mayr, 1988, 383–97).
35. Olson, 1965.
36. Mayr, 1954.
37. When Edison's light bulb was just starting to become a household fixture, the *New York Times* warned that electric light could cause blindness (Moore and Simon, 2000, 6).
38. Only lichens can make a living on bare rocks in extremely dry climates.
39. Margulis and Sagan, 1995, 146–47.
40. The complex food chain in ecological systems, for example, is rich in biodiversity but is a huge waste of energy, as compared with the alternative of self-sufficient photosynthesis. But still, the biological exuberance on Earth has barely touched the potential provided by the biochemical energy of the sun and the Earth (Bagemihl, 1999, 252–55). The genetic structure is also full of redundancy; Ockham's Razor does not apply to biology (Keller, 2000, 112).
41. Jacob, 1974, 309.
42. Lynch and Conery, 2003.
43. Buckingham and Coffman, 1999, 57.
44. Predictions such as this are typical of the linear extrapolation pitfall, which has accounted for many bad predictions up to the present day.
45. Keynes, 1930.
46. Kass, 1999.
47. Quinn, 1999. In fact, people who make their living in small businesses such as restaurants, lawn care, and construction may already be leading a life similar to what he describes. But still, this is a "dare to be different" attitude that fits some people's taste.
48. See Dennett, 2003, 2. Also see Juan Forero, "Leaving the Wild, and Rather Liking the Change," in the *New York Times*, May 11, 2006.
49. For example, even technology trailblazers such as Jaron Lanier expressed nostalgia about the good old days when computer user interfaces were much more responsive, forgetting that today's computers are asked to perform far more complex functions than was the case fifteen years ago (Kurzweil, 2005, 436).
50. Quoted in Dyson, 1997, 192.

51. The protein is used by all animals, plants, and fungi as a common housing frame for their DNA molecules (Silver, 2006, 213).
52. Ormerod, 2005.
53. In addition to creating a risk-sharing platform for investors and an organizational structure for managers, the company is a basis for taxation and regulation for the government, and it creates an organizational layer that benefits society in many ways.
54. See article "The Business of Survival" in *Economist*, December 18, 2004, 104–5.
55. Berliner, 1976.
56. Jacob, 1977.
57. "Natural" in this case means free market–originated as opposed to governmentally mandated.
58. Dyson, 1997, 19–20.
59. Hoffer, 1952, 126.
60. Petroski, 1992.
61. This could be a form of ignorant wishful thinking or an unconscious attempt to not ruffle feathers.
62. Kurzweil, 2005, 324.
63. Wendell, 2000, 11.

Chapter 10

1. Randolph Nesse, quoted in Konner, 2002, 226.
2. J. B. S. Haldane, quoted in Dyson, 1997, 226.
3. Grove, 1996, 117.
4. Joseph Conrad, quoted in Robin, 2004, 51.
5. Austin, 1998, 608.
6. Konner, 2002, 205.
7. A five-year study by scientists (led by Stephen Porter) at Dalhousie University in Halifax found that pleasant events were more difficult to recall than unhappy ones.
8. Schacter, 2001.
9. Watson, 1996, 3.
10. Fair, 1969, 91.
11. Maslow, 1968, 60. Also recall the famous remark by a woman upon hearing about Darwin's man-evolved-from-apes theory: "I wish it would not become known."
12. Becker, 1973, 74, 86–87.
13. Minsky, 1986, 46. The last point echoes Keynes's famous quip that the "new" policies of today's politicians merely reflect the ideas of some long-dead economists. Indeed, "few discoveries are more irritating than those which expose the pedigree of ideas" (Lord Acton, quoted in Hayek, 1944, 3).
14. Indigenous South Americans are the descendants of people who moved farthest from the African homeland. More than two-thirds of these

people have the novelty-seeking gene; this is the highest in any group. By comparison, only a quarter of Africans and Europeans have the gene (Burnham and Phelan, 2000, 88–90).

15. After eating from the tree, Adam hides from God and confesses, "I am afraid, because I am naked" (Robin, 2004, 1).
16. Henig, 2004, 5.
17. Thus people might drink only bottled water and eat organic food while engaging in risky sports, and they might consider nuclear power to be far riskier to themselves than smoking and handguns.
18. Joseph Stalin recognized this by saying, "America's primary weapons . . . are stockings, cigarettes, and other merchandise," although he failed to see their power. (Quoted in Mandelbaum, 2002, 14.)
19. In an old Jewish story, the rabbi of Krakow interrupted his prayers one day with a wail to announce that he had just seen the death of the rabbi of Warsaw two hundred miles away. The Krakow congregation, though saddened, was of course much impressed with the visionary powers of their rabbi. A few days later, travelers brought the news that the old rabbi was alive and well. The news caused snickering, but a few undaunted disciples came to the defense of their rabbi, admitting that he may have been wrong on the specifics, but "nevertheless, what vision!" See Hirschman, 1977, 117.
20. Bailwy, 2005, 242.
21. Martin, 2000, 305.
22. Easterbrook, 2003, xiii.
23. John Adams, "Risk and Morality: Three Framing Devices." In Ericson and Doyle (eds.), 2003, 100–101.
24. Scott Kilman, "Seed Firms Bolster Crops Using Traits of Distant Relatives." *Wall Street Journal,* October 31, 2006.
25. Brown and Duguid, 2000b.
26. Smith, 2000, 18.
27. Telushkin, 1994, 174. The "Love your neighbor" command was made famous by Jesus (Matthew 22:39), but he was simply quoting the Torah.
28. Lovejoy et al., 1935. The Five Ages include the ages of Gold, Silver, Bronze, Heroes, and Iron. Hesiod's story was developed further by Aratus and Ovid in Greek.
29. Popper, 1945, vol. 1, 37.
30. Jonathan Swift, quoted in Conway and Siegelman, 2005, 255.
31. Arthur C. Clarke famously said, "Every revolutionary idea evokes three stages of reactions: At first people say, 'It's completely impossible.' Then they say, 'Maybe it's possible to do it, but it would cost too much.' Finally they say, 'I always thought it was a good idea.'" Arthur Schopenhauer's original version is: "New truths go through three stages. First they are ridiculed, second they are violently opposed, and then, finally, they are accepted as self-evident."
32. Frankl, 1997, 25–29.
33. Arendt, 1958, 239.
34. Nietzsche, 1976, 179–80.

35. The word *robot* was coined by Josef Capek for the automatons in the play *R.U.R.* It derives from the Czech word *robata,* meaning "slave labor."
36. Freeman, 1995, 132.
37. Boulding, 1978, 259.
38. Homer, *The Iliad.* Quoted in Sagan and Druyan, 1992, 98.
39. Quoted in Shermer, 2004, 253.
40. Golems, the animated beings created out of inanimate matter (e.g. clay), are powerful but not intelligent. If commanded to perform a task, they will take the instructions literally.
41. Quoted in Barrow, 1998, 57.
42. Thayer, 1976.
43. Quoted in Fukuyama, 2002, 148.
44. The term "deep ecology" was coined by the Norwegian philosopher Arne Naess in the early 1970s, in contrast to what he called "shallow environmentalism," which only tries to remedy the environmental damages done by industrial societies.
45. Burch, 2001.
46. Quote attributed to Saint Bonaventura, a thirteenth-century theologian (de Waal, 1996, 97).
47. Of course, there are many conflicting interpretations of events in the lives of such individuals. One interpretation of Jesus's behavior is that throughout his life he had developed the illusion of his own invincibility. He had faith in his ability to summon up supernatural forces until the last minute of his life, when he cried out from the cross, "Eli, Eli, lama sabachthani"—"My God, my God, why hast thou forsaken me?" Some in the crowd thought he was calling to the prophet Elijah and gazed around to see if Elijah would come and save him, but Jesus merely uttered, "I thirst" (Humphrey, 1996, 99; Solomon and Higgins, 1998, 17; Flexner and Flexner, 2000, 21).

    An alternate interpretation of the quotation is one of an *incipit*—that he was reciting the opening line of Psalm 22 and deliberately skipped the rest as well as Psalm 23 that follows. In this interpretation, since everyone present knew Psalm 23, which states, "The Lord is my shepherd . . . Though I walk through the Valley of the Shadow of Death, I will fear no evil, for Thou art with me," the dying Jesus was being powerfully creative: while he himself did not speak Psalm 23, he made the crowd think it and turn it into *their* prayer (Hart, 2001, 103).
48. Sapolasky, 1994.
49. Dyson, 2003.
50. Dray, 2005.
51. Ward, 2001, 7.
52. Such limiting presumptions are based on the perception that the specific natural environment we experience on Earth today is eternally fixed. Modern science has shown again and again that what is eternal for a single person's lifetime is not for the history of the Earth.
53. For various philosophical and scientific arguments and counterarguments for the existence of God, see Shermer, 2000, 91–109.

54. See Galton, 2001, 9–10; Fernández-Armesto, 2004, 152.
55. Conway Morris, 2003, 319.
56. Lynn, 1996, 14–15.
57. Hitler wrote in *Mein Kampf*, "If nature does not wish that weaker individuals should mate with stronger, she wishes even less that a superior race should intermingle with an inferior."
58. This includes the criticism of Robert Wright's claim that "the secret of life" is its nonzero logic. Similar criticism is applied to any scientific theory that claims to provide the basis of reality.
59. Pinker, 2002, 157.
60. The German principle of *Vorsorgeprinzip* is the forebear of PP. http://www.extropy.org/proactionaryprinciple.htm.
61. Baudrillard, 1994, 17.
62. Appleyard, 1999, 151.
63. http://www.extropy.org/proactionaryprinciple.htm.
64. Garreau, 2005, 182.
65. Glover, 2000.
66. John Milton, *Paradise Lost*, Book 8.
67. Shattuck, 1996, 28–29.
68. Shattuck, 1996, 33.
69. More, 2004.

Chapter 11

1. Quoted in Hughes, 2004, 75.
2. Paul Virilio makes a similar statement—that "the capacity for war is the capacity for movement" (quoted in Hayles, 1999, 117).
3. http://members.tripod.com/spacetimenow/section6.html.
4. Wilson, 2007.
5. Dawkins, 2003.
6. Natalie Angier, "One Thing They Aren't: Maternal." *New York Times*, May 9, 2006.
7. Nietzsche, 1996a, 8.
8. Howard R. Gruber, *Darwin on Man: A Psychological Study of Scientific Creativity (University of Chicago Press, 1974)* as quoted in Wilson, 1978, 224.
9. Peoples and the ecosystems they occupy are self-regulating. Those who managed to cause permanent damage to their environment destroyed their own way of life. And those who stumbled upon environmentally friendly superstitions or taboos were able to survive.
10. Ruse and Wilson, 1986, 186.
11. Sharon Begley, "Researchers Seek Roots of Morality in Biology, with Intriguing Results." *Asian Wall Street Journal*, June 14, 2004.
12. Nietzsche wrote in 1887, "[M]y ideas are so indescribably strange and dangerous that only much later will anybody be ready for them." In this

sense, he was a more of a prophet than a thinker in the conventional sense (Friedman, 1995, 198–200).

13. The Silver Rule, which is the negatively stated form of the Golden Rule, takes on such wordings as "Do not do unto others what you would not have them do unto you," or "One should not treat others in ways that one would not like to be treated." This version neutralizes Confucius's criticism of the Golden Rule when he asked questions like, "Shall the masochist inflict pain on his neighbor?" Other variants are also based on the "do unto others" formula, but with often opposite results. The so-called Brazen (or Brass) Rule is "Do unto others as they do unto you." The Iron Rule elevates ruthlessness—"Do unto others as you like, before they do it unto you." These descriptions were most famously presented by Carl Sagan—see his book *Billions and Billions* ("The Rules of the Game"), and http://www.freeonlineinformation.com/rulesofthegame.htm; see also http://www.webniaga.com/j3f2c/index.php/site-map/articles/84-philosophy/78-imagine-this-1.
14. Nozick, 1974, 45–46.
15. The field of neurotheology attempts to explain how we can have religious feelings, and how the understanding of natural laws and the magnificence of cosmic evolution can stimulate transcendent religious feelings.
16. Stark, 2005, 14. To keep perspective, however, monasteries were dedicated to spiritual pursuits; science and natural philosophy were marginal to this enterprise. Even liberal leaders such as Albert the Great and Thomas Aquinas would never allow reason to prevail over revelation (Lindberg, 1992, 156, 234).
17. Rubenstein, 2003, 191.
18. Raimondo, 2000, 325–27.
19. Popper, 1994.
20. Noble, 1998, 206–7.
21. Friedman, 1995, 109.
22. Kass, 2003, 5.
23. Kurzweil, 1999, 62.
24. Kundera, 1984, 156.
25. Murchie, 1978, 622–23.
26. Midgley, 1992, 163–64.
27. Ruth A. Roland, "The Linguist of Ancient and Medieval Days," quoted in http://www.songsouponsea.com/Promenade/Court.html.
28. The *New York Times* reported that the Russians were trying to persuade a Cuban heiress to lend some of her monkeys for further experiments. See http://www.scotsman.com/news/world/stalin-s-half-man-half-ape-super-warriors-1-686693.
29. Regis, 1990, 92.
30. Shermer, 2004, 211.
31. Quoted in Allen, 2004, 163.
32. Heinberg, 1999, 139.
33. Hughes, 2004, 157.
34. International Theological Commission, 2004, 84.

35. Stock, 2003, 2.
36. Fukuyama, 1992, 64.
37. Sandel, 2002.
38. International Theological Commission, 2004.
39. Quoted in John Byrne, "Peter Drucker: Why His Ideas Still Matter." *Businessweek,* November 28, 2005.
40. Silver, 2006, 191–209.
41. These Kantian categories represent what Willem Drees calls two types of theology, one reflecting the hidden spiritual dimension beyond reality and the other concerning moral issues such as justice (Drees, 1991, 176).

Chapter 12

1. Science fiction with truly strange nonhuman narratives and alien values exists, but these works tend to be neglected and regarded as minor. This inherent limitation is unlikely to change.
2. Schnaars, 1989. For wildly pessimistic forecasts, see Goertzel and Bugaj, 2006, 1–7.
3. Traditional linear growth is achieved by accumulating production inputs of labor, capital, and land. It is also made possible by increasing trade and specialization, which result in better production resource allocation. Growth after the Industrial Revolution adds the dynamics of technological innovation and competitive "creative destruction."
4. The concept of technological singularity powered by superintelligence was first suggested by mathematician John von Neumann in the mid-1950s and popularized more recently by Vernor Vinge and Ray Kurzweil.
5. It is no wonder that the greatest task of top business leaders is not to provide leadership, but to develop tomorrow's leaders to replace the aging current generation.
6. Ullman, 1997, 116–17.
7. Alman Service called this the Law of Evolutionary Potential (Service, 1960, 97).
8. Brooks, 2004, 263.
9. Dyson, 1979, 230.
10. Quoting Stanley Greenspan: "A youngster must learn to use the greeting only with those for whom it is appropriate. Teaching him some general principle, such as 'Greet everyone who lives within three blocks of our house,' won't work; he can't stop to ask people their addresses. Nor will 'Greet everyone you see' suffice; he might give a warm smile to a thief or kidnapper. Nor can we count on 'Greet only our friends and members of our family'; there are many old chums and distant relatives he hasn't met. Even if he could learn a set of rote rules, by the time he decided whether to say hello, the person would be gone.

    "Instead, through countless encounters in his early years, the child works out the problem for himself. As he goes about his daily life, he eventually comes to associate saying hello with a particular emotion—the warmth of seeing someone he or his family knows" (Greenspan, 1997, 23).

11. One of the best "realistic" plans, which was endorsed by Arthur C. Clarke, is that of Marshall T. Savage in *The Millennial Project: Colonizing the Galaxy in Eight Easy Steps* (1994).
12. Erikson, 1964, 115.
13. See, for example, the works of Ray Kurzweil and Hans Moravec.
14. Kurzweil, 1999, 118, 281.
15. Conway Morris, 2003, 17.
16. Delbrück, 1986, 52.
17. Masson and McCarthy, 1995.
18. Gray, 2001, 27–29.
19. Bailey, 2005, 169–70.
20. The voting system in a democracy is a conventionalized, or ritualized, competition with the votes as the "canine teeth" and threat display within an artificial universe of unchanging institutions that ensure continuity. See Tiger and Fox, 1971, 41–43.
21. Webber, 1947, 130–32.
22. Finer, 1997, 15–16.
23. Stanislaw Lem speculated that this is the capability of a billion-year-old civilization, which no longer employs instrumental technologies but plays with natural laws.

Chapter 13

1. Bernal, 1929, 5. There are rare exceptions—for example, in Olaf Stapledon's science-fiction novels written in the 1930s and 1940s.
2. Dostoyevsky, 1992, 22–24.
3. Mayr and Diamond, 2001.
4. John F. Kennedy, speech in Dublin, Ireland, on June 28, 1963.
5. Paul and Cox, 1996, 338.
6. Becker, 1973, 74.
7. Woody Allen's famous quip.
8. Ronald Reagan, "A Time for Choosing," speech on October 27, 1964.

BIBLIOGRAPHY

Note: This is a substantially abridged bibliography, leaving out well-known classics and many works that influenced the author's thinking but may not be essential for readers. The full version of bibliography can be found at http://www.transhumanpotential.com.

Abbott, Edwin. *Flatland: A Romance in Many Dimensions.* New York: Dover, 1952.

Ackoff, Russell. *The Art of Problem Solving: Accompanied by Ackoff's Fables.* New York: John Wiley & Sons, 1978.

Adams, Robert McCormick. *Paths of Fire: An Anthropologist's Inquiry into Western Technology*. Princeton, NJ: Princeton University Press, 1996.

Agar, Nicholas. *Humanity's End: Why We Should Reject Radical Enhancement.* Cambridge, MA: MIT Press (A Bradford Book), 2010.

Ainslie, George, and John Monterosso. "A Marketplace in the Brain?" *Science* 306, no. 5695 (2004): 421–23.

Albus, James S. *Peoples' Capitalism: The Economics of the Robot Revolution.* Kensington, MD: New World Books, 1976.

Alexander, Christopher. *The Phenomenon of Life: The Nature of Order (Book One)*. Berkeley: Center for Environmental Structure, 1980.

Alexander, Denis, and Robert S. White. *Science, Faith, and Ethics: Grid or Gridlock?* Peabody, MA: Hendrickson Publishers, 2006.

Alkon, Daniel L. "Memory Storage and Neural Systems." *Scientific American* 261, no. 1 (1989): 42–50.

Allen, Anita L. *The New Ethics: A Guided Tour of the Twenty-First Century Moral Landscape*. New York: Miramax Books, 2004.

Allman, William F. *The Stone Age Present: How Evolution Has Shaped Modern Life—from Sex, Violence, and Language to Emotions, Morals, and Communities.* New York: Simon & Schuster, 1994.

Amar, Akhil R. *America's Constitution: A Biography.* New York: Random House, 2005.

Anderson, Walter Truett. *To Govern Evolution: Further Adventures of the Political Animal.* Boston: Harcourt Brace Jovanovich, 1987.

Angell, Ian. *The New Barbarian Manifesto: How to Survive the Information Age.* London: Dover, 2000.

Ansell Pearson, Keith. *Viroid Life: Perspectives on Nietzsche and the Transhuman Condition*. London: Routledge, 1997.

Appleyard, Brian. *Brave New Worlds: Staying Human in the Genetic Future.* London: HarperCollins, 1999.

——. *How to Live Forever or Die Trying: On the New Immortality*. London: Charles Scribner's Sons, 2007.

Arendt, Hannah. *The Human Condition.* Chicago: University of Chicago Press, 1998.

Armstrong, Karen. *A History of God: The 4000-Year Quest of Judaism, Christianity and Islam.* New York: Alfred A. Knopf, 1993.

Arnason, Johann P., S. N. Eisenstadt, Bjorn Wittrock, eds. *Axial Civilizations and World History*. Leiden, the Netherlands: Brill Academic Publishers, 2005.

Arora, Ashish, Ralph Landau, and Nathan Rosenberg, eds. *Chemicals and Long-term Economic Growth: Insights from the Chemical Industry.* New York: Wiley-Interscience, 2000.

Arrison, Sonia. *100 Plus: How the Coming Age of Longevity Will Change Everything, from Careers and Relationships to Family and Faith.* New York: Basic Books, 2011.

Arrow, K., et al. "Determining Benefits and Costs for Future Generations." *Science* 341, no. 6144 (26 July 2013), 349–50.

Arrow, Kenneth. *Essays in the Theory of Risk-Bearing.* Amsterdam: North-Holland, 1971.

Aslan, Reza. *No God but God: The Origins, Evolution, and Future of Islam.* New York: Random House, 2005.

Atkins, Peter. *Galileo's Finger: The Ten Greatest Ideas of Science.* New York: Oxford University Press, 2003.

Atkinson, William I. *Nanocosm.* New York: AMACOM American Management Association, 2005.

Atran, Scott. *In God We Trust*. Oxford: Oxford University Press, 2002.

Austin, James H. *Zen and the Brain.* Cambridge, MA: MIT Press, 1999.

Axelrod, Robert. *The Evolution of Cooperation*. New York: Basic Books, 1984.

Ayres, Ian. *Super Crunchers.* New York: Bantam Dell, 2007.

Backhouse, Roger. *The Ordinary Business of Life: A History of Economics from the Ancient World to the Twenty-first Century*. Princeton: Princeton University Press, 2002.

Backman, Mark. *Rhetoric and the Rise of Self-consciousness.* Woodbridge, CT: Ox Bow Press, 1991.

Bagemihl, Bruce. *Biological Exuberance: Animal Homosexuality and Natural Diversity.* New York: St. Martin's Press, 1999.

Bailey, James. *After Thought: The Computer Challenge to Human Intelligence.* New York: Basic Books, 1996.

Bailey, Ronald. *Liberation Biology: The Scientific and Moral Case for the Biotech Revolution.* Amherst, NY: Prometheus Books, 2005.

——. "Transhumanism: The Most Dangerous Idea? Why Striving to Be More Than Human Is Human." *Reason.* August 24, 2004.

Bailey, Ronald, ed. *The True State of the Planet.* New York: Free Press, 1995.

Bailyn, Bernard. *The Ideological Origins of the American Revolution*. Cambridge, MA: Belknap Press, 1992.

Bainbridge, William Sims. "The Transhuman Heresy." *Journal of Evolution & Technology* 14, issue 2, 2005.

Bainbridge, William Sims, and Mihail C. Roco, eds. *Managing Nano-Bio-Info-Cogno Innovations: Converging Technologies in Society*. The Netherlands: Springer, 2006.

Bak, Per. *How Nature Works: The Science of Self-organized Criticality.* New York: Springer-Verlag, 1996.

Baldi, Pierre. *The Shattered Self: The End of Natural Evolution*. Cambridge, MA: MIT Press, 2001.

Ball, Philip. *Critical Mass: How One Thing Leads to Another.* New York: Farrar, Straus and Giroux, 2004.

Ballard, J. G. *A User's Guide to the Millennium: Essays and Reviews*. New York: Picador USA, 1997.

Barash, David P. *Whisperings Within*. New York: Penguin Books, 1981.

Barbour, Ian G. *Religion and Science: Historical and Contemporary Issues.* New York: HarperCollins, 1997.

Barlow, Connie, ed. *Evolution Extended: Biological Debates on the Meaning of Life.* Cambridge, MA: MIT Press, 1994.

Barnes, B., D. Bloor, and J. Henry. *Scientific Knowledge: A Sociological Analysis*. London: Athlone Press, 1996.

Baron-Cohen, Simon. *The Essential Difference: The Truth about the Male and Female Brain*. New York: Basic Books, 2003.

Barr, Stephen M. "The Miracle of Evolution." *First Things*. February 2006.

Barrat, James. *Our Final Invention: Artificial Intelligence and the End of the Human Era*. New York: Thomas Dunne Books, 2013.

Barrett, William. *Death of the Soul: From Descartes to the Computer.* New York: Doubleday, 1986.

Barrow, John. *The Artful Universe: The Cosmic Source of Human Creativity.* Boston: Back Bay Books, 1996.

———. *Impossibility: The Limits of Science and the Science of Limits*. Oxford: Oxford University Press, 1998.

Barrow, John, and Frank Tipler. *The Anthropic Cosmological Principle.* Oxford: Oxford University Press, 1988.

Barzun, Jacques. *From Dawn to Decadence: 500 Years of Western Cultural Life, 1500 to the Present*. New York: Harper Perennial, 2001.

Baudrillard, Jean. *The Illusion of the End*. Translated by Chris Turner. Oxford: Polity Press, 1994.

Baum, Eric. *What Is Thought?* Cambridge, MA: MIT Press, 2004.

Bayly, C. A. *The Birth of the Modern World, 1780–1914: Global Connections and Comparisons*. Malden, MA: Blackwell Publishers, 2004.

Bazerman, Max H., and Michael D. Watkins. *Predictable Surprises: The Disasters You Should Have Seen Coming, and How to Prevent Them*. Cambridge, MA: Harvard Business School Press, 2004.

Becker, Ernest. *The Denial of Death.* New York: Free Press, 1973.

Behe, Michael. *Darwin's Black Box: The Biochemical Challenge to Evolution.* New York: Free Press, 1996.

Beniger, James R. *The Control Revolution: Technological and Economic Origins of the Information Society.* Cambridge, MA: Harvard University Press, 1986.

Bentley, Jerry H. *Old World Encounters: Cross Cultural Contacts and Exchanges in Pre-modern Times.* New York: Oxford University Press, 1993.

Benton, Michael J. *When Life Nearly Died: The Greatest Mass Extinction of All Time.* New York: Thames & Hudson, 2003.

Berger, Peter L. *The Sacred Canopy: Elements of a Sociological Theory of Religion.* New York: Anchor, 1990.

Bergling, Kurt. *Moral Development: The Validity of Kohlberg's Theory.* Stockholm: Almqvist & Wiksell, 1981.

Bergson, Henri. *Creative Evolution.* Translated by A. Mitchell. London: Macmillan, 1964.

——. *The Two Sources of Morality and Religion.* Translated by R. Ashley Audra and Cloudesley Brereton. New York: Henry Holt and Company, 1932.

Berliner, Joseph S. *The Innovation Decision in Soviet Industry.* Cambridge, MA: MIT Press, 1976.

Bernal, J. D. *The World, the Flesh & the Devil: An Enquiry into the Future of the Three Enemies of the Rational Soul.* Bloomington: Indiana University Press, 1969.

Bernstein, Peter L. *Against the Gods: The Remarkable Story of Risk.* New York: John Wiley & Sons, 1996.

Bernstein, William J. *The Birth of Plenty: How the Prosperity of the Modern World Was Created.* New York: McGraw-Hill, 2004.

Berry, Wendell. *Life Is a Miracle: An Essay against Modern Superstition.* Washington, D.C.: Counterpoint Press, 2000.

Bird, Richard J. *Chaos and Life: Complexity and Order in Evolution and Thought.* New York: Columbia University Press, 2003.

Birner, Jack, and Rudy van Zijp, eds. *Hayek, Co-ordination and Evolution: His Legacy in Philosophy, Politics, Economics, and the History of Ideas.* London: Routledge, 1994.

Birx, James. *Man's Place in the Universe: An Introduction to Scientific Philosophical Anthropology.* Arcade, NY: Tri-County Publications, 1977.

Bjorklund, David F., and Anthony D. Pellegrini. *The Origins of Human Nature: Evolutionary Developmental Psychology.* Washington, D.C.: American Psychological Association, 2002.

Blackmore, Susan. *The Meme Machine.* New York: Oxford University Press, 1999.

Blainey, Geoffrey. *A Short History of the World.* Chicago: Ivan R. Dee, 2002.

Bloch, Marc. *The Historian's Craft.* Translated by Peter Putnam. Manchester: Manchester University Press, 1992.

Bloch, Maurice. *Prey into Hunter: The Politics of Religious Experience.* Cambridge: Cambridge University Press, 1991.

Bloland, Sue Erikson. "Fame: The Power and Cost of a Fantasy." *Atlantic Monthly,* November (1999): 51–62.

Bloom, Allan. *The Closing of the American Mind.* New York: Simon & Schuster, 1987.

Bloom, Howard. *Global Brain: The Evolution of Mass Mind from the Big Bang to the 21st Century.* New York: John Wiley & Sons, 2000.

——. *The Lucifer Principle: A Scientific Expedition into the Forces of History.* New York: The Atlantic Monthly Press, 1997.

Bloom, Paul. *How Children Learn the Meanings of Words.* Cambridge, MA: MIT Press, 2000.

Boas, Franz. *The Mind of Primitive Man.* New York: Free Press, 1963.

Bobbitt, Philip. *The Shield of Achilles: War, Peace, and the Course of History.* New York: Alfred A. Knopf, 2002.

Boden, Margaret A, ed. *The Philosophy of Artificial Life.* Oxford: Oxford University Press, 1996.

Boeke, Kees. *Cosmic View: The Universe in Forty Jumps.* New York: John Day Company, 1957. http://www.vendian.org/mncharity/cosmicview/.

Boghossian, Paul A. *Fear of Knowledge: Against Relativism and Constructivism.* Oxford: Oxford University Press, 2006.

Bool, F. H., J. R. Kist, J. L. Locher, and F. Wierda. *M.C. Escher: His Life and Complete Graphic Work.* New York: Harry N. Abrams, Inc., 1982.

Boorsook, Paulina. *Cyberselfish: A Critical Romp Through the Terribly Libertarian Culture of High Tech.* New York: Public Affairs, 2001.

Boorstin, Daniel J. *The Seekers: The Story of Man's Continuing Quest to Understand His World.* New York: Random House, 1998.

Borges, Jorge Luis. *A Personal Anthology.* New York: Grove Press, 1967.

Bostrom, Nick. "A History of Transhumanist Thought." *Journal of Evolution & Technology* 14, issue 1, 2005b.

——. *Transhumanist Values.* http://www.transhumanism.org, 2005a.

Bostrom, Nick, et al. *The Transhumanist FAQ.* http://www.transhumanism.org/resources/faq.html.

Boulding, Kenneth. *Ecodynamics: A New Theory of Social Evolution.* Beverly Hills: Sage Publications, 1978.

Bourdieu, Pierre. *Homo Academicus.* Stanford: Stanford University Press, 1988.

Bourke, Joanna. *An Intimate History of Killing: Face-to-face Killing in Twentieth-century Warfare.* New York: Basic Books, 1999.

Bowler, Peter J. *Evolution: The History of an Idea.* Berkeley: University of California Press, 1984.

Boyer, Pascal. *The Naturalness of Religious Ideas: A Cognitive Theory of Religion.* Berkeley: University of California Press, 1994.

——. *Religion Explained: The Evolutionary Origins of Religious Thought.* New York: Basic Books, 2001.

Brague, Remi. *Eccentric Culture: A Theory of Western Civilization.* South Bend, IN: St. Augustine's Press, 2002.

——. *The Wisdom of the World: The Human Experience of the Universe in Western Thought.* Translated by Teresa Lavender Fagan. Chicago: University of Chicago Press, 2003.

Brand, Stewart. *The Clock of the Long Now: Time and Responsibility.* New York: Basic Books, 1999.

Brain, Marshall. Manna: Two Visions of Humanity's Future. Raleigh, NC: BYG Publishing, 2012.

Braudel, Fernand. *The Perspective of the World: Civilization and Capitalism 15th–18th Century, Vol. 3.* Translated by Sian Reynolds. New York: Harper & Row, 1984.

Breasted, J. H. *The Conquest of Civilization.* New York: Literary Guild, 1938.

Bremmer, Ian. *The J Curve: A New Way to Understand Why Nations Rise and Fall.* New York: Simon & Schuster, 2006.

Brinton, Crane, John B. Christopher, Robert Lee Wolff. *A History of Civilization.* Englewood Cliffs, NJ: Prentice-Hall, 1975.

Brockman, John, ed. *The New Humanists: Science at the Edge.* New York: Barnes & Noble, 2003.

———. *The Third Culture: Beyond the Scientific Revolution.* New York: Simon & Schuster, 1995.

Broderick, Damien. *The Spike: How Our Lives Are Being Transformed by Rapidly Advancing Technologies.* New York: Forge, 2001.

Bronson, Po. *What Should I Do with My Life? The True Story of People Who Answered the Ultimate Question.* New York: Random House, 2002.

Brooke, John Hedley. *Science and Religion: Some Historical Perspectives.* Cambridge: Cambridge University Press, 1991.

Brooks, David. *On Paradise Drive: How We Live Now (and Always Have) in the Future Sense.* New York: Simon & Schuster, 2004.

Brooks, Rodney. *Flesh and Machines: How Robots Will Change Us.* New York: Pantheon, 2002.

Brooks, Rodney A., and Pattie Maes, eds. *Artificial Life IV.* Cambridge, MA: MIT Press (A Bradford Book), 1994.

Brothers, Leslie. *Friday's Footprint: How Society Shapes the Human Mind.* Oxford: Oxford University Press, 1997.

Brower, Kenneth. *The Starship and the Canoe.* New York: HarperCollins, 1983.

Brown, Aaron. *The Poker Face of Wall Street.* New York: John Wiley & Sons, 2006.

Brown, Cynthia Stokes. *Big History: From the Big Bang to the Present.* New York: New Press, 2007.

Brown, Donald E. *Human Universals.* Philadelphia: Temple University Press, 1991.

Brown, John S., and Paul Duguid. "A Response to Bill Joy and the Doom-and-Gloom Technofuturists." *Industry Standard,* April 13, 2000a.

———. *The Social Life of Information.* Cambridge, MA: Harvard Business School Press, 2000b.

Brownstone, David F., and Irene Franck. *Timelines of War: A Chronology of Warfare from 100,000 B.C. to the Present.* New York: Little, Brown and Company, 1996.

Bruner, Jerome. *Acts of Meaning.* Cambridge, MA: Harvard University Press, 1990.

Brunschwig, Jacques, and Geoffrey E. R. Lloyd. *A Guide to Greek Thought: Major Figures and Trends*. Translated under the direction of Catherine Porter. Cambridge, MA: Harvard University Press, 2003.

Brynjolfsson, Erik, and Andrew McAfee. *Race Against the Machine: How the Digital Revolution Is Accelerating Innovation, Driving Productivity, and Irreversibly Transforming Employment and the Economy*. Cambridge, MA: Digital Frontier Press, 2011.

Bryson, Kenneth A. *Persons and Immortality*. Amsterdam: Rodopi, 1999.

Bucke, Richard Maurice. *Cosmic Consciousness: A Study on the Evolution of the Human Mind*. Bedford, MA: Applewood Books, 2001.

Buckingham, Marcus, and Curt Coffman. *First, Break All the Rules: What the World's Greatest Managers Do Differently*. New York: Simon & Schuster, 1999.

Burch, Greg. "Progress, Counter-Progress, and Counter-Counter-Progress." 2001. http://gregburch.net/progress.html.

Burger, William C. *Perfect Planet, Clever Species: How Unique Are We?* Amherst, NY: Prometheus Books, 2002.

Burgess, Anthony. *A Clockwork Orange*. New York: Norton, 1988.

Burke, James, and Robert Ornstein. *The Axemaker's Gift: A Double-edged History of Human Culture*. New York: Grosset/Putnam, 1995.

Burnham, John. *How Superstition Won and Science Lost*. New Brunswick, NJ: Rutgers University Press, 1987.

Burnham, Terry, and Jay Phelan. *Mean Genes: From Sex to Money to Food—Taming Our Primal Instincts*. Cambridge, MA: Perseus Books, 2000.

Burrows, William E. *The Survival Imperative: Using Space to Protect Earth*. New York: Forge, 2006.

Burt, Austin, and Robert Trivers. *Genes in Conflict: The Biology of Selfish Genetic Elements*. Cambridge, MA: Belknap Press, 2006.

Bury, J. B. *Idea of Progress: An Inquiry into Its Origins and Growth*. New York: Dover Publications, 1955.

Bush, Vannevar. "As We May Think." *Atlantic Monthly*, July 1, 1945.

Buss, David M. *The Murderer Next Door: Why the Mind Is Designed to Kill*. New York: The Penguin Press, 2005.

Butler, Declan. "2020 Computing: Everything, Everywhere." *Nature* 440 (2006): 402–5.

Butterworth, Brian. *The Mathematical Brain*. London: Macmillan, 1999.

Callahan, Daniel. *What Price Better Health?: Hazards of the Research Imperative*. Berkeley: University of California Press, 2003.

Calvin, William. *The Ascent of Mind: Ice Age Climates and the Evolution of Intelligence*. New York: Bantam, 1990.

Cameron, J. M. "The Theory and Practice of Autobiography." In *Language, Meaning and God*, edited by B. Davies. London: Geoffrey Chapman, 1987.

Campbell, Joseph. *The Hero with a Thousand Faces*. New York: MJF Books, 1996 (originally published 1949).

——. *The Inner Reaches of Outer Space: Metaphor as Myth and as Religion*. New York: HarperCollins, 1988.

Camus, Albert. *The Myth of Sisyphus and Other Essays.* Translated by Justin O'Brien. New York: Vintage International, 1991.

Cannon, Walter B. *The Wisdom of the Body.* New York: W. W. Norton, 1939.

Cantillon, Richard. *Essay on the Nature of Commerce in General.* New Brunswick, NJ: Transaction Publishers, 2001 (originally published 1755).

Cantor, Norman F. *In the Wake of the Plague: The Black Death and the World It Made.* New York: Free Press, 2001.

Capra, Fritjof. *The Tao of Physics: An Exploration of the Parallels between Modern Physics and Eastern Mysticism.* Boston: Shambhala, 2000.

Carey, James R. *Longevity: The Biology and Demography of Life Span.* Princeton, NJ: Princeton University Press, 2003.

Carr, Nicholas. *Does IT Matter?* Cambridge, MA: Harvard Business School Press, 2004.

Carroll, Sean B. *Endless Forms Most Beautiful: The New Science of Evo Devo and the Making of the Animal Kingdom.* New York: W. W. Norton, 2005.

Carse, James P. *Finite and Infinite Games.* New York: Free Press, 1986.

Carter, Rita. *Exploring Consciousness.* Berkeley: University of California Press, 2002.

——. *Mapping the Mind.* Berkeley: University of California Press, 1998.

Carter, Vernon Hill, and Tom Dale. *Top Soil and Civilization.* Norman: University of Oklahoma Press, 1974.

Cartwright, Frederick F. *Disease and History.* New York: Barnes & Noble, 1991.

Castells, Manuel. *The Rise of the Network Society.* Malden, MA: Blackwell Publishers, 1996.

Casti, John. *Complexification: Explaining a Paradoxical World through the Science of Surprise.* New York: Harper Perennial, 1995.

Chaisson, Eric J. *Cosmic Evolution: The Rise of Complexity in Nature.* Cambridge, MA: Harvard University Press, 2001.

——. *Epic of Evolution: Seven Ages of the Cosmos.* New York: Columbia University Press, 2006.

Chalmers, David J. *The Conscious Mind: In Search of a Fundamental Theory.* Oxford: Oxford University Press, 1997.

Chandler, Alfred D., Jr. *The Visible Hand: The Managerial Revolution in American Business.* Cambridge, MA: Belknap Press, 1977.

Chi, Michelene T. H., Robert Glaser, and M. J. Farr. *The Nature of Expertise.* Hillsdale, NJ: Erlbaum, 1988.

Chiles, James R. *Inviting Disaster: Lessons from the Edge of Technology.* New York: HarperBusiness, 2001.

Churchland, Patricia S. *Neurophilosophy: Toward a Unified Science of the Mind/Brain.* Cambridge, MA: The MIT Press, 1986.

Churchland, Paul M. *The Engine of Reason, the Seat of the Soul: A Philosophical Journey into the Brain.* Cambridge, MA: MIT Press, 1995.

Cialdini, Robert B. *Influence.* New York: Quill, 1993.

Clark, Andy. *Being There: Putting Brain, Body, and World Together Again.* Cambridge, MA: MIT Press, 1997.

Clark, Gregory. *A Farewell to Alms: A Brief Economic History of the World.* Princeton, NJ: Princeton University Press, 2007.

Clarke, Arthur C. *The City and the Stars.* New York: Signet, 1957.

——. *Profiles of the Future.* New York: Harper & Row, 1962.

Clausewitz, Carl von. *On War.* Translated by Michael Howard and Peter Paret. New York: Alfred A. Knopf, 1993 (originally published 1832).

Clifford, William K. "The Ethics of Belief." *Contemporary Review.* 1877. http://www.xs4all.nl/~maartens/logic/Clifford/WKCliffordEthicsOfBelief.htm_.

Coates, Joseph F., and Jennifer Jarratt. *What Futurists Believe.* Bethesda, MD: World Future Society, 1989.

Cohen, Daniel. *How the World Will End.* New York: McGraw-Hill, 1973.

Colinvaux, P. *Why Big Fierce Animals Are Rare.* Princeton: Princeton University Press, 1978.

Colling, Richard G. *Random Designer: Created from Chaos to Connect with the Creator.* Bourbonnais, IL: Browning Press, 2004.

Collins, Francis S. *The Language of God: A Scientist Presents Evidence for Belief.* New York: Free Press, 2006.

Collins, Randall. *The Sociology of Philosophies: A Global Theory of Intellectual Change.* Cambridge, MA: Harvard University Press, 1998.

Conquest, Robert. *Reflections on a Ravaged Century.* New York: W. W. Norton, 1999.

Conway, Flo, and Jim Siegelman. *Dark Hero of the Information Age: In Search of Norbert Wiener, the Father of Cybernetics.* New York: BasicBooks, 2005.

Conway Morris, Simon. "The Cambrian 'Explosion': Slow Fuse or Megatonnage?" *Proceedings of the National Academy of Sciences* 97 (2000): 4426–29.

——. *Life's Solution: Inevitable Humans in a Lonely Universe.* Cambridge: Cambridge University Press, 2003.

Cook, Francis H. *Hua-Yen Buddhism: The Jewel Net of Indra.* University Park: The Pennsylvania State University Press, 1977.

Cowan, Ruth Schwartz. *More Work for Mother: The Ironies of Household Technology from the Open Hearth to the Microwave.* New York: Basic Books, 1983.

Cowen, Tyler. *Average Is Over: Powering America Beyond the Age of the Great Stagnation.* New York: Dutton, 2013.

Crain, William. *The Theories of Development: Concepts and Applications.* Englewood Cliffs, NJ: Prentice-Hall, 1992.

Crick, Francis. *The Astonishing Hypothesis: The Scientific Search for the Soul.* New York: Simon & Schuster, 1994.

Crosby, Alfred W. *Ecological Imperialism: The Biological Expansion of Europe, 900–1900.* Cambridge: Cambridge University Press, 1986.

——. *The Measure of Reality: Quantification in Western Europe, 1250–1600.* Cambridge: Cambridge University Press, 1997.

Crump, Thomas. *A Brief History of Science: As Seen through the Development of Scientific Instruments.* New York: Carroll & Graf, 2002.

Csikszentmihalyi, Mihaly. *The Evolving Self: A Psychology for the Third Millennium.* New York: HarperCollins, 1993.

——. *Flow: The Psychology of Optimal Experience.* New York: Harper & Row, 1990.

Cummins, Denise. *The Other Side of Psychology: How Experimental Psychologists Find Out about the Way We Think and Act.* New York: St. Martin's Press, 1995.

Cziko, Gary. *Without Miracles: Universal Selection Theory and the Second Darwinian Revolution.* Cambridge, MA: MIT Press, 1995.

Damasio, Antonio. *Descartes' Error: Emotion, Reason, and the Human Brain.* New York: Putnam, 1994.

——. *The Feeling of What Happens: Body and Emotion in the Making of Consciousness.* New York: Harcourt Brace, 1999.

——. *Looking for Spinoza: Joy, Sorrow, and the Feeling Brain.* Boston: Harcourt Brace, 2003.

Daniels, Ted, ed. *A Doomsday Reader: Prophets, Predictors, and Hucksters of Salvation.* New York: New York University Press, 1999.

Darwin, Charles Galton. *The Next Million Years.* Garden City, NY: Dolphin Books, 1952.

Darwin, Charles R. *The Autobiography of Charles Darwin,* edited by Nora Barlow. London: St James's Place, 1958.

——. *The Descent of Man, and Selection in Relation to Sex.* Princeton, NJ: Princeton University Press, 1981 (originally published 1871).

——. *The Origin of Species by Means of Natural Selection.* Reprint of 1872 edition. New York: Modern Library, 1936.

Davies, E. Brian. *Science in the Looking Glass: What Do Scientists Really Know?* Oxford: Oxford University Press, 2003.

Davies, Paul. *Are We Alone?: Philosophical Implications of the Discovery of Extraterrestrial Life.* New York: BasicBooks, 1995.

——. *The Mind of God: The Scientific Basis for a Rational World.* New York: Simon & Schuster, 1992.

Davies, Paul, and John Gribbin. *The Matter Myth: Dramatic Discoveries That Challenge Our Understanding of Physical Reality.* New York: Simon & Schuster, 1992.

Davies, Philip J., and Reuben Hersh. *Descartes' Dream: The World According to Mathematics.* Boston: Houghton Mifflin, 1986.

Dawkins, Richard. *The Blind Watchmaker.* London: Longmans, 1986.

——. *A Devil's Chaplain: Reflections on Hope, Lies, Science, and Love.* Boston: Houghton Mifflin, 2003.

——. *Extended Phenotype: The Gene as the Unit of Selection.* San Francisco: Freeman, 1982.

——. *The Selfish Gene.* Oxford: Oxford University Press, 1976.

Day, Clarence Burton. *The Philosophers of China: Classical and Contemporary.* Secaucus, NJ: Citadel Press, 1978.

Day, William. *Genesis on Planet Earth: The Search for Life's Beginning.* East Lansing, MI: House of Talos, 1979.

de Becker, Gavin. *The Gift of Fear: Survival Signals That Protect Us from Violence.* Boston: Little, Brown and Company, 1997.

de Condorcet, Antoine-Nicolas. *Sketch for a Historical Picture of the Progress of the Human Mind*, translated by June Barraclough. London: Weidenfeld & Nicolson, 1955 (originally published 1795).

de Duve, Christian. *Vital Dust: The Origin and Evolution of Life on Earth.* New York: Basic Books, 1995.

De Landa, Manuel. *A Thousand Years of Nonlinear History.* New York: Zone Books, 1997.

de Sousa, Ronald. *The Rationality of Emotion.* Cambridge, MA: MIT Press, 1987.

de Waal, Frans B. M. *Chimpanzee Politics: Power and Sex among Apes.* New York: Harper Collins, 1983.

——. *Good Natured: The Origins of Right and Wrong in Humans and Other Animals.* Cambridge, MA: Harvard University Press, 1996.

——. *Primates and Philosophers: How Morality Evolved.* Princeton, NJ: Princeton University Press, 2006.

de Waal, Frans B. M., and Frans Lanting. *Bonobo: The Forgotten Ape.* Berkeley: University of California Press, 1998.

Deacon, Terrence W. *The Symbolic Species: The Coevolution of Language and the Brain.* New York: W. W. Norton, 1997.

Delbrück, Max. *Mind from Matter: An Essay on Evolutionary Epistemology.* Palo Alto, CA: Blackwell Scientific Publications, 1986.

Dembski, William A. *The Design Inference: Eliminating Chance through Small Probabilities.* Cambridge: Cambridge University Press, 1998.

——. *Intelligent Design.* Downers Grove, IL: InterVarsity Press, 1999.

Dennett, Daniel C. *Darwin's Dangerous Idea: Evolution and the Meanings of Life.* New York: Simon & Schuster, 1995.

——. *Freedom Evolves.* New York: Viking, 2003.

Denton, Michael. *Evolution: A Theory in Crisis.* Bethesda, MD: Adler & Adler, 1986.

——. *Nature's Destiny: How the Laws of Biology Reveal Purpose in the Universe.* New York: Free Press, 1998.

Deutsch, David. *The Fabric of Reality: The Science of Parallel Universes—and Its Implications.* New York: Penguin Books, 1997.

Devlin, Keith. *The Math Gene: How Mathematical Thinking Evolved and Why Numbers Are Like Gossip.* New York: Basic Books, 2001.

Dewdney, A. K. *Beyond Reason: Eight Great Problems That Reveal the Limits of Science.* New York: John Wiley & Sons, 2004.

Diamandis, Peter, and Steven Kotler. *Abundance: The Future Is Better Than You Think.* New York: Free Press, 2012.

Diamond, Jared. *Guns, Germs, and Steel: The Fates of Human Societies.* New York: Norton, 1999.

——. *The Third Chimpanzee: The Evolution and Future of the Human Animal.* New York: Harper Collins, 1992.

——. *Why Is Sex Fun? The Evolution of Human Sexuality.* New York: Basic Books, 1997.

Dienstag, Joshua Foa. *Pessimism: Philosophy, Ethic, Spirit.* Princeton, NJ: Princeton University Press, 2006.

Dobzhansky, Theodosius. "Nothing in Biology Makes Sense Except in the Light of Evolution." *American Biology Teacher* 35 (March 1973): 125–29.

Donald, Merlin. *A Mind So Rare: The Evolution of Human Consciousness*. New York: W. W. Norton, 2001.

Dostoyevsky, Fyodor. *Notes from the Underground*. New York: Dover Publications, 1992 (originally published 1864).

Douglas, Mary, and Aaron Wildavsky. *Risk and Culture: An Essay on the Selection of Technological and Environmental Dangers*. Berkeley: University of California Press, 1982.

Dray, Philip. *Stealing God's Thunder: Benjamin Franklin's Lightning Rod and the Invention of America*. New York: Random House, 2005.

Drees, Willem. *Beyond the Big Bang: Quantum Cosmologies and God*. La Salle, IL: Open Court, 1991.

Drexler, Eric. *Engines of Creation*. New York: Anchor Press, 1986.

Dreyfus, Hubert L. *Being-in-the-World: A Commentary on Heidegger's Being and Time, Division I*. Cambridge, MA: MIT Press, 1990.

——. *On the Internet*. New York: Routledge, 2001.

——. *What Computers Still Can't Do: A Critique of Artificial Reason*. Cambridge, MA: MIT Press, 1992.

Drucker, Peter. *The Age of Discontinuity: Guidelines to Our Changing Society*. New York: Harper & Row, 1968.

——. *The Ecological Vision: Reflections on the American Condition*. New Brunswick, NJ: Transaction Publishers, 1993.

Dubos, René. *A God Within*. New York: Charles Scribner's Sons, 1972.

Dunbar, Robin I. M. *Grooming, Gossip, and the Evolution of Language*. Cambridge, MA: Harvard University Press, 1997.

——. *The Human Story: A New History of Mankind's Evolution*. London: Faber and Faber, 2004.

du Noüy, Pierre Lecomte. *Human Destiny*. New York: Longmans, Green and Co., 1947.

Durant, Will. *The Greatest Minds and Ideas of All Time*. New York: Simon & Schuster, 2002.

Durant, Will, and Ariel Durant. *The Lessons of History*. New York: Simon & Schuster, 1968.

Dyson, Esther, et al. "A Magna Carta for the Knowledge Age." http://www.ifla.org/documents/libraries/net/magna.txt.

Dyson, Freeman. *Disturbing the Universe*. New York: Harper & Row, 1979.

——. "The Future Needs Us!" *New York Review of Books* 50, no. 2 (2003). http://www.nybooks.com/articles/16053.

——. *Imagined Worlds*. Cambridge, MA: Harvard University Press, 1997.

——. *Infinite in All Directions*. New York: Harper & Row, 1988.

——. *Origins of Life*. Cambridge: Cambridge University Press, 1985.

——. *The Sun, the Genome, and the Internet: Tools of Scientific Revolutions*. Oxford: Oxford University Press, 1999.

Dyson, George. *Darwin among the Machines: The Evolution of Global Intelligence*. Reading, MA: Perseus Books, 1997.

Eames, Charles, and Ray Eames. *The Power of Ten: A Flipbook*. New York: W. H. Freeman, 1998. (Short film with the same name available at http://powersof10.com/.)

Easterbrook, Gregg. *The Progress Paradox: How Life Gets Better While People Feel Worse*. New York: Random House, 2003.

Eckhardt, William. "War-Related Deaths Since 3000 B.C." *Bulletin of Peace Proposals* 22, no. 4 (1991): 437–43.

Edelman, Gerald M. *Bright Air, Brilliant Fire: On the Matter of the Mind*. New York: Basic Books, 1992.

——. *Neural Darwinism: The Theory of Neuronal Group Selection*. New York: Basic Books, 1987.

Edelman, Gerald M., and Giulio Tononi. *A Universe of Consciousness: How Matter Becomes Imagination*. New York: Basic Books, 2000.

Edgerton, David. *The Shock of the Old: Technology and Global History since 1900*. Oxford: Oxford University Press, 2007.

Edmonds, Eric V., and Nina Pavcnik. "Child Labor in the Global Economy." *Journal of Economic Perspectives* 19, no. 1 (2005): 199–220.

Einstein, Albert. *Ideas and Opinions*. Translated by Sonja Bargmann. New York: Wings Books, 1954.

Eisenstadt, Shmuel N., ed. *The Origins of Diversity of Axial Age Civilization*. Albany: State University of New York Press, 1986.

Eldredge, Niles. *Why We Do It: Rethinking Sex and the Selfish Gene*. New York: W. W. Norton, 2004.

Eldredge, Niles, and Steven Jay Gould. "Punctuated Equilibria: An Alternative to Phyletic Gradualism." In *Models in Paleobiology*, edited by T. J. M. Schopf and J. M. Thomas: 82–115. San Francisco: Freeman, Cooper, 1972.

Eliade, Mircea. *Shamanism: Archaic Techniques of Ecstasy*. Princeton, NJ: Princeton University Press, 1964.

Ericson, Richard V., and Aaron Doyle, eds. *Risk and Morality*. Toronto: University of Toronto Press, 2003.

Erikson, Erik. *Insight and Responsibility*. New York: W. W. Norton, 1964.

Esfandiary, F. M. *Are You a Transhuman?: Monitoring and Stimulating Your Personal Rate of Growth in a Rapidly Changing World*. New York: Warner Books, 1989.

Etcoff, Nancy L. *The Survival of the Prettiest: The Science of Beauty*. New York: Doubleday, 1999.

Ettinger, Robert C. *Man into Superman*. New York: St. Martin's Press, 1972.

——. *The Prospect of Immortality*. New York: Doubleday, 1964.

Fair, Charles M. *The Dying Self*. Middletown, CT: Wesleyan University Press, 1969.

Fawcett, William. *You Did What?: Mad Plans and Great Historical Disasters*. New York: Harper, 2004.

Fernández-Armesto, Felipe. *Humankind: A Brief History*. New York: Oxford University Press, 2004.

——. *Ideas That Changed the World*. New York: DK Publishing, 2003.

Ferris, Timothy. *Coming of Age in the Milky Way*. New York: William Morrow and Company, 1988.

——. *The Mind's Sky: Human Intelligence in a Cosmic Context*. New York: Bantam, 1992.

Feyerabend, Paul. *Against Method*. London: Verso, 1975.

Finer, Sammy E. *The History of Government from the Earliest Times*. 3 vols. New York: Oxford University Press, 1997.

Fisher, Helen. *Why We Love: The Nature and Chemistry of Romantic Love*. New York: Henry Holt and Company, 2004.

Fleck, Ludwik. *Genesis and Development of a Scientific Fact*. Translated by Thaddeus J. Trenn and Robert K. Merton. Chicago: University of Chicago Press, 1979 (originally published 1935).

Fletcher, Joseph. *The Ethics of Genetic Control*. New York: Doubleday Anchor, 1974.

Flexner, Stuart Berg, and Doris Flexner. *The Pessimist's Guide to History: An Irresistible Compendium of Catastrophes, Barbarities, Massacres and Mayhem from the Big Bang to the New Millennium*. New York: Harper, 2000.

Flynn, Daniel J. *Intellectual Morons: How Ideology Makes Smart People Fall for Stupid Ideas*. New York: Crown Forum, 2004.

Fodor, Jerry. *The Mind Doesn't Work That Way: The Scope and Limits of Computational Psychology*. Cambridge, MA: MIT Press, 2000.

Ford, Martin. *The Lights in the Tunnel: Automation, Accelerating Technology and the Economy of the Future*. CreateSpace Independent Publishing Platform: 2009.

Foucault, Michel. *The Order of Things: An Archaeology of the Human Sciences*. Translated by A. M. Smith. New York: Vintage Books, 1973.

Fowler, James. *Stages of Faith: The Psychology of Human Development and the Quest for Meaning*. New York: HarperCollins, 1995.

Frank, Robert H. *Passions within Reason: The Strategic Role of Emotions*. New York: W. W. Norton, 1988.

Frankl, Viktor E. *Man's Search for Meaning*. Boston: Beacon Press, 1992 (originally published 1942).

——. *Man's Search for Ultimate Meaning*. Cambridge, MA: Perseus Books, 1997.

Frederic, Harold. *The Damnation of Theron Ware*. Greenwich, CT: Fawcett Publications, Inc., 1962 (originally published 1896).

Freeman, Derek. *Margaret Mead and Samoa: The Making and Unmaking of an Anthropological Myth*. Cambridge, MA: Harvard University Press, 1983.

Freeman, Walter J. *How Brains Make Up Their Mind*. Berkeley: University of California Press, 2000.

——. *Societies of Brains: A Study in the Neuroscience of Love and Hate*. Hillsdale, NJ: Lawrence Erlbaum Associates, 1995.

Freud, Sigmund. *Beyond the Pleasure Principle*. Translated by James Strachey. New York: Bantam Books, 1967 (originally published 1920).

——. *Civilization and Its Discontents*. Translated by Joan Riviere. New York: Dover Publications, 1994 (originally published 1930).

——. *The Future of an Illusion*. Translated by W. D. Robson-Scott. New York: Anchor Books, 1964 (originally published 1927).

Friedman, Benjamin M. *The Moral Consequences of Economic Growth*. New York: Alfred A. Knopf, 2005.

Friedman, David. *The Machinery of Freedom: A Guide to Radical Capitalism*. La Salle, IL: Open Court, 1978.

Friedman, Milton. *Capitalism and Freedom*. Chicago: University of Chicago Press, 1962.

Friedman, Milton, and Rose Friedman. *Free to Choose: A Personal Statement*. New York: Harcourt Brace & Company, 1980.

Friedman, Richard Elliott. *The Disappearance of God*. Boston: Little, Brown and Company, 1995.

Friend, David et al., eds. *The Meaning of Life: Reflections in Words and Pictures on Why We Are Here*. Boston: Little, Brown and Company, 1991.

Fromkin, David. *The Way of the World: From the Dawn of Civilizations to the Eve of the Twenty-first Century*. New York: Alfred A. Knopf, 1999.

Fry, Iris. *The Emergence of Life on Earth: A Historical and Scientific Overview*. New Brunswick, NJ: Rutgers University Press, 2000.

Fukuyama, Francis. *The End of History and the Last Man*. New York: Free Press, 1992.

——. *Our Posthuman Future*. New York: Farrar, Straus and Giroux, 2002.

Futuyma, Douglas J. *Science on Trial: The Case for Evolution*. New York: Pantheon, 1983.

Gaddis, John Lewis. *The Landscape of History: How Historians Map the Past*. Oxford: Oxford University Press, 2002.

Galison, Peter. *Image and Logic: A Material Culture of Microphysics*. Chicago: University of Chicago Press, 1997.

Galton, David. *In Our Own Image: Eugenics and the Genetic Modification of People*. London: Little, Brown and Company, 2001.

Gamow, George. *One Two Three ... Infinity: Facts and Speculations of Science*. New York: Bantam, 1979.

Gardenfors, Peter. *How Homo Became Sapiens: On the Evolution of Thinking*. Oxford: Oxford University Press, 2004.

Gardner, Howard. *Multiple Intelligence: The Theory in Practice*. New York: Basic Books, 1993.

Gardner, James. *The Intelligent Universe*. Pompton Plains, NJ: New Page Books, 2007.

Garreau, Joel. *Radical Evolution: The Promise and Peril of Enhancing Our Minds, Our Bodies—and What It Means to Be Human*. New York: Doubleday, 2005.

Gasenbeek, Bert, and Babu Gogineni, eds. "International Humanist and Ethical Union 1952–2002: Past, present and future." http://www.iheu.org.

Gaylin, Willard. *Hatred: The Psychological Descent into Violence*. New York: Public Affairs, 2003.

Gazzaniga, Michael S. *The Ethical Brain*. New York: Dana Press, 2005.

——. *The Mind's Past*. Berkeley: University of California Press, 1998.

——. *The Social Brain: Discovering the Networks of the Mind*. New York: Basic Books, 1985.

Gelernter, David. *Machine Beauty: Elegance and the Heart of Technology*. New York: Basic Books, 1999.

Gell-Mann, Murray. *The Quark and the Jaguar: Adventures in the Simple and the Complex*. New York: W. H. Freeman, 1994.

Gellner, Ernest. *Plough, Sword, and Book: The Structure of Human History*. Chicago: University of Chicago Press, 1988.

——. *Postmodernism, Reason and Religion*. New York: Routledge, 1992.

Gigerenzer, Gerd. *Adaptive Thinking: Rationality in the Real World*. Oxford: Oxford University Press, 2000.

Gilovich, Thomas. *How We Know What Isn't So: The Fallibility of Human Reason in Everyday Life*. New York: Free Press, 1991.

Gilovich, Thomas, Dale Griffin, Daniel Kahneman, eds. *Heuristics and Biases: The Psychology of Intuitive Judgment*. Cambridge: Cambridge University Press, 2002.

Gimpel, Jean. *The Cathedral Builders*, translated by Teresa Waugh. New York: Grove Press, 1983.

Gintis, Herbert. "The Hitchhiker's Guide to Altruism: Genes, Culture, and the Internalization of Norms." *Journal of Theoretical Biology* 220, no. 4 (2003): 407–18.

Giovannoli, Joseph. *The Biology of Belief: How Our Biology Biases Our Beliefs and Perceptions*. Rosetta Press, Inc., 1999.

Glad, John. *Future Human Evolution: Eugenics in the Twenty-first Century*. Schuylkill Haven, PA: Hermitage Publishers, 2006.

Glover, Jonathan. *Humanity: A Moral History of the Twentieth Century*. New Haven: Yale University Press, 2000.

——. *What Sort of People Should There Be?* New York: Penguin, 1984.

Goertzel, Ben, and Stephan Vladimir Bugaj. *The Path to Posthumanity: 21st Century Technology and Its Radical Implications for Mind, Society, and Reality*. Bethesda, MD: Academica Press, 2006.

Goldberg, Elkhonon. *The Executive Brain: Frontal Lobes and the Civilized Mind*. Oxford: Oxford University Press, 2001.

Goldhill, Simon. *Love, Sex, and Tragedy: How the Ancient World Shapes Our Lives*. Chicago: University of Chicago Press, 2004.

Goldstein, Rebecca. *Incompleteness: The Proof and Paradox of Kurt Gödel*. New York: W. W. Norton, 2005.

Gonzalez, Guillermo, and Jay Richards. *The Privileged Planet: How Our Place in the Cosmos Is Designed for Discovery*. Washington, D.C.: Regnery Publishing, 2004.

Goodall, Jane. *The Chimpanzees of Gombe*. Cambridge, MA: Harvard University Press, 1986.

Goodwin, Brian. *How the Leopard Changed Its Spots: The Evolution of Complexity*. New York: Charles Scribner's Sons, 1994.

Goody, Jack. *The Domestication of the Savage Mind*. Cambridge: Cambridge University Press, 1977.

Gopnik, Alison, Andrew N. Meltzoff, and Patricia K. Kuhl. *The Scientist in the Crib: Minds, Brains, and How Children Learn*. New York: William Morrow and Company, 1999.

Gordon, Robert J. "Is U.S. Economic Growth Over? Faltering Innovation Confronts the Six Headwinds." *NBER Working Paper No. 18315*, August 2012.

Gosling, J. C. *The Greeks on Pleasure*. Oxford: Oxford University Press, 1982.

Gottlieb, Anthony. *The Dream of Reason: A History of Philosophy from the Greeks to the Renaissance*. New York: W. W. Norton, 2001.

Gould, Stephen Jay. *Full House: The Spread of Excellence from Plato to Darwin*. New York: Three Rivers Press, 1997.

——. *Wonderful Life: The Burgess Shale and the Nature of History*. New York: W. W. Norton, 1989.

Gould, Stephen Jay, and E. Vrba. "Exaptation: A Missing Term in the Science of Form." *Paleobiology* 8 (1982): 4–15.

Gould, Stephen Jay, and R. L. Lewontin. "The Spandrels of San Marco and the Panglossian Paradigm: A Critique of the Adaptationist Programme." *Proceedings of the Royal Society of London*, B 205 (1979): 581–98.

Grand, Steve. *Creation: Life and How to Make It*. Cambridge, MA: Harvard University Press, 2001.

Grant, Edward. *The Foundations of Modern Science in the Middle Ages: Their Religious, Institutional, and Intellectual Contexts*. Cambridge: Cambridge University Press, 1996.

Gray, Chris Hables. *Cyborg Citizen: Politics in the Posthuman Age*. New York: Routledge, 2001.

Green, Ronald M. *Babies by Design: The Ethics of Genetic Choice*. New Haven, CT: Yale University Press, 2008.

Greene, Brian. *The Fabric of the Cosmos: Space, Time, and the Texture of Reality*. New York: Alfred A. Knopf, 2004.

Greene, John C. *Science, Ideology, and World View: Essays in the History of Evolutionary Ideas*. Berkeley: University of California Press, 1981.

Greene, J. D., et al. "The Neural Bases of Cognitive Conflict and Control in Moral Judgment." *Neuron* 44 (2004): 389–400.

Greenfield, Susan, ed. *The Human Mind Explained: The Control Centre of the Living Machine*. London: Cassell, 1996.

Greenspan, Stanley, I. *The Growth of the Mind and the Endangered Origins of Intelligence*. Reading, MA: Addison Wesley Longman, 1997.

Greenstein, George. *The Symbiotic Universe: Life and the Cosmos in Unity*. New York: William Morrow & Co., 1988.

Gregory, Richard L., et al. *The Oxford Companion to the Mind*. Oxford: Oxford University Press, 1998.

Gress, David. *From Plato to NATO*. New York: Free Press, 1998.

Gribbin, John. *Deep Simplicity: Chaos, Complexity, and the Emergence of Life*. London: Penguin, 2004.

Griffin, Donald. *Animal Minds*. Chicago: University of Chicago Press, 1992.

——. *Animal Thinking*. Cambridge, MA: Harvard University Press, 1984.

Grinnell, Frederick. *The Scientific Attitude*. New York: Guilford Press, 1992.

Grossman, Dave. *On Killing: The Psychological Cost of Learning to Kill in War and Society*. Boston: Little, Brown and Company, 1995.

Grossman, Gene M., and Alan B. Krueger. "Economic Growth and the Environment." *Quarterly Journal of Economics* 110, no. 2 (1995): 353–77.

Grove, Andrew. *Only the Paranoid Survive: How to Exploit the Crisis Points That Challenge Every Company*. New York: Currency, 1996.

Guigley, Carroll. *Evolution of Civilizations*. Indianapolis, IN: Liberty Fund, 1979.

Guth, Alan. *The Inflationary Universe: The Quest for a New Theory of Cosmic Origins*. Cambridge, MA: Perseus Books, 1997.

Guthrie, Stewart Elliott. *Faces in the Clouds: A New Theory of Religion*. Oxford: Oxford University Press, 1993.

Gutkind, Lee. *Almost Human*. New York: W. W. Norton, 2007.

Guttmann, Allen. *From Ritual to Record: The Nature of Modern Sports*. New York: Colombia University Press, 1978.

Habermas, Jurgen. *The Future of Human Nature*. Cambridge: Polity Press, 2003.

Haeckel, Ernst. *The Riddle of the Universe at the Close of the Nineteenth Century.* New York: Harper & Brothers, 1900.

Hafner, Katie, and Matthew Lyon. *Where Wizards Stay Up Late: The Origins of the Internet*. New York: Simon & Schuster, 1996.

Haldane, J. B. S. *Daedalus, or, Science and the Future.* London: Kegan Paul, Trench, Trubner and Co., 1923. http://www.cscs.umich.edu/~crshalizi/Daedalus.html.

——. *The Inequality of Man, and Other Essays.* London: Chatto & Windus, 1932.

Hamer, Dean. *The God Gene: How Faith Is Hardwired into Our Genes*. New York: Doubleday, 2004.

Hamilton, William D. "A Review of *Dysgenics: Genetic Deterioration in Modern Populations*." *Annals of Human Genetics* 64, issue 4 (2000): 363–74.

Hansell, Gregory R., and William Grassie, eds. *H+/-: Transhumanism and Its Critics.* Bloomington, IN: Xlibris, 2011.

Hanson, Robin. "Economic Growth Given Machine Intelligence." 1998. http://hanson.gmu.edu/aigrow.pdf.

Haraway, Donna. "A Cyborg Manifesto: Science, Technology, and Socialist-Feminism in the Late Twentieth Century." In *Simians, Cyborgs and Women: The Reinvention of Nature.* New York: Routledge, 1991: 149–81.

Harold, Franklin M. *The Way of the Cell: Molecules, Organism, and the Order of Life.* New York: Oxford University Press, 2001.

Harré, Rom. *Problems of Scientific Revolution: Progress and Obstacles to Progress in the Sciences.* Oxford: Clarendon Press, 1975.

Harris, John. *Enhancing Evolution: The Ethical Case for Making Better People.* Princeton: Princeton University Press, 2007.

Harris, Marvin. *Cows, Pigs, Wars and Witches: The Riddles of Culture.* New York: Random House, 1974.

Harris, Sam. *The End of Faith: Religion, Terror, and the Future of Reason.* New York: W. W. Norton, 2004.

Harrison, Edward. *Masks of the Universe: A Physicist's Remarkable Portrayal of Mankind's Search for Meaning in the Universe.* New York: Collier Books, 1985.

Hart, Jeffrey. *Smiling through the Cultural Catastrophe: Toward the Revival of Higher Education*. New Haven, CT: Yale University Press, 2001.

Hart, Michael H. *The 100: A Ranking of the Most Influential Persons in History*. New York: Citadel Press, 1978.

Harth, Erich. *The Creative Loop: How the Brain Makes a Mind*. Reading, MA: Addison-Wesley Publishing Company, 1993.

——. *Dawn of a Millennium: Beyond Evolution and Culture*. Boston: Little, Brown and Company, 1990.

Haught, John F. *Deeper Than Darwin: The Prospect for Religion in the Age of Evolution*. Boulder, CO: Westview Press, 2004.

——. *God After Darwin: A Theology of Evolution*. Boulder, CO: Westview Press, 1999.

Hauser, Marc D. *Moral Minds: How Nature Designed Our Universal Sense of Right and Wrong*. New York: Ecco, 2006.

——. *Wild Minds: What Animals Really Think*. New York: Henry Holt and Company, 2000.

Hawking, Stephen W. *A Brief History of Time: From the Big Bang to Black Holes*. New York: Bantam Books, 1988.

Hawkins, Jeff. *On Intelligence: How a New Understanding of the Brain Will Lead to the Creation of Truly Intelligent Machines*. New York: Henry Holt and Company, 2004.

Hawley, Amos. *Human Ecology: A Theoretical Essay*. Chicago: University of Chicago Press, 1986.

Hayek, Friedrich A. *The Road to Serfdom*. Chicago: University of Chicago Press, 1944.

——. *Studies in Philosophy, Politics and Economics*. New York: Simon & Schuster, 1967.

Hayles, Katherine. *How We Become Post-human: Virtual Bodies in Cybernetics, Literature, and Informatics*. Chicago: University of Chicago Press, 1999.

Hegel, G. W. F. *The Phenomenology of Spirit*. Translated by A. V. Miller. Oxford: Oxford University Press, 1977.

Heinberg, Richard. *Cloning the Buddha: The Moral Impact of Biotechnology*. Wheaton, IL: Quest Books, 1999.

Helminiak, Daniel A. *Spiritual Development: An Interdisciplinary Study*. Chicago: Loyola University Press, 1987.

Henderson, Lawrence J. *The Fitness of the Environment*. Boston: Beacon Press, 1959 (originally published 1913).

Henig, R. M. *Pandora's Baby: How the First Test Tube Babies Sparked the Reproductive Revolution*. Boston: Houghton Mifflin, 2004.

Herman, Arthur. *The Idea of Decline in Western History*. New York: Free Press, 1997.

Hilgard, E. R. *Divided Consciousness: Multiple Controls in Human Thought and Action*. New York: John Wiley & Sons, 1977.

Hirschman, Albert O. *The Passions and the Interests: Political Arguments for Capitalism Before Its Triumph*. Princeton: Princeton University Press, 1977.

Hitchens, Christopher. *God Is Not Great: How Religion Poisons Everything*. New York: Twelve, 2007.

Hoffer, Eric. *The Ordeal of Change*. New York: Harper & Row, 1963.

——. *The True Believer.* New York: Harper & Row, 1952.

Hoffman, Donald. *Visual Intelligence: How We Create What We See*. New York: W. W. Norton, 1998.

Hofstadter, Douglas R. *Gödel, Escher, Bach: An Eternal Golden Braid.* New York: Vintage Books, 1979.

——. *Metamagical Themas.* New York: Basic Books, 1985.

Hofstadter, Douglas R., and Daniel C. Dennett, eds. *The Mind's I: Fantasies and Reflections on Self and Soul.* New York: Bantam Books, 1981.

Holloway, Carson. *The Right Darwin?: Evolution, Religion, and the Future of Democracy.* Dallas: Spence Publishing Company, 2008.

Holmes, Richard. *Acts of War: The Behavior of Men in Battle*. New York: Free Press, 1989.

Hook, Sidney. *The Hero in History*. Boston: Beacon Press, 1943.

Hopkins, Jasper. *Nicholas of Cusa on Learned Ignorance: A Translation and an Appraisal of De Docta Ignorantia*. Minneapolis: Arthur J. Banning Press, 1985.

Hopkins, Thomas J. *The Hindu Religious Tradition*. Encino, CA: Dickenson Publishing, 1971.

Horgan, John. *The End of Science: Facing the Limits of Knowledge in the Twilight of the Scientific Age*. Reading, MA: Helix Books, 1996.

Hrdy, Sarah Blaffer. *Mother Nature: Maternal Instincts and How They Shape the Human Species*. New York: Ballantine, 2000.

Hubbard, Barbara Marx. *Conscious Evolution: Awakening the Power of Our Social Potential*. Novato, CA: New World Library, 1997.

Hughes, Howard. *Sensory Exotica: A World Beyond Human Experience*, Cambridge, MA: MIT Press, 1999.

Hughes, James. *Citizen Cyborg*. Cambridge, MA: Westview Press, 2004.

——. "The Compatibility of Religious and Transhumanist Views of Metaphysics, Suffering, Virtue, and Transcendence in an Enhanced Future." 2007. http://ieet.org/archive/20070326-Hughes-ASU-H + Religion.pdf.

Humphrey, Nicholas. *Leaps of Faith: Science, Miracles, and the Search for Supernatural Consolation*. New York: Basic Books, 1996.

——. *The Mind Made Flesh: Essays from the Frontiers of Psychology and Evolution*. Oxford: Oxford University Press, 2002.

Huxley, Aldous. *Brave New World*. New York: Perennial Classics, 1998 (originally published 1932).

——. *The Perennial Philosophy.* New York: Harper Colophon Books, 1970 (originally published 1945).

Huxley, Julian. *Religion without Revelation*. New York: Mentor Books, 1957 (originally published 1928).

Huxley, Thomas H. *Evolution and Ethics and Other Essays*. London: Macmillan, 1894.

Ierley, Merritt. *Wondrous Contrivances: Technology at the Threshold*. New York: Clarkson Potter, 2002.

The Immortality Institute. *The Scientific Conquest of Death: Essays on Infinite Lifespans.* Buenos Aires: LibrosEnRed, 2004.

International Theological Commission. "Communion and Stewardship: Human Persons Created in the Image of God." http://www.catholicculture.org/docs/doc_view.cfm?recnum=6664#3a.

Jacob, Francois. "Evolution and Tinkering." *Science* 196, no. 4295 (1977): 1161–66.

——. *The Logic of Life: A History of Heredity.* Translated by Betty E. Spillmann. New York: Pantheon, 1974.

Jacob, Margaret C. *The Cultural Meaning of the Scientific Revolution.* Philadelphia: Temple University Press, 1988.

Jacobs, Jane. *Dark Age Ahead.* New York: Random House, 2004.

James, William. *The Varieties of Religious Experience.* New York: Touchstone, 1997 (originally published 1902).

——. *The Will to Believe, and Other Essays in Popular Philosophy and Human Immortality.* New York: Dover Publications, 1956 (originally published 1896).

Jantsch, Erich. *The Self-Organizing Universe: Scientific and Human Implications of the Emerging Paradigm of Evolution.* New York: Pergamon Press, 1980.

Jaspers, Karl. *The Origin and Goal of History.* London: Routledge and Kegan Paul, 1953.

——. *The Great Philosophers.* New York: Harcourt, Brace & World, 1955.

Jastrow, Robert. *Until the Sun Dies.* New York: W. W. Norton, 1977.

Jaynes, Julian. *The Origin of Consciousness in the Breakdown of the Bicameral Mind.* Boston: Houghton Mifflin, 1976.

Johnson, George. *Fire in the Mind.* New York: Alfred A. Knopf, 1995.

Johnson, Phillip. *Darwin on Trial.* Downers Grove, IL: InterVarsity Press, 1991.

Johnson, Steve. *Emergence: The Connected Lives of Ants, Brains, Cities, and Software.* New York: Charles Scribner's Sons, 2001.

——. *Mind Wide Open: Your Brain and the Neuroscience of Everyday Life.* New York: Charles Scribner's Sons, 2005.

Joseph, Rhawn. *The Transmitter to God: The Limbic System, the Soul, and Spirituality.* San Jose, CA: California University Press, 2000.

Joseph, Rhawn, ed. *NeuroTheology: Brain, Science, Spirituality, Religious Experience.* San Jose, CA: California University Press, 2003.

Joy, Bill. "Why the Future Doesn't Need Us: Our Most Powerful 21st-Century Technologies—Robotics, Genetic Engineering, and Nanotech—Are Threatening to Make Humans an Endangered Species." *Wired*, April 2000.

Judson, H. F. *The Eighth Day of Creation: The Makers of the Revolution in Biology.* New York: Simon & Schuster, 1979.

Kaczynski, Theodore. "The Unabomber Manifesto: Industrial Society and Its Future." *Washington Post*, September 19, 1995.

Kahneman, Daniel. *Thinking, Fast and Slow.* New York: Farrar, Straus & Giroux, 2013.

Kahneman, Daniel, Ed Diener, and Norbert Schwarz, eds. *Well-Being: The Foundations of Hedonic Psychology.* New York: Russell Sage Foundation, 1999.

Kaku, Michio. *Parallel Worlds: A Journey Through Creation, Higher Dimensions, and the Future of the Cosmos.* New York: Doubleday, 2005.

Kaplan, Robert. *Warrior Politics: Why Leadership Demands a Pagan Ethos.* New York: Random House, 2002.

Kass, Leon. *The Beginning of Wisdom: Reading Genesis.* New York: Free Press, 2003.

——. "L'Chaim and Its Limits: Why Not Immortality?" *First Things.* May 2001.

——. *Life, Liberty, and the Defense of Dignity.* Washington, D.C.: The AEI Press, 2002.

——. "Moral Meaning of Genetic Technology." *Commentary* 108, no. 2 (1999): 32–38.

Kass, Leon, ed. *Reproduction and Responsibility: The Regulation of New Biotechnologies.* Washington, D.C.: President's Council on Bioethics, 2004.

Kass, Leon, and James Q. Wilson. *The Ethics of Human Cloning.* Washington, D.C.: The AEI Press, 1998.

Kauffman, Stuart A. *At Home in the Universe: The Search of the Laws of Self-Organization and Complexity.* Oxford: Oxford University Press, 1995.

——. *Investigations.* Oxford: Oxford University Press, 2000.

Keeley, Lawrence H. *War Before Civilization: The Myth of the Peaceful Savage.* Oxford: Oxford University Press, 1997.

Keller, Evelyn Fox. *The Century of the Gene.* Cambridge, MA: Harvard University Press, 2000.

Keller, Laurent, ed. *Levels of Selection in Evolution.* Princeton, NJ: Princeton University Press, 1999.

Kelly, Kevin. *Out of Control: The Rise of Neo-biological Civilization.* Reading, MA: Addison-Wesley, 1994.

——. *What Technology Wants.* New York: Viking, 2010.

Kessler, Andy. *How We Got Here: A Slightly Irreverent History of Technology and Markets.* New York: HarperCollins, 2005.

Kevles, Daniel J. *In the Name of Eugenics: Genetics and the Uses of Human Heredity.* Cambridge, MA: Harvard University Press, 1997.

Keynes, John Maynard. "Economic Possibilities for Our Grandchildren." In *Essays in Persuasion.* New York: W. W. Norton, 1991 (originally published 1930).

Kierkegaard, Søren. *Fear and Trembling.* Translated by Walter Lowrie. Princeton, NJ: Princeton University Press, 1941 (originally published 1843).

Kiki, Albert Maori. *Kiki: Ten Thousand Years in a Lifetime: A New Guinea Autobiography.* New York: Frederick A. Praeger, 1968.

Kindleberger, Charles P. *Manias, Panics, and Crashes: A History of Financial Crises.* New York: John Wiley & Sons, 1978.

Kingdon, Jonathan. *Self-Made Man: Human Evolution from Eden to Extinction?* New York: John Wiley & Sons, 1993.

Kirsch, Jonathan. *A History of the End of the World: How the Most Controversial Book in the Bible Changed the Course of Western Civilization.* New York: HarperSanFrancisco, 2006.

Kirshner, Robert. *The Extravagant Universe: Exploding Stars, Dark Energy, and the Accelerating Cosmos.* Princeton, NJ: Princeton University Press, 2002.

Kitcher, Philip. *Living with Darwin: Evolution, Design, and the Future of Faith.* New York: Oxford University Press, 2006.

Kitto, Humphrey Davy Findley. *The Greeks.* New York: Penguin, 1951.

Knoll, Andrew. *Life on a Young Planet: The First Three Billion Years of Evolution on Earth.* Princeton, NJ: Princeton University Press, 2003.

Knuth, Donald E. *Things a Computer Scientist Rarely Talks About.* Stanford, CA: Center for the Study of Language and Information, 2001.

Koestler, Arthur. *The Sleepwalkers: A History of Man's Changing Vision of the Universe.* New York: Penguin, 1991 (originally published 1959).

Kohlberg, Laurence. *The Philosophy of Moral Development: Moral Stages and the Idea of Justice.* San Francisco: Harper & Row, 1981.

Konner, Melvin. *The Tangled Wing: Biological Constraints on the Human Spirit.* New York: Henry Holt and Company, 2002.

Kosco, Bart. *Fuzzy Thinking.* New York: Hyperion, 1993.

Koyre, Alexandre. *From the Closed World to the Infinite Universe.* Baltimore, MD: John Hopkins University Press, 1957.

Krauss, Lawrence. *Atom: An Odyssey from the Big Bang to Life on Earth ... and Beyond.* Boston: Little, Brown and Company, 2001.

Kuhn, Thomas S. *The Structure of Scientific Revolution.* Chicago: University of Chicago Press, 1996 (originally published 1962).

Kundera, Milan. *The Unbearable Lightness of Being: A Novel.* Translated by Michael Henry Heim. New York: Harper & Row, 1984.

Kurzweil, Ray. *The Age of Spiritual Machines: When Computers Exceed Human Intelligence.* New York: Penguin Books, 1999.

———. *The Singularity Is Near: When Humans Transcend Biology.* New York: Viking, 2005.

Kuxe, Kjell, and Luigi F. Agnati, eds. *Volume Transmission in the Brain.* New York: Raven Press, 1991.

Kwok, Man-Ho, et al., translation. *Tao Te Ching.* New York: Barnes & Noble Books, 1994.

Laidler, Keith. *The Harmonious Universe: The Beauty and Unity of Scientific Understanding.* Amherst, NY: Prometheus Books, 2004.

Lakoff, George, and Mark Johnson. *Philosophy in the Flesh: The Embodied Mind and Its Challenge to Western Thought.* New York: Basic Books, 1999.

Landes, David S. *The Wealth and Poverty of Nations: Why Some Are So Rich and Some So Poor.* New York: Norton, 1998.

Lane, Nick. *Oxygen: The Molecule That Made the World.* Oxford: Oxford University Press, 2000.

———. *Power, Sex, Suicide: Mitochondria and the Meaning of Life.* Oxford: Oxford University Press, 2005.

Lanier, Jaron. "A Future That Loves Us: An Optimistic One Thousand Year Scenario." http://www.advanced.org/jaron/lovely/default.htm.

———. "One Half of a Manifesto." *Edge.* http://www.edge.org/documents/archive/edge74.html.

Larson, Edward J. *Evolution: The Remarkable History of a Scientific Theory.* New York: Random House, 2004.

Last, Jonathan. *What to Expect When No One's Expecting: America's Coming Demographic Disaster*. San Francisco: Encounter Books, 2013.

LaTorra, Michael. "Trans-Spirit: Religion, Spirituality, and Transhumanism." *Journal of Evolution and Technology* 14, issue 1 (2005): 41–55.

Laurent, Clint. *Tomorrow's World: A Look at the Demographic and Socio-economic Structure of the World in 2032*. New York: John Wiley & Sons, 2013.

Layard, Richard. *Happiness: Lessons from a New Science*. New York: Penguin, 2005.

Leakey, Richard, and Roger Lewin. *The Sixth Extinction: Patterns of Life and the Future of Humankind*. New York: Anchor Books, 1995.

Lederman, Leon, and Christopher Hill. *Symmetry and the Beautiful Universe*. Amherst, NY: Prometheus Books, 2005.

LeDoux, Joseph E. *The Emotional Brain: The Mysterious Underpinnings of Emotional Life*. New York: Simon & Schuster, 1996.

———. *The Synaptic Self: How Our Brains Become Who We Are*. New York: Penguin Books, 2002.

Leslie, John. *The End of the World: The Science and Ethics of Human Extinction*. New York: Routledge, 1996.

LeVay, Simon. *The Sexual Brain*. Cambridge, MA: MIT Press, 1994.

Levitt, Norman. *Prometheus Bedeviled: Science and the Contradictions of Contemporary Culture*. New Brunswick, NJ: Rutgers University Press, 1999.

Levy, Neil. *Neuroethics: Challenges for the 21st Century*. New York: Cambridge University Press, 2007.

Lewis, Bernard. *What Went Wrong? The Clash Between Islam and Modernity in the Middle East*. New York: Harper Perennial, 2002.

Lewis, C. S. *The Problem of Pain*. New York: HarperCollins, 2001.

Lewis, Thomas, Fari Amini, and Richard Lannon. *A General Theory of Love*. New York: Random House, 2000.

Lin, Yotang, ed. *The Wisdom of China and India*. New York: The Modern Library, 1942.

Lindberg, David. *The Beginnings of Western Science: The European Scientific Tradition in Philosophical, Religious, and Institutional Context, 600 B.C. to A.D. 1450*. Chicago: University of Chicago Press, 1992.

Lloyd, Seth. *Programming the Universe: A Quantum Computer Scientist Takes On the Cosmos*. New York: Alfred A. Knopf, 2006.

Loevinger, Jane. *Ego Development*. San Francisco, CA: Jossey-Bass, 1976.

Lorenz, Konrad. *On Aggression*. Translated by Marjorie Kerr Wilson. New York: Harcourt, Brace & World, 1963.

Lovejoy, Arthur O. *The Great Chain of Being: A Study of the History of an Idea*. Cambridge, MA: Harvard University Press, 1936.

Lovejoy, Arthur, George Boas, et al. *Primitivism and Related Ideas in Antiquity*. Baltimore, MD: Johns Hopkins University Press, 1935.

Lowie, Robert H. *Are We Civilized? Human Culture in Perspective*. New York: Harcourt, Brace and Company, 1929.

Lucretius. "Does Neuroscience Refute Ethics?" http://www.mises.org/story/1943.

Lurquin, Paul F. *The Origins of Life and the Universe*. New York: Columbia University Press, 2003.

Lykken, David T. *The Antisocial Personalities*. Hillsdale, NJ: Lawrence Erlbaum Associates, 1995.

——. *Happiness: The Nature and Nurture of Joy and Contentment*. New York: St. Martin's Press, 2000.

Lynch, Michael, and John S. Conery. "The Origins of Genome Complexity." *Science* 302, no. 5649 (2003): 1401–4.

Lynn, Richard. *Dysgenics: Genetic Deterioration in Modern Populations*. Westport, CT: Praeger, 1996.

——. *Dysgenics: A Reassessment*. Westport, CT: Praeger, 2001.

MacHale, Des. *Wit*. Kansas City, MO: Andrews McMeel Publishing, 2003.

Mallove, Eugene. *The Quickening Universe: Cosmic Evolution and Human Destiny*. New York: St. Martin's Press, 1987.

Mandelbaum, Michael. *The Ideas That Conquered the World: Peace, Democracy, and Free Markets in the Twenty-first Century*. New York: PublicAffairs, 2002.

Mandler, George. *Mind and Body: Psychology of Emotion and Stress*. New York: W.W. Norton, 1984.

Manson, Neil A., ed. *God and Design: The Teleological Argument and Modern Science*. New York: Routledge, 2003.

Margulis, Lynn, and Rene Fester, eds. *Symbiosis as a Source of Evolutionary Innovation*. Cambridge, MA: MIT Press, 1991.

Margulis, Lynn, and Dorion Sagan. *What Is Life?* New York: Peter Nevraumont, Inc., 1995.

Markley, O. W., and Willis W. Harman, eds. *Changing Images of Man*. New York: Pergamon Press, 1982.

Marar, Ziyad. *The Happiness Paradox*. London: Reaktion Books, 2003.

Marcus, Gary. *The Birth of the Mind: How a Tiny Number of Genes Creates the Complexities of Human Thought*. New York: Basic Books, 2003.

Martenson, Chris. *The Crash Course: The Unsustainable Future of Our Economy, Energy, and Environment*. New York: John Wiley & Sons, 2011.

Martin, James. *After the Internet: Alien Intelligence*. Washington, D.C.: Capital Press, 2000.

Maslow, Abraham. *Toward a Psychology of Being*. New York: Van Nostrand Reinhold Company, 1968.

Masson, Jeffrey Moussaieff, and Susan McCarthy. *When Elephants Weep: The Emotional Lives of Animals*. New York: Delacorte Press, 1995.

Maturana, Humberto R., and Francisco Varela. *The Tree of Knowledge: The Biological Roots of Human Understanding*. Boston: Shambhala, 1987.

Maurice, Charles, and Charles Smithson. *The Doomsday Myth: 10,000 Years of Economic Crises*. Stanford, CA: Hoover Institution Press, 1984.

Maxwell, Kenneth. *The Sex Imperative: An Evolutionary Tale of Sexual Survival*. New York: Plenum Press, 1994.

May, Robert. "Biological Diversity in a Crowded World: Past, Present and Likely Future." *Current Science* 82, no. 11 (2002): 1325–31.

Maynard Smith, John, and Eörs Szathmáry. *The Major Transitions in Evolution.* Oxford: Oxford University Press, 1998.

——. *The Origins of Life: From the Birth of Life to the Origin of Language.* Oxford: Oxford University Press, 1999.

Mayr, Ernst. "Change of Genetic Environment and Evolution." In J. Huxley, A. C. Hardy, and E. B. Ford, eds., *Evolution as a Process*, 157–80. London: Allen and Unwin, 1954.

——. *Toward a New Philosophy of Biology.* Cambridge, MA: Harvard University Press, 1988.

——. *What Evolution Is.* New York: Basic Books, 2001.

Mazlish, Bruce. *The Fourth Discontinuity: The Co-evolution of Humans and Machines.* New Haven, CT: Yale University Press, 1993.

McGrath, Alister. *The Twilight of Atheism: The Rise and Fall of Disbelief in the Modern World.* New York: Doubleday, 2004.

McIntosh, Steve. *Evolution's Purpose: An Integral Interpretation of the Scientific Story of Our Origins.* New York: Select Books, 2012.

McKibben, Bill. *The End of Nature.* New York: Random House, 1989.

——. *Enough: Staying Human in an Engineered Age.* New York: Times Books, 2003.

McLuhan, Marshall, and Fiore Quentin. *The Medium Is the Message.* New York: Random House, 1967.

McMichael, Tony. *Human Frontiers, Environments and Disease.* Cambridge: Cambridge University Press, 2001.

McNeill, John R. *Something New Under the Sun: An Environmental History of the Twentieth-Century World.* New York: W. W. Norton, 2000.

McNeill, William H. *The Rise of the West.* Chicago: University of Chicago Press, 1963.

McPhee, John. *The Control of Nature.* New York: Farrar, Straus & Giroux, 1989.

McWhorter, John. *The Power of Babel: A Natural History of Language.* New York: W. H. Freeman, 2002.

Meadows, Donella, Jorgen Randers, and Dennis Meadows. *Limits to Growth: The 30-Year Update.* Post Mills, VT: Chelsea Green Publishing Company, 2004.

Merson, John. *The Genius That Was China: East and West in the Making of the Modern World.* Woodstock, NY: The Overlook Press, 1990.

Midgley, Mary. *Beast and Man: The Roots of Human Nature.* Ithaca, NY: Cornell University Press, 1978.

——. *Science as Salvation: A Modern Myth and Its Meaning.* London: Routledge, 1992.

Miles, Jack. *God: A Biography.* New York: Alfred A. Knopf, 1995.

Miller, George A. "The Magical Number Seven, Plus or Minus Two: Some Limits on Our Capacity for Processing Information." *Psychological Review* 63 (1956): 81–97.

Miller, James, ed. *An Evolving Dialogue: Theological and Scientific Perspectives on Evolution.* Harrisburg, PA: Trinity Press International, 2001.

Miller, Henry, and Gregory Conko. *The Frankenfood Myth: How Protest and Politics Threaten the Biotech Revolution*. Westport, CT: Praeger Publishers, 2004.

Miller, James. *Singularity Rising: Surviving and Thriving in a Smarter, Richer, and More Dangerous World*. Dallas: BenBella Books, 2012.

Miller, Kenneth R. *Finding Darwin's God: A Scientist's Search for Common Ground Between God and Evolution*. New York: Cliff Street Books, 1999.

Minsky, Marvin. *The Society of Mind*. New York: Simon & Schuster, 1986.

Mintzberg, Henry. *The Rise and Fall of Strategic Planning*. New York: Free Press, 1994.

Mises, Ludwig von. *Human Action: A Treatise on Economics*. Chicago: Henry Regnery, 1996 (originally published 1949).

Mithen, Steven. *After the Ice: A Global Human History, 20,000–5000 BC*. Cambridge, MA: Harvard University Press, 2004.

——. *The Pre-history of the Mind: The Cognitive Origins of Art and Science*. London: Thames and Hudson, 1996.

Mokyr, Joel. *The Lever of Riches: Technological Creativity and Economic Progress*. New York: Oxford University Press, 1990.

Moltmann, Jurgen. *The Theology of Hope*. Translated by James W. Leitch. New York: Harper & Row, 1975.

Monod, Jacques. *Chance and Necessity: An Essay on the Natural Philosophy of Modern Biology*. Translated by Austryn Wainhouse. New York: Alfred A. Knopf, 1971.

Montaigne, Michel de. *Essays*. Translated by John M. Cohen. London: Penguin Classics, 1993.

Moore, Stephen, and Julian L. Simon. *It's Getting Better All the Time: 100 Greatest Trends of the Last 100 Years*. Washington, D.C.: Cato Institute, 2000.

Morain, Lloyd, and Mary Morain. *Humanism as the Next Step*. Amherst, NY: Humanist Press, 1998 (originally published 1954).

Moravec, Hans. *Mind Children: The Future of Robot and Human Intelligence*. Cambridge, MA: Harvard University Press, 1988.

——. *Robot: Mere Machine to Transcendent Mind*. Oxford: Oxford University Press, 1999.

More, Max. "The Extropian Principles: A Transhumanist Declaration." http://www.extropy.org/principles.htm.

——. "The Proactionary Principle." http://www.maxmore.com/proactionary.htm.

More, Max, and Natasha Vita-More, eds. *The Transhumanist Reader: Classical and Contemporary Essays on the Science, Technology, and Philosophy of the Human Future*. Malden, MA: Wiley-Blackwell, 2013.

More, Thomas. *Utopia*. Norwalk, CT: Easton Press, 1963.

Morowitz, Harold J. *Cosmic Joy and Local Pain: Musings of a Mystic Scientist*. New York: Charles Scribner's Sons, 1987.

Morris, Desmond. *The Naked Ape*. New York: McGraw-Hill, 1967.

Morris, Ian, *The Measure of Civilization: How Social Development Decides the Fate of Nations*. Princeton, NJ: Princeton University Press, 2013.

———. *Why the West Rules, for Now: The Patterns of History, and What They Reveal about the Future*. New York: Farrar, Straus & Giroux, 2010.

Morrison, Reg. *The Spirit in the Gene: Humanity's Proud Illusion and the Laws of Nature*. Ithaca, NY: Cornell University Press, 1999.

Muehlhauser, Luke. *Facing the Intelligence Explosion*. Berkeley: Machine Intelligence Research Institute, 2013.

Mueller, John. *Retreat from Doomsday: The Obsolescence of Major War*. New York: Basic Books, 1989.

Mumford, Lewis. *The Myth of the Machine: The Pentagon of Power*. New York: Harcourt, Brace and World, 1970.

———. *Technics and Civilization*. New York: Harcourt, Brace and World, 1934.

Murchie, Guy. *The Seven Mysteries of Life: An Exploration in Science and Philosophy*. Boston: Houghton Mifflin, 1978.

Murray, Charles. "For God's Eye: The Surprising Role of Christianity in Great Human Accomplishment." *American Enterprise*, October 2003, 22–26.

———. *Human Accomplishment: The Pursuit of Excellence in the Arts and Sciences, 800 B.C. to 1950*. New York: Harper Perennial, 2003.

Myers, David. *The Pursuit of Happiness: Who Is Happy—and Why*. New York: William Morrow and Company, 1992.

Naam, Ramez. *More Than Human: Embracing the Promise of Biological Enhancement*. New York: Broadway Books, 2005.

Negroponte, Nicholas. *Being Digital*. New York: Alfred A Knopf, 1995.

Neisser, Ulric, et al. *The Rising Curve: Long-Term Gains in IQ and Related Measures*. Washington, D.C.: American Psychological Association, 1998.

Nelson, Richard R., and Sidney G. Winter. *An Evolutionary Theory of Economic Change*. Cambridge, MA: Belknap Press, 1982.

Nettle, Daniel. *Happiness: The Science behind Your Smile*. Oxford: Oxford University Press, 2005.

Neustadt, Richard, and Ernest May. *Thinking in Time: The Uses of History for Decision Makers*. New York: Macmillan, 1986.

Newberg, Andrew, Eugene G. D'Aquili, and Vince Rause. *Why God Won't Go Away*. New York: Ballantine Books, 2001.

Nietzsche, Friedrich. *Beyond Good and Evil / The Genealogy of Morals*. New York: Barnes & Noble, 1996a (originally published 1886 and 1887, respectively).

———. *Human, All Too Human: A Book for Free Spirits*. Cambridge: Cambridge University Press, 1996b (originally published 1878).

———. *The Portable Nietzsche*. Edited and translated by Walter Kaufmann. New York: Penguin Books, 1976 (originally published by Viking Press, 1954).

Nisbet, Robert. *History of the Idea of Progress*. New York: Basic Books, 1980.

Nisbett, Richard E. *The Geography of Thought: How Asians and Westerners Think Differently . . . And Why*. New York: Free Press, 2003.

Nitecki, Matthew H. *Evolutionary Progress*. Chicago: University of Chicago Press, 1988.

Noble, David. *The Religion of Technology: The Divinity of Man and the Spirit of Invention*. New York: Alfred A. Knopf, 1998.

Norretranders, Tor. *The User Illusion: Cutting Consciousness Down to Size.* Translated by Jonathan Sydenham. New York: Viking, 1998 (originally published 1991).

North, Douglass, and Robert P. Thomas. *The Rise of the Western World.* Cambridge: Cambridge University Press, 1973.

Novak, Michael. *On Two Wings: Humble Faith and Common Sense at the American Founding.* San Francisco: Encounter Books, 2002.

Nowak, Martin A. *Evolutionary Dynamics: Exploring the Equations of Life.* Cambridge, MA: Belknap Press, 2006.

Nozick, Robert. *Anarchy, State, and Utopia.* New York: Basic Books, 1974.

——. *Invariances: The Structure of the Objective World.* Cambridge, MA: Belknap Press, 2001.

Nüsslein-Volhard, Christiane. *Coming to Life: How Genes Drive Development.* San Diego: Kales Press, 2006.

O'Clery, Conor. *The Billionaire Who Wasn't: How Chuck Feeney Secretly Made and Gave Away a Fortune.* New York: PublicAffairs, 2007.

Oderberg, David S. *Moral Theory: A Non-consequentialist Approach.* Malden, MA: Blackwell Publishers, 2000.

Offer, Avner. *The Challenge of Affluence: Self-control and Well-being in the United States and Britain since 1950.* Oxford: Oxford University Press, 2006.

Olivelle, Patrick, trans. *Upanisads.* Oxford: Oxford University Press, 1996.

Olson, Mancur. *The Logic of Collective Action: Public Goods and the Theory of Groups.* Cambridge, MA: Harvard University Press, 1965.

Ormerod, Paul. *Why Most Things Fail: Evolution, Extinction and Economics.* London: Faber and Faber, 2005.

Osborne, Roger. *Civilization: A New History of the Western World.* New York: Pegasus Books, 2006.

Pagels, Heinz. *The Dreams of Reason: The Computer and the Rise of the Sciences of Complexity.* New York: Simon & Schuster, 1988.

Parker, Andrew. *In the Blink of an Eye.* Cambridge, MA: Perseus Publishing, 2003.

Patterson, Orlando. *Freedom: Freedom in the Making of Western Culture.* New York: Basic Books, 1991.

Paul, Gregory, and Earl Cox. *Beyond Humanity: CyberEvolution and Future Minds.* Rockland, MA: Charles River Media, 1996.

Pearce, Joseph Chilton. *Evolution's End: Claiming the Potential of Our Intelligence.* San Francisco: HarperSanFrancisco, 1992.

Pearson, Simon. *A Brief History of the End of the World: Apocalyptic Beliefs from Revelation to UFO Cults.* New York: Carroll & Graf, 2006.

Peirce, Charles S. *Values in a Universe of Chance: Selected Writings,* edited by Philip P. Wiener. Stanford, CA: Stanford University Press, 1958.

Penrose, Roger. *The Emperor's New Mind: Concerning Computers, Minds, and the Laws of Physics.* Oxford: Oxford University Press, 1989.

Perrin, Noel. *Giving Up the Gun: Japan's Reversion to the Sword, 1543–1879.* Boston: D. R. Godine, 1979.

Peters, Ted. *Playing God: Genetic Determinism and Human Freedom.* New York: Routledge, 1997.

Petroski, Henry. *To Engineer Is Human: The Role of Failure in Successful Design.* New York: Vintage Books, 1992.

Pico, Richard. *Consciousness in Four Dimensions.* New York: McGraw-Hill, 2002.

Pinker, Steven. *The Blank Slate: The Modern Denial of Human Nature.* New York: Viking Press, 2002.

——. *The Language Instinct.* New York: HarperCollins, 1994.

Polanyi, Michael. *Personal Knowledge: Towards a Post-critical Philosophy.* Chicago: University of Chicago Press, 1974 (originally published 1958).

Polkinghorne, John. *Reason and Reality: The Relationship between Science and Theology.* Philadelphia: Trinity Press International, 1991.

Ponting, Clive. *A Green History of the World: The Environment and the Collapse of Great Civilizations.* New York: St. Martin's Press, 1991.

Popper, Karl R. *The Myth of Framework: In Defence of Science and Rationality.* London: Routledge, 1994.

——. *The Open Society and Its Enemies.* London: Routledge, 1945.

Postman, Neil. *Technopoly: The Surrender of Culture to Technology.* New York: Alfred A. Knopf, 1992.

Postrel, Virginia. *The Future and Its Enemies: The Growing Conflict over Creativity, Enterprise, and Progress.* New York: Free Press, 1999.

Potts, Richard. *Humanity's Descent: The Consequences of Ecological Instability.* New York: Avon Books, 1997.

President's Council on Bioethics. *Beyond Therapy: Biotechnology and the Pursuit of Happiness.* Washington, D.C.: President's Council on Bioethics, 2003.

Prigogine, Ilya, and Isabelle Stengers. *Order Out of Chaos: Man's New Dialogue with Nature.* Toronto: Bantam Books, 1984.

Primack, Joel R., and Nancy Ellen Abrams. *The View from the Center of the Universe: Discovering Our Extraordinary Place in the Cosmos.* New York: Riverhead, 2006.

Prothero, Donald R. *Evolution: What the Fossils Say and Why It Matters.* New York: Columbia University Press, 2007.

Quinn, Daniel. *Beyond Civilization: Humanity's Next Great Adventure.* New York: Harmony Books, 1999.

Raimondo, Justin. *An Enemy of the State: The Life of Murray N. Rothbard.* Amherst, NY: Prometheus Books, 2000.

Ramachandran, V., and S. Blakeslee. *Phantoms in the Brain: Human Nature and the Architecture of the Mind.* New York: William Morrow, 1998.

Ramachandran, V., and W. Hirstein. "The Science of Art." *Journal of Consciousness Studies* 6, no. 6–7 (1999): 15–51.

Raup, David M. *Extinction: Bad Genes or Bad Luck?* New York: W. W. Norton, 1991.

Rawls, John. *A Theory of Justice.* Cambridge, MA: Belknap Press, 1971.

Rees, Dai, and Steven Rose, eds. *The New Brain Sciences: Perils and Prospects.* Cambridge: Cambridge University Press, 2004.

Rees, Martin. *Our Final Hour: A Scientist's Warning: How Terror, Error, and Environmental Disaster Threaten Humankind's Future in this Century—on Earth and Beyond.* New York: Basic Books, 2003.

Regis, Ed. *Great Mambo Chicken and the Transhuman Condition: Science Slightly over the Edge*. Reading, MA: Addison-Wesley, 1990.

Reiss, Steven. *Who Am I?: The 16 Basic Desires That Motivate Our Actions and Define Our Personality*. New York: Putnam, 2000.

Richards, Jay W., ed. *Are We Spiritual Machines?: Ray Kurzweil vs. the Critics of Strong AI.* Seattle: Discovery Institute, 2002.

Richerson, Peter J., and Robert Boyd. *Not by Genes Alone: How Culture Transformed Human Evolution*. Chicago: University of Chicago Press, 2004.

Ridley, Mark, ed. *The Cooperative Gene: How Mendel's Demon Explains the Evolution of Complex Beings*. New York: Free Press, 2001.

——. *Evolution, 2nd edition*. Oxford: Oxford University Press, 1997.

——. *Evolution, 3rd edition*. Malden, MA, and Oxford: Wiley-Blackwell (Blackwell Publishing), 2003.

Ridley, Matt. *Nature via Nurture: Genes, Experience, and What Makes Us Human*. New York: HarperCollins, 2003.

——. *The Origins of Virtue: Human Instincts and the Evolution of Cooperation*. New York: Penguin Books, 1996.

Rifkin, Jeremy. *The Biotech Century: Harnessing the Gene and Remaking the World.* New York: Jeremy P. Tarcher / Putnam, 1998.

Rizzuto, Ana-Maria. *The Birth of the Living God: A Psychoanalytic Study.* Chicago: The University of Chicago Press, 1979.

Robin, Corey. *Fear: The History of a Political Idea*. Oxford: Oxford University Press, 2004.

Rogers, Everett M. *Diffusion of Innovations*. New York: Free Press, 1995.

Ronan, Colin A. *The Shorter Science and Civilization in China: An Abridgement of Joseph Needham's Original Text.* Cambridge: Cambridge University Press, 1978.

Root-Bernstein, Robert S. *Discovering: Inventing and Solving Problems at the Frontiers of Scientific Knowledge.* Cambridge, MA: Harvard University Press, 1989.

Rose, Hilary, and Steven Rose, eds. *The Political Economy of Science*. London: Macmillan, 1976.

Rose, Michael R. *Evolutionary Biology of Aging*. New York: Oxford University Press, 1994.

Rose, Steven. *Lifelines: Biology, Freedom, Determinism*. London: Allen Lane, 1997.

Rosner, Hillary. "Body and Soul: The First Great Bioethics Debate Began 2,000 Years Ago with a Clash between the Scalpel and the Cross." *Wired,* 10.01 (2002).

Rothschild, Michael. *Bionomics: Economy as Ecosystem*. New York: Henry Holt and Company, 1990.

Roughgarden, Joan. *Evolution and Christian Faith: Reflections of an Evolutionary Biologist*. Washington, D.C.: Island Press, 2006.

Roughley, Neil, ed. *Being Humans: Anthropological Universality and Particularity in Transdisciplinary Perspectives.* New York: Walter de Gruyter, 2000.

Rubenstein, Richard E. *Aristotle's Children: How Christian, Muslims, and Jews Rediscovered Ancient Wisdom and Illuminated the Middle Ages.* New York: Harcourt, 2003.

Ruse, Michael. *Monad to Man: The Concept of Progress in Evolutionary Biology.* Cambridge, MA: Harvard University Press, 1997.

Ruse, Michael, and E. O. Wilson. "Moral Philosophy as Applied Science." *Philosophy* 61, no. 236 (1986): 173–92.

Russell, Bertrand. *Religion and Science.* New York: Henry Holt and Company, 1935.

Russell, Peter. *The Global Brain: Speculations on the Evolutionary Leap to Planetary Consciousness.* Los Angeles: Jeremy P. Tarcher, Inc., 1983.

Russell, Robert John. *Cosmology, Evolution, and Resurrection Hope: Theology and Science in Creative Mutual Interaction.* Telford, PA: Pandora Press, 2006.

Sacks, Oliver W. *Awakenings.* New York: Harper Perennial, 1990 (fourth edition).

Sagan, Carl. *The Demon-Haunted World: Science as a Candle in the Dark.* New York: Random House, 1995.

Sagan, Carl, and Ann Druyan. *Shadows of Forgotten Ancestors: A Search for Who We Are.* New York: Random House, 1992.

Sandel, Michael J. *The Case Against Perfection: Ethics in the Age of Genetic Engineering.* Cambridge, MA: Harvard University Press, 2007.

——. "What's Wrong with Enhancement." 2002. The President's Council on Bioethics. http://www.bioethics.gov/background/sandelpaper.html.

Sapolsky, Robert M. *The Trouble with Testosterone: And Other Essays on the Biology of the Human Predicament.* New York: Charles Scribner's Sons, 1997.

Satinover, Jeffrey. *The Quantum Brain: The Search for Freedom and the Next Generation of Man.* New York: John Wiley & Sons, 2001.

Savage, Marshall T. *The Millennial Project: Colonizing the Galaxy in Eight Easy Steps.* Boston: Little, Brown and Company, 1994.

Schacter, Daniel. *The Seven Sins of Memory.* Boston: Houghton Mifflin, 2001.

Schnaars, Steven P. *Megamistakes: Forecasting and the Myth of Rapid Technological Change.* New York: Free Press, 1989.

Schneider, Stephen H., et al., eds. *Scientists Debate Gaia: The Next Century.* Cambridge, MA: MIT Press, 2004.

Schrodinger, Erwin. *What Is Life? The Physical Aspect of the Living Cell.* Cambridge: Cambridge University Press, 1944.

Schumpeter, Joseph. *Capitalism, Socialism, and Democracy.* London: George Allen & Unwin Ltd., 1943.

Schwartz, Peter. *The Art of the Long View: Planning for the Future in an Uncertain World.* New York: Currency Doubleday, 1996.

Scigliano, Eric. "Through the Eye of an Octopus: An Exploration of the Brainpower of a Lowly Mollusk," *Discover* 24, no. 10 (October 2003).

Seabright, Paul. *The Company of Strangers: A Natural History of Economic Life.* Princeton, NJ: Princeton University Press, 2004.

Searle, John R. *The Rediscovery of the Mind.* Cambridge, MA: MIT Press, 1992.

Seife, Charles. *Decoding the Universe: How the New Science of Information Is Explaining Everything in the Cosmos, from Our Brains to Black Holes.* New York: Viking, 2006.

Service, Elman. "The Law of Evolutionary Potential." In *Evolution and Culture,* edited by Marshall D. Sahlins and Elman Service, 93–122. Ann Arbor: University of Michigan Press, 1960.

Shanks, Niall. *God, the Devil, and Darwin.* New York: Oxford University Press, 2004.

Shattuck, Roger, *Forbidden Knowledge: From Prometheus to Pornography.* New York: St. Martin's Press, 1996.

Shelley, Mary. *Frankensteinor, the Modern Prometheus.* New York: Quality Paperback Book Club, 1994 (originally published 1818).

Shermer, Michael. *How We Believe: The Search for God in an Age of Science.* New York: W. H. Freeman, 2000.

———. *The Science of Good and Evil: Why People Cheat, Gossip, Care, Share, and Follow the Golden Rule.* New York: Henry Holt and Company, 2004.

Shorter, Edward. *A History of Psychiatry: From the Era of the Asylum to the Age of Prozac.* New York: John Wiley & Sons, 1997.

Silver, Brian, L. *The Ascent of Science.* New York: Oxford University Press, 1998.

Silver, Lee. *Challenging Nature: The Clash of Science and Spirituality at the New Frontiers of Life.* New York: Ecco, 2006.

———. "Future Stories of Human Evolution." 2005. http://www.leemsilver.net/SilverArticles/05EnglesbergRevised.pdf.

———. *Remaking Eden: Cloning and Beyond in a Brave New World.* New York: Avon Books, 1997.

Simon, Herbert. *The Sciences of the Artificial.* Cambridge, MA: MIT Press, 1969.

Simon, Julian. *The Ultimate Resource 2.* Princeton, NJ: Princeton University Press, 1998.

Simon, Robert I. *Bad Men Do What Good Men Dream: A Forensic Psychiatrist Illuminates the Darker Side of Human Behavior.* Washington, D.C.: American Psychiatric Association, 1996.

Simon, Yves R. *The Tradition of Natural Law: A Philosopher's Reflections.* Bronx, NY: Fordham University Press, 1992.

Sinclair, Upton. *The Jungle.* New York: Doubleday, Page & Co., 1906.

Singer, Peter. *Practical Ethics.* Cambridge: Cambridge University Press, 1993.

———. "Shopping at the Genetic Supermarket." In *Asian Bioethics in the 21st Century,* edited by Song, S. Y., Y. M. Koo, and D. R. J. Macer, 143–56. Tsukuba: Eubios Ethics Institute, 2003.

Smith, Adam. *The Theory of Moral Sentiment.* Amherst, NY: Prometheus Books, 2000 (originally published 1759).

Smolin, Lee. *The Life of the Cosmos.* Oxford: Oxford University Press, 1997.

———. *Time Reborn: From the Crisis in Physics to the Future of the Universe.* Boston: Houghton Mifflin Harcourt, 2013.

Snell, Bruno. *The Discovery of Mind: The Greek Origins of European Thought.* Translated by T. G. Rosenmeyer. New York: Harper, 1960.

Snow, Charles P. *The Two Cultures and the Scientific Revolution.* New York: Cambridge University Press, 1959.

Solomon, Robert. *Love: Emotion, Myth, and Metaphor.* New York: Doubleday, 1981.

Solso, Robert L., ed. *Mind and Brain Sciences in the 21st Century.* Cambridge, MA: MIT Press, 1997.

Sorrentino, Richard M., and E. Tory Higgins, eds. *Handbook of Motivation and Cognition: Foundations of Social Behavior*, vols. 1–3. New York: The Guilford Press, 1986, 1990, 1996.

Sowell, Thomas. *Knowledge and Decisions*. New York: Basic Books, 1980.

Spetner, Lee M. *Not by Chance!* New York: The Judaica Press, 1997.

Sproul, Barbara. *Primal Myths: Creation Myths around the World.* San Francisco: HarperSanFrancisco, 1979.

Stamenov, Maksim I., and Vittorio Gallese, eds. *Mirror Neurons and the Evolution of Brain and Language*. Philadelphia, PA: John Benjamins Publishing Company, 2002.

Stanford, Craig B. *The Hunting Ape: Meat-Eating and the Origins of Human Behavior.* Princeton, NJ: Princeton University Press, 1999.

Stark, Rodney. *One True God: Historical Consequences of Monotheism.* Princeton, NJ: Princeton University Press, 2001.

———. *The Rise of Christianity*. Princeton, NJ: Princeton University Press, 1996.

———. *The Victory of Reason: How Christianity Led to Freedom, Capitalism, and Western Success*. New York: Random House, 2005.

Steinhardt, Paul J., and Neil Turok. *Endless Universe: Beyond the Big Bang.* New York: Doubleday, 2007.

Stent, Gunther S. *The Coming of the Golden Age: A View of the End of Progress.* Garden City, NY: Natural History Press, 1969.

Stevenson, Betsey, and Justin Wolfers. "Economic Growth and Subjective Well-Being: Reassessing the Easterlin Paradox." *Brookings Papers on Economic Activity* 1 (2008): 1–87.

Stevenson, Leslie, and David Haberman. *Ten Theories of Human Nature.* New York: Oxford University Press, 1998.

Stock, Gregory. *Metaman: The Merging of Humans and Machines into a Global Superorganism.* New York: Simon & Schuster, 1993.

———. *Redesigning Humans: Choosing Our Genes, Changing Our Future.* Boston: Houghton Mifflin, 2003.

Strauch, Barbara. *The Primal Teen: What the New Discoveries about the Teenage Brain Tell Us about Our Kids*. New York: Doubleday, 2003.

Strauss, Leo. *Natural Right and History*. Chicago: University of Chicago Press, 1953.

Strogatz, Steven. *Sync: The Emerging Science of Spontaneous Order*. New York: Theia, 2003.

Sunstein, Cass. *Why Societies Need Dissent.* Cambridge, MA: Harvard University Press, 2003.

Sutherland, Stuart. *Irrationality: Why We Don't Think Straight!* New Brunswick, NJ: Rutgers University Press, 1994.

Sutton, Robert I. *Weird Ideas That Work.* New York: Free Press, 2001.

Suzuki, Daisetz T. *Zen and Japanese Culture.* New York: MJF Books, 1995.

Swenson, Rod. "Thermodynamics, Evolution, and Behavior." In *The Encyclopedia of Comparative Psychology*, edited by Greenberg, G. and M. Haraway. New York: Garland, 1997.

Swimme, Brian, and Thomas Berry. *The Universe Story: From the Primordial Flaring Forth to the Ecozoic Era.* New York: HarperSanFrancisco, 1994.

Tagore, Rabindranath. *The Religion of Man.* Boston: Beacon Press, 1970 (originally published 1931).

Tainer, Joseph A. *The Collapse of Complex Societies.* Cambridge: Cambridge University Press, 1988.

Taleb, Nassim Nicholas. *Fooled by Randomness: The Hidden Role of Chance in Life and in the Markets.* New York: Texere, 2001.

Tarnas, Richard. *The Passion of the Western Mind: Understanding the Ideas That Have Shaped Our World View.* New York: Harmony Books, 1991.

Teilhard de Chardin, Pierre. *The Future of Man.* New York: Harper Torchbooks, 1969.

——. *Man's Place in Nature.* New York: Harper & Row, 1966.

——. *The Phenomenon of Man.* New York: Harper Torchbooks, 1965.

Telushkin, Joseph. *Jewish Wisdom: Ethical, Spiritual, and Historical Lessons from the Great Works and Thinkers.* New York: William Morrow and Company, 1994.

Templeton, John Marks, and Kenneth Seeman Giniger, eds. *Spiritual Evolution: Scientists Discuss Their Beliefs.* Philadelphia: Templeton Foundation Press, 1998.

Tenner, Edward. *Why Things Bite Back: Technology and the Revenge of Unintended Consequences.* New York: Alfred A. Knopf, 1996.

Thayer, Lee. "The Functions of Incompetence." In *Vistas in Physical Reality*, edited by Ervin Laszlo, 171–187. New York: Plenum Press, 1976.

Thomas, Lewis. *The Fragile Species.* New York: Collier Books, 1993.

Thompson, D'Arcy Wentworth. *On Growth and Form.* New York: Dover Publications, 1992 (originally published 1917).

Tiger, Lionel. *Optimism: The Biology of Hope.* New York: Simon & Schuster, 1979.

Tiger, Lionel, and Robin Fox. *The Imperial Animal.* New York: Holt, Rinehart and Winston, 1971.

Tilly, Charles. *Democracy.* Cambridge: Cambridge University Press, 2007.

Tipler, Frank. *The Physics of Immortality.* New York: Doubleday, 1994.

Toffler, Alvin. *Future Shock.* New York: Bantam Books, 1970.

——. *The Third Wave.* New York: Bantam Books, 1980.

Toulmin, Stephen, and June Goodfield. *The Discovery of Time.* New York: Harper & Row, 1965.

Tuan, Yi-Fu. *Cosmos and Hearth: A Cosmopolite's Viewpoint*. Minneapolis: University of Minnesota Press, 1996.

Turchin, Peter. *War and Peace and War: The Life Cycles of Imperial Nations*. New York: Pi Press, 2005.

Turchin, Valentin F. *The Phenomenon of Science*. Translated by Brand Frentz. New York: Columbia University Press, 1977.

Turnbull, Colin M. *The Mountain People*. New York: Simon & Schuster, 1972.

Turner, Mark. *The Literary Mind*. New York: Oxford University Press, 1996.

Turner, Frederick. "An Ecopoetics of Beauty and Meaning." In *Biopoetics: Evolutionary Explorations in the Arts*, edited by B. Cooke and F. Turner, 119–37. Lexington, KY: ICUS, 1999.

Ullman, Ellen. *Close to the Machine: Technophilia and Its Discontents*. San Francisco: City Lights Books, 1997.

van Loon, Hendrik Willem. *The Story of Mankind*. New York: Boni & Liveright, 1921.

Varela, F. J., Evan Thompson, and Eleanor Rosch. *The Embodied Mind: Cognitive Science and Human Experience*. Cambridge, MA: MIT Press, 1991.

Venter, Craig L. *Life at the Speed of Light: From the Double Helix to the Dawn of Digital Life*. New York: Viking, 2013.

Vertosick, Frank T., Jr. *The Genius Within: Discovering the Intelligence of Every Living Thing*. New York: Harcourt, 2002.

Vinge, Vernor. "The Coming Technological Singularity: How to Survive in the Post-Human Era." VISION-21 Symposium, NASA, 1993.

Vizi, E. S. *Non-synaptic Interaction between Neurons: Modulation of Neurochemical Transmission*. New York: John Wiley & Sons, 1984.

Walker, Gabrielle. *Snowball Earth: The Story of the Great Global Catastrophe That Spawned Life as We Know It*. New York: Random House, 2003.

Walker, Mark. "Genetic Virtue." 2003. http://ieet.org/index.php/IEET/more/walker20031119.

Walter, Dave. *Today Then: America's Best Minds Look 100 Years into the Future on the Occasion of the 1893 World's Columbian Exposition*. Helena, MT: American & World Geographic Publishing, 1992.

Wang, Hao. *Reflections on Kurt Gödel*. Cambridge, MA: MIT Press, 1987.

Warburton, D.M., ed. *Pleasure: The Politics and the Reality*. New York: John Wiley & Sons, 1994.

Ward, Keith. *God, Chance & Necessity*. Oxford: Oneworld Publications, 1996.

Ward, Peter. *Future Evolution*. New York: Times Books, 2001.

Ward, Peter, and Donald Brownlee. *Rare Earth: Why Complex Life Is Uncommon in the Universe*. New York: Copernicus Books, 1999.

Watson, James, and Andrew Berry. *DNA: The Secret of Life*. New York: Alfred A. Knopf, 2003.

Watson, Lyall. *Dark Nature: A Natural History of Evil*. New York: HarperCollins, 1996.

Watts, Duncan J. *Small Worlds: The Dynamics of Networks between Order and Randomness*. Princeton, NJ: Princeton University Press, 1999.

Weinberg, Steven. *The First Three Minutes*. New York: Basic Books, 1976.

Weiner, Robert Paul. *Creativity and Beyond.* Albany: State University of New York Press, 2000.

Whitehead, Alfred N. *Process and Reality.* New York: Free Press, 1979 (originally published 1929).

——. *Science and the Modern World.* New York: Free Press, 1967 (originally published 1925).

Whitrow, G. J. *Time in History: Views of Time from Prehistory to the Present Day.* Oxford: Oxford University Press, 1989.

Wiener, Norbert. *The Human Use of Human Beings: Cybernetics and Society.* Garden City, NY: Doubleday Anchor Books, 1954.

Wilber, Ken. *Eye to Eye: The Quest for the New Paradigm.* Garden City, NY: Anchor Books, 1983.

Wilson, David S. *Darwin's Cathedral: Evolution, Religion, and the Nature of Society.* Chicago: University of Chicago Press, 2002.

——. *Evolution for Everyone: How Darwin's Theory Can Change the Way We Think About Our Lives.* New York: Delacorte Press, 2007.

Wilson, Edward O. *Consilience: The Unity of Knowledge.* New York: Alfred A. Knopf, 1998.

——. *On Human Nature.* Cambridge, MA: Harvard University Press, 1978.

Wilson, James. *The Moral Sense.* New York: Free Press, 1993.

Wilson, Sloan. *The Man in the Gray Flannel Suit.* London: Cassell, 1956.

Winner, Langdon. "Are Humans Obsolete?" (2003). http://www.rpi.edu/~winner/AreHumansObsolete.html.

——. *The Whale and the Reactor: A Search for Limits in an Age of High Technology.* Chicago: University of Chicago Press, 1986.

Witham, Larry. *The Measure of God: Our Century-Long Struggle to Reconcile Science and Religion.* New York: HarperSanFrancisco, 2005.

Wolf, Gary. "The Church of the Non-believers." *Wired,* November 2006: 182–93.

Wolfe, Alan. *Moral Freedom: The Search for Virtue in a World of Choice.* New York: W. W. Norton, 2001.

Wolpert, Lewis. *The Unnatural Nature of Science: Why Science Does Not Make (Common) Sense.* Cambridge, MA: Harvard University Press, 1993.

World Medical Organization. "Declaration of Helsinki." *British Medical Journal* 313, no. 7070 (1996): 1448–49.

Wrangham, Richard, and Dale Peterson. *Demonic Males: Apes and the Origins of Human Violence,* Boston: Houghton Mifflin, 1996.

Wright, John C. *The Golden Age: A Romance of the Far Future.* New York: Tor, 2002.

Wright, Robert. *NonZero: The Logic of Human Destiny.* New York: Pantheon, 2000.

——. *Three Scientists and Their Gods: Looking for Meaning in an Age of Information.* New York: Times Books, 1988.

Young, Dudley. *Origins of the Sacred: The Ecstasies of Love and War.* New York: St. Martin's Press, 1991.

Young, Louise. *The Unfinished Universe.* New York: Oxford University Press, 1986.

Young, Peyton H. *Individual Strategy and Social Structure: An Evolutionary Theory of Institutions*. Princeton, NJ: Princeton University Press, 1998.

Zerzan, John. *Elements of Refusal.* Seattle: Left Bank Books, 1988.

Zimbardo, Philip. *The Lucifer Effect: Understanding How Good People Turn Evil.* New York: Random House, 2007.

Zimmer, Carl. *Soul Made Flesh: The Discovery of the Brain and How It Changed the World*. New York: Free Press, 2004.

Zimmerman, Michael. "The Singularity: A Crucial Phase in Divine Self-Actualization?" *Cosmos and History: The Journal of Natural and Social Philosophy* 4, no. 1–2 (2008).

Zubko, Andy. *Treasury of Spiritual Wisdom: A Collection of 10,000 Inspirational Quotations.* San Diego: Blue Dove Press, 2004.

INDEX

C

E

H

I

M

N

O

P

S

T

U

V

W

X

Y

Z